A Personal History of Nuclear Medicine

Henry N. Wagner, Jr.

A Personal History of Nuclear Medicine

 Springer

Henry N. Wagner, Jr., MD, PhD
Professor of Environmental Health Sciences,
Johns Hopkins Bloomberg School of Public Health;
Professor Emeritus of Medicine and Radiology,
Johns Hopkins School of Medicine
Baltimore, MD, USA

A catalogue record for this book is available from the British Library.

Library of Congress Control

ISBN-10: 1-85233-972-1   eISBN: 1-84628-072-9
ISBN-13: 978-1-85233-972-2

Printed on acid-free paper.

Printed in Singapore   (BS/KYO)

9 8 7 6 5 4 3 2 1

Springer Science+Business Media
springer.com

To Anne, our four children, their spouses and nine grandchildren
for their never-ending love and help

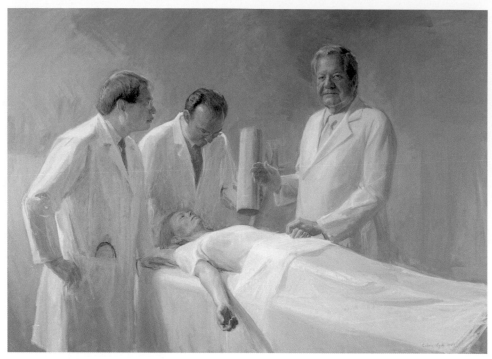

Life-size portrait of Henry N. Wagner, Jr., MD, with two colleagues that now hangs in Hopkins Nuclear Medicine Division. Portrait by Cedric Egeli.

Foreword

Each year, at the annual meeting of the Society of Nuclear Medicine, Henry Wagner summarizes his view of principal advances in the field. In *A Personal History of Nuclear Medicine*, he brings the same insight to the fifty years he has practiced, preached and breathed nuclear medicine. That same fifty years spans the era in which radioactivity has been harnessed to provide exquisite maps of physiologic function in the living human body.

Thus, the book brings the perspective of an insider, whose own contributions have been particularly influential: leader of a premier program in education and research; founding member of the American Board of Nuclear Medicine; proponent of international cooperation and the World Congress, and much more.

Because of Henry's positions and desire to meet and know colleagues throughout the world (he and his wife Anne are most gracious hosts and visitors) this autobiography is also a story of the major figures who grew the field of nuclear medicine and made the discipline into a coherent one.

The book also reflects Henry's personality: his candor and unflinching way of telling it the way he thinks it is, his punctuated use of aphorisms (some of his own making), his deep understanding of who he is, and an innocent delight in many accomplishments.

Some years ago, I suggested that Henry was a constructive troublemaker; someone who goaded us out of accepted wisdom into new, and sometimes outrageous, thinking. This volume documents his life, his philosophy, and his role in the coming of age for a remarkable medical specialty.

S. James Adelstein
Chappaquiddick
July 2005

Acknowledgment

I would like to acknowledge the inspiration and help of William G. Myers; the assistance of Judy Buchanan and Anne Wagner for reviewing the manuscript; Hiroshi Ogawa for his assistance, and Melissa Morton, Eva Senior and Robert Maged for their help.

Contents

Introduction

"There is a history in all men's lives."
—Shakespeare, *Henry V*

"The history of science is the history of scientists."
—John Lukacs

"How can man perform that long journey who has not conceived whither he is bound?"
—Henry David Thoreau

In September 2003, the National Institutes of Health (NIH) presented to the American people the goals of the NIH for medical research in the 21st century. Dr. Elias Zerhouni, who became director of the NIH in May 2002, had been Associate Dean for Research at Johns Hopkins School of Medicine before going to the NIH as the first radiologist to head that agency. He had been trained in nuclear medicine while a resident in radiology at Hopkins.

"Molecular imaging" was to be a major focus of research in the future of the NIH. This declaration of intent by the NIH was exciting for those in nuclear medicine, because molecular imaging had been the hallmark of nuclear medicine since its beginning.

The new NIH "Roadmap" focused on (1) the presymptomatic detection of disease; (2) personalized treatment based on molecular targets; and (3) the discovery of the clinical manifestations of genetic abnormalities. These had been the goals of nuclear medicine for over half a century.

In 2002, a new institute of the National Institutes of Health, the National Institute of Biomedical Imaging and Bioengineering (NIBIB), was created with an annual budget approaching $300 million, adding to the imaging research being carried out in other institutes, especially the National Cancer Institute. Imaging sciences had become a key focus of today's biomedical research, but this had not always been the case.

Those of us who had chosen to become specialists in nuclear medicine often encountered obstacles during the development of our careers. Many of the basic principles of our new specialty had not yet achieved acceptance by the medical establishment. Anatomy, radiology, and surgery remained the foundation of medical practice.

My first encounter with nuclear medicine took place when I arrived in London in July 1957, five years after I graduated from Johns Hopkins medical school. Nuclear medicine was not then a recognized medical specialty. The general public had heard the term

Figure 1 Elias Zerhouni trained in nuclear medicine at Johns Hopkins. At present, he is head of the National Institutes of Health in Bethesda, Maryland.

"atomic medicine" and associated it with the development of the atomic bomb. The field was based on the same scientific principles that had produced the atomic bomb. There was in those days an underlying fear of anything that had to due with radiation. These negative perceptions lingered long after the end of World War II. It would take decades before nuclear medicine would find its place in medical practice and biomedical research, before nuclear medicine defined itself as a scientific and clinical discipline, and people understood what the specialty was really all about. Nuclear medicine moved medicine beyond its focus on anatomy to a new focus on "molecular medicine." More than any other specialty, it brought together structure and function. Arthur Koestler has written: "In biology, what we call structures are slow processes of long duration; what we call functions are fast processes of short duration." They are both changes in mass as a function of time.

The story of the birth and growth of nuclear medicine is one of the most fascinating in physics and medicine, an excellent example of the precept that things don't happen; people make things happen. Nuclear medicine evolved from using the tools of physics and chemistry to solve patient problems. First, political, scientific, and technological challenges had to be faced.

The "tracer" principle was invented in 1913 by Georg Hevesy. It refers to our ability to "track" molecules as they participate in chemical processes. It is as if a molecule emitted a radio signal telling us what it was doing at all times.

Hevesy was born in August 1885 in Budapest. Working with Fritz Paneth in Vienna, he invented what he called "radioactive indicators." After his chemistry experiments in 1913, in 1923 he carried out his first radioisotope studies in biological systems, first in plants and then animals. In 1925, Herman Blumgart in Boston carried out the first human tracer studies by injecting his patients with solutions of the radioactive gas radon and timing how long it took for the radioactivity to travel from the injection site in an arm vein through the heart and lungs to reach the opposite arm.

In 1934, Hevesy left Berlin for political reasons and began to work in Copenhagen with Niels Bohr, who had first proposed the structure of the atom. In 1935, Hevesy began to work with phosphorus-32, being provided the radionuclide through the mail from Ernest Lawrence's cyclotron in Berkeley, California.

Figure 2 Georg Hevesy, who was awarded the Nobel prize in 1943 for the invention of the tracer principle, the most fundamental in nuclear medicine.

Figure 3 Herman Blumgart, who carried out the first studies of the circulation with solutions of radon gas.

Figure 4 Herman Blumgart decades later.

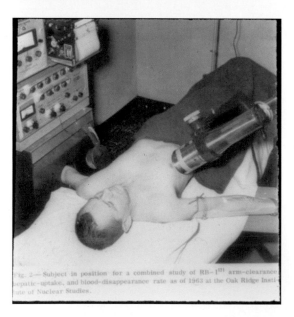

Figure 5 Measurement of the circulation time with intravenously injected tracers.

Figure 6 Forstman, discoverer of nuclear fission, and Seaborg.

Hevesy published more than 400 scientific articles and in 1943 won the Nobel prize. In 1959, he received the Atoms For Peace award by the U.S. Atomic Energy Commission. He died on July 5, 1966, in Freiburg, Germany.

In 1931, physicist Ernest Lawrence in California invented the cyclotron, which made possible the production of radionuclides not previously available. This invention was a major event along the path to nuclear medicine, occurring more than a decade before the start of the Manhattan Project, which was to build the atomic bomb and led to the invention of the nuclear reactor. The first cyclotron specifically for biomedical research was built in Cambridge, Massachusetts, by physicist Robley Evans in November 1940.

A cyclotron, which can be used to insert highly accelerated atomic particles, such as protons, into the nuclei of target molecules, can produce all of the most important radioactive elements needed for the study of living systems: radioactive oxygen, carbon, nitrogen, and fluorine (a substitute for hydrogen). Indeed, the element carbon defines organic chemistry.

Early studies in the 1940s focused on the thyroid. The fascination of the general public for this new approach to the chemistry of the living body is typified by an article in the June 4, 1963, issue of the *Wall Street Journal*, describing the construction of the cyclotron in the Physics Department at Washington University. For the first time, the economics of hospital cyclotrons were also examined.

The cyclotron was put on a back burner in biomedical research as a result of the invention of the nuclear reactor during World War II. In December 1938, Hahn and Strassman in Germany discovered fission, a process by which uranium atoms could be split into smaller elements. In December 1942, Enrico Fermi and his colleagues in Chicago built the first nuclear reactor as part of the Manhattan Project. Compared to the cyclotron, the nuclear reactor was able to provide a far wider source of radioactive elements and compounds at much lower cost. Fermi graduated from the University of Pisa in 1922 and subsequently studied in Gottingen, Germany, and the University of Florence, and then for 12 years taught at the University of Rome. When he learned that he was to receive the Nobel prize in Physics in 1938, he used the occasion to sail directly from Stockholm to New York. When the Manhattan Project began in 1942, Fermi was responsible for the study of chain reactions and plutonium research in the Metallurgical Laboratory of the University of Chicago. On December 2, 1942, he and his colleagues carried out the first production of a self-sustained nuclear chain reaction, which subsequently led to the production of the atomic bomb.

The invention of the nuclear reactor, which was a product of the Manhattan District Project of World War II, made large quantities of useful radioactive elements available to scientists and physicians throughout the world. The project was started by President Franklin Roosevelt shortly after he received a letter from Albert Einstein on August 2, 1939. Einstein had been told by E. Fermi and L. Szilard that "the element uranium may be turned into an important source of energy in the immediate future . . . that extremely powerful bombs of a new type may thus be constructed . . . You may think it desirable to have some permanent contact maintained between your administration and the group of physicists working on the chain reaction in America."

Ernest Lawrence had invented the cyclotron to make possible bombardment of atomic nuclei with high-energy sub-atomic particles, but in 1934, Frederick Joliot and Irene Curie made the startling discovery that practically every chemical element could be made radioactive by particle bombardment. Bombardment with high energy particles, such as protons, was possible in a cyclotron, because progressively high voltages of electricity could be produced conveniently, making it possible to produce hundreds of

Figure 7 Strassman and Wagner at Mainz in 1969.

isotopes of different elements, including carbon, nitrogen, and oxygen, which are of enormous importance in living systems. Indeed, carbon defines organic chemistry, the chemistry of life. Lawrence and his colleagues recognized immediately the great biomedical potential of the cyclotron.

Most of the Nobel prize winning discoveries in physics that provide the infrastructure of nuclear medicine were made at the time of a worldwide economic depression. In 1939, my parents took our family to the New York World's Fair in Flushing Meadow, N.Y. We were greatly impressed by the exhibit of "Man-made Lightning" at the General Electric Pavilion. A Van de Graaff generator could generate voltage up to 50,000 watts to produce an impressive 10-foot bolt of "lightning" that was spell-binding. The very next year, a group of six British scientists, called the Tizard mission, led by Henry Tizard, were sent by Winston Churchill to enlist the aid of American scientists in developing new technologically based weapons, which he believed was the key to winning the war spreading throughout Europe. They brought with them the results of all the top secret work on radar going on in England, and hastily set up headquarters in the Shoreham Hotel in Washington, D.C. On their voyage across the Atlantic, physicist John Cockcroft was asked to give a lecture on board ship. Because the work on radar was top secret, he chose to speak on atomic energy, which he believed was a safe topic "still considered years away from being realized and of no possible importance to the war." In his lecture, he stated that the energy in a cup of water could blow a fifty-thousand ton battleship one foot out of the sea.

Few people in the field of nuclear medicine know of the important relationships between the brilliant physicists who worked on both the development of radar and nuclear energy. The book *Tuxedo Park*, (a "must" read for everyone in the field of nuclear medicine), written in 2002 by Jennet Conant, the granddaughter of James B. Conant, President of Harvard University from 1933 to 1953 and Chairman of the National Defense Research Committee from 1941 to 1946, relates these remarkable connections between the physicists who developed radar and subsequently directed their attention and creativity to the nuclear physics foundations of nuclear medicine. The late Hal Anger was among these physicists. He had several key inventions related to radar prior to his directing his attention to nuclear instrumentation in 1948, inventing the well counter in 1951, the first of a series of basic instruments in the infant field of nuclear chemistry and medicine.

Even before the beginning of World War II, the Danish physicist Niels Bohr had lectured extensively in the United States about the destructive potential of the energy that might be released by nuclear fission. A report in Newsweek stated that atomic energy might create "an explosion that would make the forces of TNT or high-power bombs seem like firecrackers." Bohr's fears were matched by those of the Hungarian physicist Leo Szilard, who in 1939 was working with Nobel laureate Enrico Fermi on uranium fission at Columbia University.

Szilard told of his work to his 60 year old mentor, Albert Einstein, who decided immediately that the U.S. government should be warned of the possibility of making an atomic bomb, and wrote on August 2, 1939, to President Franklin Roosevelt. Szilard solicited funds to support his research on uranium from the financier tycoon and amateur physicist, Alfred Loomis, who, beginning in 1926, had built a personal research laboratory in Tuxedo Park, New York. Loomis subsequently contributed financially and helped

Ernest O. Lawrence to construct a cyclotron for the production of radioactive isotopes for research in both biomedicine and physics. With the help of Loomis and his many connections, Lawrence obtained a $1 million research grant from the Rockefeller Foundation. Loomis' consuming interest at the time was recruiting the brightest physicists to help develop advanced weapons for what he believed was certain to be a war in which the United States would become involved.

Loomis's lab would be hastily shuttered in 1940 and its research transferred to the newly established Rad Lab at MIT: "It is hard to believe that in only a few years, that bright circle (the physicists in Loomis's laboratory at Tuxedo Park) would not only build a radar system that would alter the course of the war, but would go on to create a weapon that would change the world forever."

Ernest Lawrence first visited Tuxedo Park in 1936 "to see the lab." Five years before, he had become famous for building the first cyclotron, using a radio frequency oscillator to accelerate deuterons at high speeds to bombard target atoms. As Lawrence's colleague and another Nobel prize winner, Luis Alvarez, wrote: "Lawrence had developed a new way of doing what came to be called 'big science', and that development stemmed from his ebullient nature plus his scientific insight and his charisma; he was more the natural leader than any man I've met." With the help of Arthur Loomis, Lawrence received a breathtaking $1.15 million from the Rockefeller Foundation to build a 60-inch cyclotron, far bigger than the 7-inch and 30-inch machines that had been built previously. This was long before the National Institutes of Health was even dreamed of. Nearly all scientific research was privately supported. In Loomis's words: "It was obvious from the very beginning, when he (Lawrence) was building (radioactive) isotopes, that it opened up methods for making medical measurements as well as chemical and physical measurements." After spending an enormous amount of time generating the funds, a 184-inch cyclotron was finally on the drawing board, when, on September 1, 1939, Germany invaded Poland.

Ernest's brother, John, had been in England to give a lecture on the use of P-32 to treat leukemia, and was to return on the ship, *Athenia*. Ernest heard a radio report that the Athenia had been torpedoed by a German submarine and was sinking off Scotland. It was 6 hours before he received word that all Americans on board had been rescued by a British destroyer.

In November 1939, Loomis moved to the Claremont Hotel in Oakland in order to carry out microwave experiments that Lawrence helped him design to complement his work on radar in Boston. The klystron tube had been invented by a physicist at Stanford, William Hanen, with the help of a former roommate, Russell Varian and his brother Sigurd. They were all working on the design of a radar device for navigating and detecting planes. These important advances were picked up for development by the Sperry Gyroscope Company. The 37-inch cyclotron was operating in the same building. Lawrence and his talented group were continuing to make plans for what eventually turned out to be the 184-inch cyclotron. On November 9, it was announced that Lawrence had won the Nobel prize for physics for his invention and development of the cyclotron.

When Ernest Lawrence returned to Berkeley after a visit to Loomis in 1939, he excitedly told his colleague Luis Alvarez of "his adventures on Wall Street with Loomis." When Loomis asked Lawrence to help him recruit for the new radar laboratory in MIT, to be opened after the closure of Loomis's laboratory in Tuxedo Park, Lawrence recommended two of his best students in Berkeley, Luis Alvarez and Edwin McMillan, both of whom would subsequently receive the Nobel prize. They began to work on radar a year and a month before Pearl Harbor. On February 7, 1941, Alvarez and his colleagues detected an airplane 2 miles away. The head of the laboratory, Lee DuBridge, exclaimed: "We've done it, boys."

The success in Britain and the United States on the development of radar changed the course of World War II, saved tens of thousands of lives, and subsequently revolutionized air travel, navigation, and weather forecasting. The enormous value of radar was clear in 1940 when Britain was subjected to the Blitz by the German Luftwaffe. The British could only survive and prevail because of the invention of radar, which had occurred several years before, based on the original work of Dr. Robert A. Watson-Watt, then head of Britain's Radio Research Laboratory. His work led to the establishment of a chain of Radio Detection and Ranging (RADAR) stations along the south and east coasts of England to detect enemy planes and ships.

While this work on radar was progressing, Fermi and Szilard at Columbia University were working on the possibility if obtaining a chain reaction, based on the discovery of deuterium by another Nobel laureate, Harold Urey. Before he left to work on radar, in Berkeley, Ed McMillan discovered uranium-239. His work was taken up by Glenn Seaborg and Emelio Segre, who subsequently showed that another product of uranium bombardment with deuterons was the new element, plutonium-239. They too would be among the many of Lawrence's disciples to receive the Nobel prize; McMillan with Seaborg in 1951 for their discovery of plutonium and his discovery in 1940 of neptunium; Alvarez in 1968 for his work in high energy physics.

Lawrence helped recruit every physicist of consequence in the country—many of them his former students—who were on the brink of exciting careers in nuclear physics to go to the Radiation Laboratory at MIT in Cambridge, Massachusetts, and work on the development of radar. According to Conant: "In each case, they dropped what they were doing and came for the simple reason that Lawrence had asked them to . . . Roping Lawrence into the radar project had been a stroke of brilliance . . . The Manhattan Project had not yet come into being. Here were all these unemployed nuclear physicists." Lawrence picked Lee DuBridge, a protégé and Chairman of the Physics Department at the University of Rochester, to direct the radar project, and he continued modifying and enlarging his 37-inch cyclotron. By the fall of 1941, Lawrence was convinced that every effort should be made to build an atomic bomb using either uranium-235 or plutonium-239. As Nobel laureate, Arthur Compton wrote in his memoir, *Atomic Quest*, the unique contribution of Lawrence was "a feasible proposal for making a bomb. No one else ever proposed the possibility. He came forward with what he felt could be carried through, and had something tangible to take hold of."

Although Ernest himself devoted all his efforts to physics, he appointed his brother, John Lawrence, to be Director of the University's Medical Physics Laboratory. The first application of a radioisotope in clinical medicine was the use of phosphorus 32 to treat certain blood disorders, including leukemia and polycythemia vera.

With most of the world, I heard about the atomic bombing of Hiroshima on August 6, 1945. I was aboard a three-masted, full-rigged training ship, *Danmark*, of the U.S. Coast Guard, that had fled to the United States at the beginning of World War II instead of returning to its homeport in Denmark. We sailed under a bridge spanning the Thames

River in New London, Connecticut, and docked at the dock of the Coast Guard Academy. I was one of 100 first year cadets who had entered the Academy in June 1945 after I had finished the first year of college at Johns Hopkins University in Baltimore. The news of the bombing of Hiroshima and Nagasaki was a tremendous shock, greater than the invasion of France on D-Day and the saturation incendiary bombing of Tokyo and other Japanese cities. The atomic bombings led to the sudden surrender of the Japanese within days.

The public had been kept in the dark about the development of the atomic bomb during the two and a half years of its development by the Manhattan Project. Some secrets had leaked out, but most people had never even heard of "radioactivity," a word that was for decades to incite fear in the minds of people all over the world. "Radioactivity" would hang as a cloud over the lives of those of us who chose to dedicate our professional lives to developing the "peaceful uses of atomic energy" in biology and medicine.

Radioactive elements, especially carbon-14, were key products of the Manhattan Project, and could be produced in large quantities by the newly invented nuclear reactors. They would provide the world with new tools for chemical and biomedical research. Radioactive "tracers" were able to "broadcast" their presence in "radiolabeled" molecules as they participated in the "chemistry of life". Being able to measure the chemical processes in every part of the body of living organisms would revolutionize biology and medicine. The radionuclides, chiefly carbon-14 and phosphorus-32, led to the birth of biochemistry.

Martin D. Kamen started working at the radiation laboratory of Dr. Ernest Lawrence at the University of California in Berkeley in 1937. He discovered carbon-14 but had the misfortune of suspicions arising from a dinner he had with two officials from the Russian consulate in 1944. He was fired by the University of California at Berkeley. He spent decades trying to prove his innocence. With the help of friends, he became Professor of Biochemistry at Washington University in St. Louis in 1945. He moved to Brandeis University in 1975, and was influential in the founding of the Universisty of California in San Diego in 1957. In 1996, he won the prestigious Enrico Fermi Award given by the U.S. Department of Energy. Among his discoveries was that the oxygen produced by the process of photosynthesis originates from water molecules, not from carbon dioxide as had been previously thought.

The American government made the decision after the war to make radioactive tracers available to qualified scientists all over the world. Before radioactive tracers could be used in human beings, the patients had to be convinced that it was safe to have "radioactivity" injected into their veins as part of the diagnostic process or medical treatment. Fear was understandable.

"Fallout" was another cause of fear. It can occur when radioactive debris that has accumulated in the atmosphere after the testing of atomic bombs falls to the earth. Radioactive particles are sucked up in millions of tons of earth, rising to altitudes greater than 40,000 feet, attaching themselves to vapor and dust that would be carried around the world because of the winds and rotation of the earth, and then falling back to earth as rain. The potential carcinogenic effects of fallout were described in newspapers all over the world. Especially fearful was that radioactive particles are invisible and cannot be detected by the natural senses. Another fear was environmental contamination from accidents during shipments of radioactive materials to hospitals and research laboratories around the country. Nuclear power plants were being built all over the country, which increased concerns about the possibility of accidents resulting in huge areas of contamination. Some feared (erroneously) that nuclear power plants could explode in the same way as atomic bombs. The greatest fear was "proliferation" of nuclear weapons by hostile countries.

Nuclear reactors at universities could also lead to nuclear weapons. Even today, five university nuclear reactors—the University of Wisconsin, Oregon State, Washington State, Purdue, the University of Florida—are fueled with weapons-grade uranium. More than 99% of naturally-occurring uranium is U-238, not suitable fuel for bombs. U-235, which makes up about 0.7% of naturally-occurring uranium, splits easily and can be used for making atomic bombs. The Department of Energy has spent large amounts of money to develop low-grade uranium fuel for university and other reactors. By July 30, 2004, 39 of 105 research reactors all over the world were to have been converted to U-235. Energy Secretary Spencer Abraham tried to have all of these reactors converted to U-235 by 2014.

Since World War II, proliferation of nuclear weapons has hung over the heads of everyone in the world. Some believed that the developing knowledge of the relationship between brain chemistry and behavior might help us to better understanding of the emotions of fear, rage, and insecurity that plague the human race.

Since the Cold War ended in December 1991, the greatest fear has been nuclear terrorism that could end civilization as we know it today. Those who have benefited professionally from the peaceful uses of nuclear energy have an obligation to help diminish the potential danger that could result from misuse of nuclear reactors used in research and in providing the necessary radioactive tracers on which our specialty is based. We must help face the challenge of keeping the world's nuclear materials out of the hands of the world's most dangerous people.

The pioneers of "atomic medicine" had to confront all these fears. Only their understanding, dedication, persistence, and ingenuity made success possible. They were able to convince their colleagues and the public of the benefits that radioactive materials can provide in medical diagnosis and treatment. They had to educate their colleagues about the "tracer principle," and its potential role in the practice of medicine and biomedical research.

We can see the spirit of the times right after World War II in the book, *From Hiroshima to the Moon*, by Daniel Lang. He quoted Dr. Willard F. Libby, a commissioner of the civilian U.S. Atomic Energy Commission, charged in 1946 with directing and controlling atomic energy, including atomic bomb production. Libby did not reassure the public when he said:

"In the event of a thermonuclear attack on the United States, a large fraction of the bombs would explode high above the earth, so that fallout of radioactivity would be minimized by the enemy's attempt to maximize the blast and thermal effects." This hardly made people feel better!

Would nuclear medicine have reached the widespread use in health care that exists today if the atomic bomb had not been developed by the expenditure of billions of dollars of government money? My answer is "yes," but the process would have taken far longer. Support by the U.S. government in promoting "peaceful uses of atomic energy" in medicine and other scientific fields played a major role in the development and growth of nuclear medicine all over the world. Most of the support for research in nuclear medicine at Hopkins came from the National Institutes of Health (NIH) Over the past decades, the Department of Energy (successor to the Atomic Energy Commission) has played a major role in development of instruments and radionuclides as part of intra- and extra-mural AEC programs. The NIH has emphasized support of biomedical research, while AEC research provided the tools. The efforts of both government agencies—the AEC (now called Department of Energy) and the NIH—have been synergistic. An example is the Human Genome Project.

After I had finished college, medical school, a three-year residency in internal medicine at Johns Hopkins, and two years as a Clinical Associate at the National Institutes of Health in Bethesda, Maryland, Professor Mac Harvey, Chairman of the Department of Medicine at Hopkins, told me that I had been selected for the highly desirable position of Chief Resident in medicine on the Osler Medical Service at Johns Hopkins Hospital.

The Osler residency was the first modern residency in the United States, begun in 1890 with assistant residents and a chief resident in each specialty. In 1897, an internship was added when Johns Hopkins medical school graduated its first class. Osler established the sleep-in-residency system where "house staff" physicians lived in the Administrative Building of the Hospital. The house staff lived an almost monastic life, many with rooms on the third floor of the building overlooking a large statue of Christ in the lobby. It was said jokingly that the house staff could look down on God, just as God looked down on them. Susequently, when administrators took over the house staff quarters which became offices, an elevator was soon installed.

Osler introduced the clinical clerkship, having third and fourth year medical students work on the wards. They would "follow a case day by day, hour by hour." Patients welcomed the house staff without whom they could not be cared for efficiently and effectively. Unlike today, in those days there was no scheduled time off. When the patients did not require immediate care and did not present specific problems, one could "sign out" to one's house staff colleague and spend a few hours at home.

A colleague of mine, Dr. Wilbur Mattison, had also been selected for the position of Chief Resident in medicine, but since there could be only one chief resident at a time, Professor Harvey said: "You and Wilbur decide who will go first." We literally flipped a coin. The result determined that I would go second, thereby giving me a free year before returning from the NIH to the Chief Residency at Hopkins. I decided to go to Hammersmith Hospital in London in 1957 to work under the direction of Professor Russell Fraser, head of endocrinology, the most exciting field in internal medicine at that time.

After my year at Hammersmith Hospital, I returned to Johns Hopkins Hospital. On August 24, 1867, Johns Hopkins, a Baltimore merchant, who provided the funds and inspiration for the founding of Johns Hopkins University and Hospital, wrote: ". . . It will be your duty, hereafter, to provide for the erection, upon other ground, of suitable buildings for the reception, maintenance and education of orphan colored children . . . It will be your special duty to secure for the service of the Hospital surgeons and physicians of the highest character and greatest skill . . . The Active Staff . . . shall regularly practice a hospital-based specialty." Johns Hopkins was among the earliest hospitals to have a full-time faculty. The Hospital and School of Nursing began operations in 1889, and the medical school, closely linked to the Hospital opened in 1893. Today, greatly expanded in size, the Hospital is still at this site, despite occasional temptations to follow other hospitals to the more affluent suburbs of Baltimore.

Two years before I went to Hammersmith Hospital, the Medical Research Council of the United Kingdom had built a cyclotron dedicated to biomedical research. Soon after I arrived, I recognized immediately the potential that radioactive isotopes could play in medicine. They could be measured by radiation detectors directed from outside of the patient's body. These new techniques might help solve many problems of patients that I had seen since my graduation from medical school five years before. One of the physicians at Hammersmith who was active in the use of radioiodine in diagnosis and therapy beginning in 1969 was Dr. A.W.D. Goolden. I often saw patients with him, as well as with Professor Fraser. Goolden subsequently published an article in 1971 on the use of technetium-99m for the routine assessment of thyroid function.

The "tracer principle" was to become the focus of my professional life for the next half century. After a year at Hammersmith, I returned to Johns Hopkins as Chief Resident in medicine, and then joined the full time faculty of internal medicine at Hopkins in 1958, with the goal of establishing a nuclear medicine division with John McAfee. We visualized the division as a joint effort of radiology and internal medicine. I still wonder why internal medicine never viewed nuclear medicine as an important part of internal medicine.

Beginning in those early days, which subsequently extended to almost half a century in the field of nuclear medicine, I felt that I was walking up the upward-moving escalator of nuclear medicine, an escalator powered by the discovery of radioactivity, the

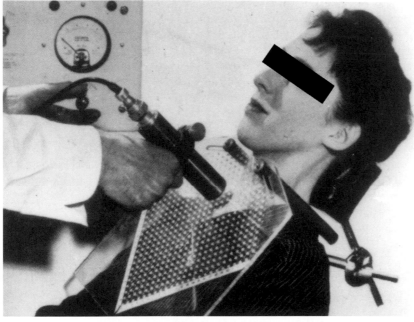

Figure 8 Measurement of the accumulation of radioactive iodine with a Geiger-Mueller counter placed at different points indicated by a plastic grid over the patient's neck.

cyclotron, nuclear reactor, radiochemistry, rectilinear scanner, Anger camera, computer, positron emission tomography (PET), single photon emission computed tomography (SPECT), PET/CT, and SPECT/CT. The combining of PET and SPECT with CT (computed tomography) brought anatomy and biochemistry together.

In 1958, when I told Professor Harvey, Chairman of Internal Medicine, that I wanted to work full time at Hopkins on the application of "radioisotopes" in medicine, he recommended that I consider an alternative, that is, to join Dr. Lawrence Shulman in the field of arthritis and rheumatology. At the time, I thought this was a curious recommendation, but in retrospect I believe that he knew of the work going on at that time in the laboratory of Dr. Dewitt Stetten at the NIH. In the summer of 1957, a young biochemist named Marshall Nirenberg had just come to the NIH and with his colleagues in the National Institute of Arthritis and Metabolic Diseases carried out research that was to win the Nobel prize for his work in molecular biology. He and his colleagues discovered that RNA consisted of chains of four nucleotide bases that served as templates for the synthesis of proteins containing 20 kinds of amino acids.

When political leaders such as Senator Lister Hill and Congressman John Fogarty responded to NIH director James Shannon's request for funds to "fight arthritis," they didn't realize at the time that they were helping to found molecular biology, a principal component of modern "molecular" medicine. Nirenberg received the Nobel prize for his work in 1968. The great accomplishments of investigators at the NIH were the result of Shannon's vision that clinical progress would come only through fundamental research.

I had no knowledge of this exciting work in molecular biology at that time, so I stuck with my plan to join John McAfee to co-found the Division of Nuclear Medicine at Hopkins. This new division was a combination of a new Division in Radiology, directed by John, and one from Internal Medicine, directed by me. My mental image at that time was that I was standing with one foot in each of two rowboats, one being Radiology, the other Internal Medicine, hoping that I would not fall in the water. We faced many hurdles over the next half century, all of them taking place against the background of the Cold War with the Soviet Union, the arising Red Chinese dragon, the rebuilding of Europe, the resurrection of Germany and Japan, the Korean, Vietnamese and Iraqi wars, and the tragedy of September 11, 2001.

My professional and personal life for the past 55 years has depended on the love, companionship, intelligence, and wonderful personality of my wife, Anne. We married on February 3, 1951, and began the spartan life that we lived during my last year of medical school, the house staff days at Hopkins, and subsequent the two years at the NIH. We were fortunate that we were able to enjoy those days without ever reflecting on how things would be better in the future.

When we moved to a two bedroom apartment at 120 Center Drive on the grounds of the NIH, we believed that our living conditions were luxurious compared to our three rooms on the 2nd floor of a row house at 1900 McElderrry Street across from the Woman's Clinic at Johns Hopkins. There was very little likelihood that I would be called to the Clinical Center during the night, as I had been almost every night when I was on the house staff at Hopkins.

After 2 years at the NIH and one year at Hammersmith Hospital in England, we returned to Baltimore, and lived for 10 months in the "Compound" on Monument Street across the street from the main building of the Hopkins Hospital. "Broadway Apartments" was the name of the rows of two-story dwellings owned by Hopkins. The two acres of green lawn enclosed in a high chain-link fence was a great playground for the children, all of whom were under 6 years of age. We on the married house staff enjoyed the proximity to the Hospital and the congeniality of other young married couples.

After 10 months living in the "Compound," Anne, our four children and I moved to 3410 Guilford Terrace to a row house built during World War I. Each three-story house in the block was different. Our next door neighbor was Paul Menton and his family; Paul was famous as sports editor of the Evening Sun for decades. Many of the people in the neighborhood were elderly, but beginning in the 1970s the neighborhood attracted doctors, lawyers, stock brokers and other professionals, including Dr. John Walton, chairman of the education department and President of the Baltimore City school board. Johns Hopkins University was only a few blocks away. Walton said: "I think it (our neighborhood) compares favorably with Georgetown." We decided to purchase our house as soon as it was shown to us by our realtor, who really understood what we wanted. We belonged to the Baltimore Protective and Improvement Association, which (among other activities) managed to block the granting of a liquor license for the dining room in the Marylander apartment house nearby until 1966 when the opposition ceased under the condition that there be no stand-up bar or cocktail lounge on the premises and that liquor would be served only at meals.

One of our neighbors, John Young, a retired stock broker said: "If it's a question of a broken curb or hole in the street, I get on the phone to City Hall. It's been my experience that if you call the right people down there, you get results."

On July 20, 1969, on her 40th birthday Anne and I, together with a friend, the late Bishop Frank Murphy, watched the first landing on the moon on television. After living 22 years on Guilford Terrace, we moved to Mt. Washington to live with Anne's parents in a carriage house remodeled by Anne's father during WWII. Our son-in-law, an architect, tripled the size of the original house before we moved in.

We had returned with our four children to the house where Anne and I had had our first date, several days after meeting on March 11, 1948 in Levering Hall on the campus of Johns Hopkins University. I was then 20 years old and Anne was 18. A great adventure lay ahead.

1

Survival of the Luckiest

In 1925, the Harvard physician Herman Blumgart injected a solution of a radioactive gas, radon, into the arm vein of a patient "to measure the velocity of the circulation." He measured the time it took for the tracer to pass through the heart and lungs and reach the opposite arm. His experiment was little noted at the time but is of great historic interest. It was the first time a physiological process had been measured with a radioactive tracer, making the measurements with an externally-placed radiation detector directed at a part of the body of a living human being.

On July 4, 1924, a year before Herman Blumgart's historic first study of the circulation with a radioactive tracer, my mother planned to accompany my 60-year-old grandmother on an overnight trip on a steamboat going down the Chesapeake Bay to visit her daughter, Alma, who lived in Crisfield, Maryland. Grandmother had arrived in Baltimore in 1885, emigrating from Germany on the *Brandenburg*, a 8,000 ton vessel which plied between Bremen and Baltimore. She was 5 feet $2^1/_2$ inches tall and weighed 125 pounds. Her maiden name was Barbara Krautblatter.

Grandmother Wagner had immigrated to Baltimore from Bavaria, Germany. Widowed soon after her arrival in Baltimore, Barbara lived with her son, my father, Henry, and his wife, Gertrude. Several times a year, she took the overnight steamer to visit her daughter, Alma, and her husband, Jim Thornton, a seafood salesman in Crisfield, Maryland, 100 miles south of Baltimore, near the mouth of the Bay.

Crisfield was founded in 1867, built on a giant mound of oyster shells, and became the seafood capital of the Bay. At its peak in the 1870s, 9,000,000 bushels of oysters were shipped every year from Crisfield. In 1884, 15,000,000 bushels were shipped. Today, the oyster and blue crab industries are in trouble because of the effect of pollution on the yields.

At the last minute, my mother decided not to join Grandmother Wagner on the boat to Crisfield, because of a head cold. This decision saved her life (and made possible mine). Grandmother boarded the tiny coal-fired steamship, the *Three Rivers*, at Pier 5 on Light Street in downtown Baltimore, and headed down the Bay, passing the shipyards of the Bethlehem Steel Company, the two-century old houses on Fells Point, the city-owned Recreation Pier, the Seven Foot Knoll light house, and the guns of Fort McHenry, where, in 1812, Francis Scott Key had viewed the "star spangled banner by the dawn's early light" and written the poem that became our national anthem.

On board the *Three Rivers* were 50 noisy, excited newsboys who delivered the *Baltimore Sun*. They were celebrating a successful year of steadily increasing newspaper sales.

Figure 9   Henry N. Wagner, Sr. (father of HNW) and his mother.

For an hour before retiring to her cabin, she had been amused by the goings-on of the boys. As she slept, the boat proceeded quietly down the broad waters of the Bay.

As the *Three Rivers* steamed past Cove Point on the western shore of the Bay just across from Crisfield, a bright full moon was shining, and most of the passengers were asleep in their cabins. Suddenly, grandmother was awakened by the cries of "fire, fire!" and smelled smoke entering her cabin on the third deck. She tried to escape through the door of her cabin but was stopped by a heavy cloud of smoke in the passageway. She turned back, terrified, not knowing what to do.

Not only do the fittest survive, but also the luckiest. Grandmother's luck was that among the newsboys was a 17-year-old newsboy, William Elkins, who played the bass horn in the newsboy band. He heard grandmother's cries for help, and broke into her cabin through the single porthole facing the deck. Throwing her arms over his shoulders, he crawled back out on deck, grabbing two life jackets. He tried unsuccessfully to launch a life raft, and then, giving up, threw a rope down the side of the boat, and descended from the third deck down into the water. Grandmother was hanging on with her arms over his shoulders. With her crying but not struggling, he swam two hundred yards with her still clinging to his back. They reached a lifeboat launched by another boat, the *Middlesex*, which had responded to the SOS and raced to the rescue of the *Three Rivers*. They were hauled aboard the *Middlesex* to join the other survivors.

My father was waiting with the crowd of anxious relatives at the dock on Light Street in the Baltimore harbor, when the *Middlesex* docked in the early hours of the morning. Grandmother and William, with blankets over their shoulders, walked solemnly off the ship with the other survivors. Two passengers had drowned, and five newsboys were missing. They were subsequently found to have drowned.

Frank Morse, leader of the newsboys' band, told a Sunpaper reporter: "Many of the boys were among the last to leave the burning boat. They threw life preservers to those struggling in the water. Some manned fire hoses and fought the flames." The five newsboys who died were buried in a semi-circle in Loudon Park, Baltimore, beneath a copper and granite shaft designed by sculptor J. Maxwell Miller. The monument features a life-sized boy in three-quarters relief holding a raised flute. A tablet contains a line from Henry Wadsworth Longfellow's poem "Hiawatha":

"They have moved a little nearer to the master of all music."

William's heroism was celebrated in an article on the front page of the *Baltimore Sun*. A few days later, my grandmother and parents-to-be invited William to a celebratory dinner at their 1919 W. Fayette St. home, where they presented him with a gold watch to express their gratitude for his heroism. This dinner and medal were noted in another article on the front page of the *Sun*, with William's photograph and the watch. A week later, a short article appeared on the inside pages of the *Sun*: careless smoking among the young newsboys had caused the fire.

Thirty-four years later, in the 1960s, I read an obituary in the *Baltimore Sun* reporting that Elkins had died, and that he had been on the *Three Rivers* when it burned. I immediately telephoned Elkin's wife, telling her that I was the grandson of the woman whom her late husband had saved. Shocked to receive the call, she told me that it was not her recently deceased husband, Andrew, who "was the hero," but it was Andrew's brother, William. She told me that "the hero," William Elkins, had died in the 1940s when he was struck by a passing car after he had stopped to assist a motorist whose car had become incapacitated by the side of a busy highway. The unlucky often fail to survive.

Three years after the fire on the *Three Rivers*, I was born at St. Joseph's Hospital in east Baltimore, one half mile north of Johns Hopkins Hospital. St. Joseph's Hospital later moved to a more prosperous Baltimore suburb, as the city of Baltimore expanded.

On May 20, 1927, eight days after I was born, my mother and I were still in St. Joseph's hospital, when Charles Lindbergh, the Lone Eagle, took off from Roosevelt Field in New York City and flew $33\frac{1}{2}$ hours non-stop to a cheering crowd of 150,000 people at LeBourget Field in Paris.

The transatlantic flight of Lindbergh's plane, *The Spirit of St. Louis*, marked the beginning of the "Age of Aviation." It is hard to believe that so many important events have taken place over the course of my lifetime.

Another hero was Joe Louis, an African-American who became heavyweight boxing champion when I was 12 years old. We kids rooted for Joe Louis when he defeated the German boxer, Max Schmeling. As we sat on the white marble steps outside our houses in west Baltimore, we didn't know whether we should root for Joe Louis because he was an American, or for Max Schmeling because he was white. Baltimore was racially segregated then—housing, movies, transportation, restaurants, churches, schools, even the morgue at Johns Hopkins Hospital.

After spending grades 5–8 at Calvert Hall Country School in Walbrook Oval, and four years of high school at Calvert Hall High School, I entered the College of Arts & Sciences on the Homewood campus of Johns Hopkins University in north Baltimore in June 1944. Little did I know then that I would be associated with Hopkins for the next 60 years with

Figure 10 Henry Wagner.

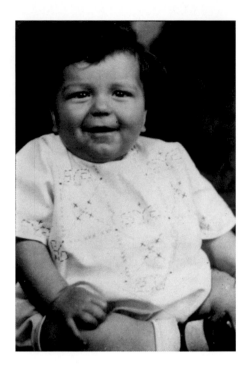

the exception of 18 months in the U.S. Coast Guard, two years at the National Institutes of Health, and one year at Hammersmith Hospital in London.

The Johns Hopkins Hospital is in one of the poorest sections of Baltimore, several miles from the main campus of the University. The admistration of the Hospital once considered moving from its historic location in east Baltimore to the suburbs, but resisted the temptation, and remained at the site of its founding. Hopkins is now the largest employer in Baltimore, and has subsequently done a lot to transform its surroundings. Today Hopkins continues to play a major role in the revitalization of east Baltimore, as well as providing out-patient care in the suburbs.

The row house at 1919 W. Fayette Street where I grew up in Baltimore was two miles west of "downtown" Baltimore. The city had been founded in 1729, 200 miles inland from the Atlantic Ocean, at a location on the Patapsco River that flowed into the Chesapeake Bay. The population of Baltimore was 25,000 in 1800, and featured its port and the developing railroads. Its shipbuilding industry brought fame during the war of 1812. In 1828, the Baltimore and Ohio Railroad, the first in the country, officially opened in Baltimore. Its inland location provided a great commercial advantage because incoming freighters could land their cargo farther inland than at other ports. Johns Hopkins himself was one of the founders of the B & O Railroad, and contributed $6 million to the founding of the University and Hospital which bear his name. By 1850, Baltimore had grown to be the third largest city in the country, with a population of more than a quarter of a million.

Our house on Fayette St. was in the parish of St. Martin's Catholic Church, one of the oldest in Baltimore, founded in 1865 on the site of a Civil War military camp, by its first pastor, John S. Foley. St. Martin's was the center of the lives and social activities of many

Figure 11 The Wagner family in 1928 (l to r, Gertrude, Mother, HNW, and Herman).

German- and Irish-American Catholics who lived in the nearby row houses with white marble steps that were kept spotlessly scrubbed. Many of our neighbors worked half a mile away at the roundhouse of the Baltimore and Ohio Railroad, where trains were repaired, now the site of a world-class railroad museum.

H.L. Mencken, the accomplished critic and journalist of the Baltimore Sunpapers and founder of the magazine, the *American Mercury*, during the "Roaring" Twenties lived near us on Hollins Street in west Baltimore, near St. Martin's Church. He wrote: ". . . the Catholic clergy of Baltimore never engage in buffooneries, and make no attempt to advertise in the newspapers, but confine themselves strictly to the proper business of their office. The congregations of the Catholic Churches in this part of southwest Baltimore grow steadily, as those of the Protestants shrink. On next Sunday morning, there

Figure 12 HNW at age 7 on the occasion of his first holy communion.

will be more worshippers and particularly more adults and more men in St. Martin's Church on Fulton Avenue than in all the religious vaudeville shows west of Eutaw Street." Today, beautiful St. Martin's Church still stands majestically, but is now in the heart of the slums of Baltimore.

In those early days, the nearby parish of Fourteen Holy Martyrs occasionally was noted by headlines in the *Sun* reporting on a sporting event: FOURTEEN HOLY MARTYRS SLAUGHTER OUR LADY OF LOURDES.

Next to St. Martin's Church is "Foley Hall," a recreation building which in those days had a gym, pool tables, and two bowling alleys. At one of the parish dances, the order of dances included waltz, two-step, Paul Jones, Spanish Boston-Varsovienne, Rye, Lanciers, and Schottische. The St. Martin's Literary and Dramatic Association occasionally sponsored dances in Foley Hall.

Young African-Americans served as "pin boys" in the bowling alleys in Foley Hall. Those of us who couldn't knock down many pins were called "pin boys' delights." The pastor of the parish was Monsignor Louis O'Donovan, ministering to 10,000 parishioners with his three assistant priests. One was T. Austin Murphy, later to become Auxiliary Bishop of Baltimore. Everyone in the parish was shocked when we heard that his brother,

Brady Murphy, an FBI agent, on his way home from work had received a telephone call telling him that a wanted criminal was in a telephone booth in the lobby of the Center Theatre in downtown Baltimore. As Agent Murphy approached the booth, the man suddenly drew a gun and killed him.

Another assistant priest at St. Martin's was Father Raymond Kelly, who was in charge of Camp St. Martin, which I attended from ages 8 to 15 with my two brothers, one older, Herman, and one younger, Albert. Every summer, we three boys spent 6 weeks at the camp, which was located at Love Point, overlooking the Chesapeake Bay. The camp had been opened in 1930, and was run by St. Martin's parish, with seminarians from St. Mary's Seminary serving as some of the counselors. The men of the parish built the camp "shacks," and women of the parish served as cooks, living in a rented house nearby. As campers, we avoided walking by the cottage at night lest we be hit by the contents of chamber pots being thrown out of the windows.

Every Memorial Day on May 30, 15 men of the parish would prepare the camp for its opening in June. Among the men were painters, carpenters, electrician, and handymen, together with youngsters, who would clean up the grounds. Among the non-Seminarian counselors were "Knotty" McCann, who subsequently became athletic director at St. Martin's High School, John "Oatey" O'Grady, who subsequently became a judge, "Jake" Lentz, who became an FBI agent, and Larry McCabe, who became a Court bailiff.

During those depression days, the cost per week to attend the camp was $5.00 for parishioners and $10.00 for those from outside the parish. After a six-week period for boys, there were two weeks of camp for girls, with women counselors. Once, having gotten in with some "unsavory" fellow campers, I told my parents, when they came for the weekly Sunday visit, that I wanted to come home at the end of my fourth week. My father asked: "Don't you love your mother?" It's nice to bring happiness to people, even if only by staying in camp.

We traveled to and from camp on the ferry boat, Philadelphia, which we called Smoky Joe. On arriving at Love Point, we all rushed from the pier at Love Point to the camp, half a mile away. We would choose a "shack" for about 10 boys, each of whom was assigned an army cot. There was only a single hand-operated pump to provide fresh drinking water, which tasted terrible because of its high iron content. There was no hot water. In the mornings, we would fill several buckets with water, place them in the sun all day, in order to have hot water each evening for a "bucket bath." An "outhouse" with four adjacent seats served males. Alongside was another enclosure with two seats for girls.

We would not sleep well the first night, but soon became totally adjusted to camp life. Although there were morning prayers every day, St. Martin's was not a religious camp, but was highly oriented toward sports, with three periods of softball, basketball and volleyball every day, followed by swimming twice a day. The teams were led by the best athletes who chose teammates once a week for the following week. It was embarrassing if you were the last to be picked.

Those campers who could swim were taken by pickup truck to the pier where the "Smoky Joe" docked. We dove from the tall pilings surrounding the docking slip, as well as from the gang plank 30 feet high from which the passengers boarded and exited the ferry boat. Those who couldn't swim walked twice a day about three blocks from the camp to a beach called "Rhodes," named after the owner of the property who was kind

Figure 13 Weekly boxing at Camp St. Martin.

enough to let the campers use the site. The water in the Chester River was perfectly clear in those days, with lots of seaweed and soft crabs. There was a weekly gathering of campers and counselors at a huge evening campfire, built from driftwood gathered by the campers from along the beach on the west side of the campgrounds. At each Thursday night campfire, one boy was awarded a loving cup, acknowledging that he had been selected as the "best camper" for that week. Both of my brothers, Herman and Albert, received this award, but I never did. I could never understand why. Many decades later, when we were middle-aged, having heard me tell this story of my disappointment so often, Albert awarded me a replica of the loving cup as a surprise.

One summer, when I was catcher during a softball game, the batter "threw his bat," and hit me on the forehead, resulting in a cut requiring several stitches. They took me to Dr. Sattlemeyer in nearly Stevensville, Maryland, to have my cut sewed up. This first encounter with medicine may have planted the seed of what would become my subsequent career.

Every Sunday, parents of the campers would visit the camp for a day, taking the Love Point ferry from Baltimore. Some would have lunch at Miss May's, a bed-and-breakfast near the ferryboat pier. Sunday afternoon would feature a visitor/counselor softball game, with all the campers and visitors as spectators. A high point of the game would be if the ball was hit over the cliff that lay beyond left and center fields. Once a center-fielder disappeared suddenly as he was going back for a fly ball to center field. He had fallen off the cliff. Fortunately, he was not hurt by the 20-foot drop. Erosion of the cliff was a big problem at the campsite. Ten feet of land would have eroded when we returned every summer. Today the camp no longer exists and the campsite has for all practical purposes disappeared because of erosion.

Attending Camp St. Martin every summer year after year helped prepare us for life. With the emphasis on sports three times every day, and later at Calvert Hall Country School and Calvert Hall High School, we were continually taught how important it was to always do our best, to strive constantly for excellence. We were taught that achievement would give meaning to our lives. We were imbued with a spirit of confidence and optimism, and constantly kept our eyes on our futures. Many campers went on to become star athletes in one of two Catholic High Schools in Baltimore: Calvert Hall and Mt. St. Joe's. Few went to Loyola, a third Catholic high school in north Baltimore, where the families had a higher social status. Most came from the lower-middle class of southwest Baltimore where St. Martin's was located. We were constantly taught by our elders to respect authority and count our "blessings." We always knew who was boss, and it wasn't us. We believed in government and authority figures. A camp rule was that whenever a counselor blew his whistle, we had to immediately stop what we were doing, and run to the whistle blower. We were far from being "goody-two-shoes," but I can still remember lying in bed before going to sleep, praying silently: "Dear God, make me a good boy." Raised as a Catholic through grammar and high school, we were very conscious of values and morality in our lives. A frequent discussion in high school was whether one could live a moral life without religion. We usually concluded that it would be very difficult. Selfishness would get out of hand. We were taught that there had to be the proper balance of freedom and order, that freedom is only possible when strong societal or religious values are accepted.

When my mother was dying at age 98, my wife whispered in her ear: "Mom, is there anything that you want?" Mother answered: "I want to be good."

My brother, Herman, was valedictorian of his class in the 8th grade at Calvert Hall Country School in 1937 and again in his senior year at Calvert Hall in 1941. For having the highest average for four years at Calvert Hall, he received a full scholarship to Johns Hopkins University. I was valedictorian at Calvert Hall Country School in 1940, and again in the senior year at Calvert Hall in 1944. Eventually, both Herman and I had Chairs named for us at Drexel University in Philadelphia and Johns Hopkins School of Medicine: the Herman Block Wagner Chair in Chemistry and the Henry N. Wagner, Jr. Chair in Nuclear Medicine. Among Herman's many inventions were sunglasses that darkened in the sunlight, and a doll that became "sun tanned," because its paint contained a compound known as dithizone.

We knew that our parent's lives were devoted to our well-being and to our achieving success in later life. Although they themselves lived happy lives, it was clear that we were the main focus of their lives. The depression of the 1930s had a great influence on children, but throughout those difficult times, we were happy with our family, friends, classmates, and neighbors. We learned to appreciate the good things in life, an appreciation which never left us. I thought that the ultimate example of luxury in those days was when

Figure 14 Football team at Calvert Hall Country School. HNW at the left of the first row.

Figure 15 The Wagner family gathered at the establishment of an endowed Chair in Nuclear Medicine at Johns Hopkins named after Henry N. Wagner, Jr.

the father of one of my girl friends gave her a nickel every night to buy an ice cream cone at our neighborhood drugstore.

Competitiveness and discipline were important in our young lives. Parents, camp counselors, or teachers never had to plead with us to do what we were told. We had to stay within clearly defined rules, or suffer the consequences. Authority figures provided feelings of security in the difficult times of the depression. We lived according to President Roosevelt's admonition: "We have nothing to fear but fear itself." As Catholics, especially Irish Catholics, we often suffered feelings of guilt, but always had the Act of Contrition to fall back on: "O my God, I am heartily (at times the word was jokingly replaced by the expression 'partly') sorry for having offended thee, and I detest all my sins because I dread the loss of heaven and the pains of hell, but most of all because they offend Thee my God who are all good, and deserving of all our love." Our respect of authority was accompanied by a touch of cynicism. I learned then, and believe now, that at least half of what I am told is probably not true, not because of any moral deficiency on the part of the person who was telling me something, but because of a human tendency to try to provide answers when a more appropriate response would be: "I don't know." Is this perhaps characteristic of scientists?

St. Martin's operated an elementary, junior and senior high schools, with over 1,000 students taught by the Sisters of Charity. There were 8 masses every Sunday. Pews rented for $50.00 per month. My father told me that if we didn't rent a pew, no priest would come to our house to administer the last rites of the church as we lay on our deathbed. The sacrament of the last rites was called "Extreme Unction," which we pronounced as "Extree-munction." Pew rents provided as much income for the parish as did the Sunday Offertory collections which were placed in the collection basket in envelopes designated for each Sunday and Holy Day of Obligation. Confessions were heard every day from 9:00 to 12:00 AM and 2:00 to 10:30 PM. None of our sins were original.

I started the first grade of school at nearby Fourteen Holy Martyrs parish, where my mother had enrolled me because I was too young (5 years old) to enter St. Martin's School, and too active to be kept at home all day. When an assistant pastor at St. Martin's, Father Manns, learned that I was attending Fourteen Holy Martyrs School, he arranged immediately for me to be transferred to St. Martin's School. My first class was music. I was given a saxophone, rather than a violin that I had hoped for. Thus ended my musical career.

An early contact with the medical profession occurred when a young intern dressed in a white coat and pants arrived at our house by ambulance to take me to Sydenham Hospital for isolation of patients with infectious diseases. I was 5 years old, and had developed scarlet fever. Visitors of patients at this infectious disease hospital had to visit with patients from a walkway outside the hospital, looking through glass windows to talk to their loved ones.

We children were very conscious of the fact that our parents were dedicated to our welfare and future success. Even though the country was in a depression, our parents managed to transfer us to Calvert Hall Country School (CHCS) after completing the fourth grade at St. Martin's. Our sister transferred to the Institute of Notre Dame, a private school up the street from where our mother was born. Two Brothers of the Christian Schools took a taxi every day from Calvert Hall in downtown Baltimore to Walbrook

Oval, where the Country School was located. Two grades, 5 and 6, and 7 and 8, were taught by one of the two brothers in the same room. Brother G. Edward taught the 5th and 6th grades in the same room, and Brother Matthew taught the 7th and 8th grades. There were about 10 boys in each grade.

Figure 16 Boys in the four grades (5–8) at Calvert Hall Country School. Henry Wagner in second row, three boys from the right; Herman Wagner, first row second from the right. Brother Matthew on the left, Brother Edward on the right. Tom Mooney, third from the right in the top row.

There were three football teams—for 80, 90, and 100 pound boys. I was eventually quarterback of all three teams. I remember one game, when as quarterback, I called for a pass at the last minute of the game, when we were ahead by a score of 7–6. The pass was intercepted; our opponents scored, and we lost by a score of 13 to 7. My father wouldn't speak to me on the ride home from the game. Television had not yet been invented, so we were always participants in sports, rather than spectators.

We looked up to the older boys who came to Walbrook Oval from downtown Calvert Hall to practice and play baseball. One of our greatest heroes was Martin Schwalenberg, who went on to become one of the most popular priests in Baltimore, and served as chaplain of the Baltimore Orioles.

On hot summer nights, we sat on the marble steps in front of our house, and talked with neighbors. Activities of the parish were the focus of the lives of most parishioners. The men and boys belonged to the Holy Name Society; the women belonged to the Sodality, Ladies of Charity, and Sewing Societies, all meeting weekly. There was a choir, a parish journal, a debating and literary club. The St. Vincent De Paul Society provided for the needs of the poor in the parish. There was the Ancient Order of Hibernians,

Figure 17 The 100-pound football team at Calvert Hall Country School. Henry Wagner at the left of the first row, next to Coach Tom Mooney.

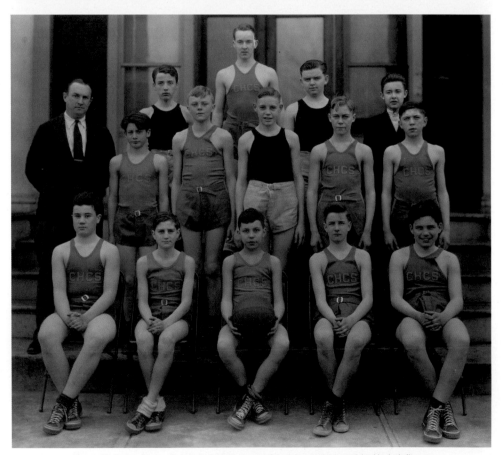

Figure 18 Herman Wagner (holding basketball), captain of the Calvert Hall Country School basketball team.

festivities in Foley Hall, including weekly Bingo games, as well as the Knights of Columbus, a social and beneficent organization that played a major role in the lives of my parents and their friends. My father became head of THE ALHAMBRA, for 4th degree Knights of Columbus. We children would often swim in the indoor pool of the K of C in downtown Baltimore. For hygienic reasons, we boys never wore bathing suits. I still don't know whether or not the girls wore bathing suits during their allotted times. St. Martin's also ran a Day Nursery for working women, two blocks from the church and next to the Bon Secours Hospital, operated by the Sisters of the Bon Secours, which often provided emergency medical care to us kids. One day, climbing over the fence around the playground of #48 public school, I cut my arm and had to be taken to Bon Secours to have it sewn up, still another favorable interaction with doctors. As children, showing great daring, we would occasionally sneak onto the grounds of the convent next to the hospital, and sample the delicious grapes that the sisters grew.

We would often drive to my other grandmother's row house on Aisquith Street in east Baltimore, going past a still-standing obelisk commemorating two young American boys, Wells and McComas, who were killed while they were still up in a tree from which they had shot the British Commanding General who led British troops into Baltimore during the Battle of North Point during the war of 1812. Up the street from my mother's childhood home was a pickle factory, where we bought pickles fished from huge wooden tanks. Also up the street from grandmother's house were St. John's Catholic Church attended by Irish-Americans and St. James Catholic Church, attended by German-Americans.

My Grandfather on my mother's side was an Irish-American Sergeant in the traffic division of the Baltimore City Police Department. During his entire life, he never went outside the city limits of Baltimore. Once he almost went to Washington, D.C., but at the last minute, the trip was cancelled. I inherited his red hair, but not his lack of interest in travel. My grandmother was a German-American, who met my grandfather after they arrived separately at the immigration pier at Locust Point in Baltimore, among hundreds of thousands of Irish and German Immigrants in the 19th Century. Baltimore was second only to New York as the point of entry of immigrants coming from Europe at that time.

After two years of high school, my mother went to work as a secretary at the Pennsylvania Railroad. After her marriage to my father, a salesman of wholesale woolens from England, she moved to west Baltimore to live with my father and his mother in a six room row house with white marble steps, a hallmark of Baltimore. In the summer, my mother would put up dark blue "blinds" in the windows, and replace them with white "blinds" in the winter. All along the red brick sidewalks outside our house were attractive gaslights, lit every evening by a lamplighter. Baltimore had been the first city in the U.S. to install gaslights.

Our row house on Fayette Street was a later design than the old house on Aisquith Street. There were windows to let in light for the middle rooms on both of the two floors, three bedrooms upstairs, and a kitchen, dining room and living room downstairs. My sister, Gertrude, now "Trudy," slept with my grandmother in the middle room, my parents in the back room facing the back alley, and we three boys in the front room facing the street. Every night as we lay in bed, we could hear the slow, laborious huffing and puffing of a steam engine a quarter of a mile away as a train tried to get traction to begin to pull a long line of freight cars carrying manufactured goods from Baltimore to the west. We

Figure 19 The Wagner family: Trudy, Herman, Albert, and Henry.

would listen for the wheels of the engine to slip, resulting in a staccato of huffs of steam, a process that would be repeated over and over again until the wheels finally held and the train moved on. That was the signal for us to go to sleep.

The back alley and a small, adjacent street called Fairmount Avenue was our playground, where we often played "step ball," "red line," and "kick the can." We also played "stick ball" with a broomstick as a bat, and a rolled up "snow ball" tray as the ball. Whenever the ice truck came down the street, we would all rush to get a piece of ice, particularly appreciated on the hot days of a Baltimore summer. In a back window of each house there would be a sign indicating to the iceman how big a piece of ice was needed that day to stock the icebox. Usually, we picked up spare chips of ice that happened to be in the wagon, but occasionally the iceman would chip off a few pieces just for us.

During his daily stroll down Fairmount Avenue, the local patrolman, Mr. Black, would always find us on our best behavior. Street "A-Rabs" would pass by leading their horse-drawn wagons, hawking vegetables, and occasionally "hard crabs."

A second encounter with Bon Secours Hospital was at age 14 when, while roller skating, I broke both bones in my left arm when I fell from holding on to the back of a moving truck on Fairmount Avenue. I was trying to impress a girlfriend standing nearby. The fracture was "compound," meaning that it had broken the skin, predisposing the site to infection and osteomyelitis. Fortunately, sulfa drugs had just been developed, and were available at Bon Secours. My cut healed uneventfully. Dr. Frank Marino, a general surgeon, and father of one of my girlfriends, had to operate to set the radius and ulna of my left arm. This was still another experience that increased my desire to become a doctor.

Eventually, in the early 1950s my parents moved from Fayette Street to a new development of row houses in Rodgers Forge, in the northern part of Baltimore, populated by the middle class. The area surrounding St. Martin's parish gradually deteriorated into one of America's best publicized slums.

From the winter of 1992 to the fall of 1993, writer David Simon and Edward Burns, a former police detective, camped out on the corner of Monroe and Fayette Streets, six houses from where I lived from the time I was born until 1951. In their book *The Corner*, published in 1997, they describe the gutted houses, drug dealers and addicts, and the poor folk who live in constant fear of crime and death.

They concluded their description of the menacing neighborhood and its inhabitants on an optimistic noting of the escape of some through the public schools, community colleges, affirmative action, and the Army.

They quote the poet, W.H. Auden, who wrote: "... the first criterion of success in any human activity, the necessary preliminary, whether to scientific discovery or artistic vision, is intensity of attention or, less pompously, love." This was true for me, and I hope it will continue to be true for those who now live in my old neighborhood.

2

So You Want To Be a Doctor

In my 1944 high school yearbook at Calvert Hall, under my picture there is written: "He plans to be a doctor." This was the first indication of a lifetime series of decisions during my climbing the "nuclear medicine" tree. At that time, Calvert Hall was in the heart of downtown Baltimore, across the street from the main branch of the Enoch Pratt Free Library, and one block away from the Remington bookstore on Charles Street, that I would often visit. I still treasure a book that I purchased in 1944: "The Advancing Front of Medicine," by George W. Gray, published in 1941. He quoted Sophocles: "Only against death shall he call for aid in vain; but from baffling maladies he hath devised escapes."

Gray concluded: "We are all indebted to research, and that means indebtedness to a small company of people, numbering only a few thousand, who have committed themselves to this kind of specialized activity." What would he think about today's NIH budget of over $25 billion. He quoted Paracelsus, who foresaw what today we call "molecular medicine." Paracelsus wrote: "The body is a conglomeration of chymical matters; when these are deranged illness results, and naught but chymical medicines may cure the same."

Outside of Calvert Hall was a parking lot, on which cars were parked, and also served as a handball court. The keys were left in these cars so they could be moved whenever a handball game was about to begin. I learned how to drive moving these cars off the court, but had to stop when my actions were discovered and the student council sentenced me to a week in "detention," an hour at the end of each school day. Detention was served in Room 205, next to the office of Brother E. James, the Principal.

During my second year of high school, on December 7, 1941, while having Sunday lunch at our kitchen table in Baltimore, we heard on the radio that the Japanese had attacked Pearl Harbor. My father often engaged in heated political discussions with his friends, listened to Fulton Lewis, Jr. and Father Coughlin on the radio, and couldn't stand Franklin or Eleanor Roosevelt. He believed that President Roosevelt was leading us into war. When we heard the news of the bombing, he told us, sadly: "we are in the war now." The titanic struggle over the next four years would leave 60 million dead and reshape the map of the world. Germany had invaded Poland two years before, and Japan had moved brutally into China and French Indochina. Roosevelt had previously stopped shipments of scrap metal and oil to Japan, and sent 50 destroyers to England under the Lend-Lease Act.

We faithfully kept up with the news of the war in the newspapers and on the radio, and in newsreels at our Saturday afternoon movies. We were thrilled by the news on

April 18, 1942, that a small force of B-25 bombers, led by Lieutenant Colonel Jimmy Doolittle, had bombed Tokyo. This news was a great morale booster. We only learned later that every aircraft but one was lost, far different from what would be broadcast widely today. We went to the movies every Saturday afternoon, regardless of what picture was being shown. We always hoped we would like the picture, hoping at least that it would not be a "love picture." Movie theaters, neighborhoods, schools, and practically everything was segregated. Most often, we went to the Capital theatre, three blocks from our house in southwest Baltimore. Occasionally we went a few blocks further away to the Lord Baltimore and Horn theaters. On special occasions, we took the streetcar to a downtown movie theater. We saw the first full-length cartoon, "Snow White and the Seven Dwarfs" at the Hippodrome. Both the Hippodrome and State Theatres had live vaudeville shows between featured films. Other theaters that we attended were the Auditorium, Stanley, Valencia or Boulevard Theaters. Every week at the Capital theatre, we looked forward to seeing the new episode of serial short features, which we called "chapters." Often, the hero would be shown falling to the bottom of a cliff. The next week we would see him grabbing the top edge of the cliff, escaping his fall.

Once a week at Calvert Hall Country School (CHCS), we had a woodworking lesson in a room filled with carpenter's tools under the direction of "our big brother," Tom Mooney, who lived in a house near to the school, and simultaneously coached all the athletic teams and ran the cafeteria. His brother was a Christian Brother at Calvert Hall. We admired the varsity baseball team from downtown Calvert Hall High School when they came to Walbrook Oval to practice or play games against other city schools. CHCS was located in a large, formerly private house. Every Friday the Brothers awarded gold cards for excellence in schoolwork that week. These were cherished by the recipients, who proudly showed them to their parents. Brother Matthew, who taught the 7th and 8th grade, played the violin at the ceremonies of awarding the gold cards. My parents expected the Wagner boys to get gold cards every week.

After graduating from the 8th grade at CHCS, I went by street car to high school at Calvert Hall College (CHC), as it was called since the time of its founding. The Christian Brothers were excellent and dedicated teachers, participating in sports, such as handball in the school yard, as well as in coaching the athletic teams. We were extremely competitive, and always felt that we had to win, or, at least, to always do our best. We were constantly reminded of the rules of the game by both our parents, and coaches. My mother was President of the Parents' Club at Calvert Hall for several years, and a few of the Brothers became her life-long friends. We were fortunate that, even in the face of the depression, mother did not have to work outside of the house. Until her death when I was 11, Grandmother Wagner's presence in the house and dedication to her grandchildren played a major role in the evolution of our personalites and values.

Long after we had graduated, Calvert Hall moved to its present large modern facility in north Baltimore, with a large gymnasium and football stadium, and parking lots for students in the upper grades. The school now attracts boys of diverse backgrounds, black and white, who must wear coats and ties or athletic jackets when on campus. In the 1940s, we attended school on the corner of Cathedral and Mulberry Streets in downtown Baltimore. The original school building has been replaced by a modern building housing the offices of the Catholic Diocese of Baltimore. Ironically, when the original stone structure of Calvert Hall was torn down, the stones were used to construct a building on the grounds of arch rival, Loyola College, in north Baltimore. When Calvert Hall was still at the downtown site, we walked every afternoon during basketball season to Turnverein Vorwaerts, a large row house built at 723 W. Fayette Street by German-American gymnasts. The building housed a gym, a bowling alley, lockers, and showers. As the home team during basketball season, we had the advantage of knowing that one of the backboards would vibrate for a time after the ball hit it. We delayed the next shot until the vibrations had stopped.

The Chaplain of Calvert Hall was Father Louis Brianceau (Father Brie), a French-American Sulpician priest, who taught at St. Mary's Seminary in Baltimore, the oldest seminary in the United States. There were 350 seminaries in those days, compared to 75 at present. Father Brie was also the Chaplain at the Spring Grove state mental hospital. When Father Brie was at Calvert Hall, he would stand at the school's entrance, greeting every boy by name as he entered the school. He would hear the confessions of the boys who wished to confess to him. Occasionally he would express disbelief that the boy who was confessing had actually committed the sin being confessed. We had to try to persuade him that: "Yes, we did." Sports, rather than girls, were our major interest. One boy confessed to Father Brie that he had committed ten sins, including two mortal and six venial, but none "original."

Every Sunday, he would travel by streetcar from St. Mary's Seminary, where he taught and lived, to Spring Grove. After services at the Hospital, he would be picked up by Gene Cochley, who would drive him to Calvert Hall, where he would celebrate benediction for the Calvert Hall Christian brothers. After Benediction at Calvert Hall on Sunday morning, he and several boys would travel by street car to a tributary of the Chesapeake Bay, called "Spring Gardens," where he kept two boats in a boatyard adjacent to the Hanover Street bridge. One of the boats was a small open boat that he took out alone on Wednesdays, and the other a wooden cabin "cruiser" that he took out with a few invited boys and a hired Captain on Sundays, heading down the Bay past Stony Creek to a landing at the pier in front of Nick Burcher's restaurant on Rock Creek. We would have an ice cream sundae, and then head back up the Bay to the dock by the Hanover Street bridge, secure the boat, and catch a streetcar back to our homes.

We occasionally took cruises of a few days on Father Brie's boat. I still have fond images of the picturesque lighthouses that we passed as we motored down the Chesapeake Bay: Baltimore Light (1908, the last of 100 lighthouses built on 74 sites on the Bay); Seven Foot Knoll (1855, now moved to the Mariners' Museum in Norfolk, Virginia); Fort Carroll (1848); Sandy Point (1882); Thomas Point (1825), and Bloody Point (1882). We had spectacular views of one of the most picturesque bridges in the world: the Chesapeake Bay bridge built in 1952, changing the character of the Eastern Shore. Today, tens of thousands of people cross the Bay to the beaches at Ocean City, Maryland. (We acquired 17 acres of land and built a house in 1968 on the shore of the Chester River, 5 miles above Chestertown, Maryland, which had been settled in 1698 as a major port 35 miles from the entrance of the Chester River into the Bay. Near our house was an old wharf, which from 1840 to 1925, was frequented by steamboats regularly carring freight and passengers from Baltimore to Chestertown and beyond.)

Father Brie told us about an episode, where he, alone and dressed in a typical Frenchman's blue working clothes, fell overboard, and began calling out: "Save me. I'm a Catholic priest." Fortunately, there was another boat nearby, and he was rescued. Once during the war he was stopped by U.S. Coast Guardsmen, who discovered that he did not have the required fire extinguisher on board. The next time Father Brie went to Calvert Hall, he "borrowed" a fire extinguisher from the chapel wall, and headed down to the Hanover Street dock. With John Hartman, who subsequently became a Baltimore physician, Father Brie set out to find the Coast Guard boat in order to show the sailors that he now had a fire extinguisher. Reaching the vicinity of the Coast Guard boat, he asked John to fetch the fire extinguisher. John replied: "I'm sorry, Father, but I left it on the dock." Immediately, Father Brie instructed everyone to lie down in the bottom of the boat so that the Guardsmen would not see them. John said later that he thought this was unwise, because the Guardsmen would be attracted to a boat headed down the Bay with nobody in it. Father Brie subsequently told John that if anything like that happened again, he would no longer be invited back. Often, in those wartime days, the boat would be boarded by Coast Guardsmen with rifles to check us out.

We regularly motored past the Bethlehem Steel shipyards in Baltimore, which were turning out one Victory ship, a 10,000 ton freighter with a gun on the stern, every week. It was the arming of these boats that cut down the enormous losses by torpedos from German U-boats. Seventy-five percent of all the German U-boat crews lost their lives during the war.

Father Brie died on August 23, 1950, at the age of 75. As a Sulpician priest educated in France, he had emigrated to the United States in 1898 to join the faculty at St. Mary's Seminary in Baltimore and taught there until his death.

My first date was taking a neighborhood policeman's daughter to the Calvert Hall senior prom. A week later, I delivered the Valedictorian address at the Calvert Hall high school graduation on June 6, 1944, at the Maryland Casualty auditorium. At that time, Calvert Hall did not have a large enough auditorium for the ceremony. My talk was entitled: "Four Corners, Four Freedoms," referring to the four buildings at the corner of Cathedral and Mulberry Streets: Calvert Hall, the main branch of the Enoch Pratt Free Library, the Baltimore Cathedral, and the Wentworth Apartments. The metaphor was that each represented one of the Four Freedoms.

The title referred to the famous "Four Freedoms" State of the Union speech delivered by President Roosevelt to Congress on January 6, 1941, a year before the attack on Pearl Harbor. In his speech, he said: "We Americans are vitally concerned with your defense of freedom. We are putting forth our energies, our resources and our organizing powers to give you (the British) the strength to regain and maintain a free world. We shall send you in ever-increasing numbers, ships, planes, tanks, and guns. That is our purpose and our pledge.

"We look forward to a world founded upon four essential human freedoms. The first is freedom of speech . . . The second is freedom of every person to worship God in his own way . . . The third is freedom from want . . . The fourth is freedom from fear."

The month we graduated from high school, on June 6, 1944, we learned that the Allied forces had landed on the beaches of Normandy in France, on the bloody road to victory. Nearly 10,000 soldiers died on the beaches, over 1,500 on Omaha beach in the bloodiest battle of the Normandy invasion.

The Germans surrendered on May 2, 1945, while I was a freshman at Johns Hopkins University, founded in 1876, the centennial year of our country. Shortly after the University was founded, the Johns Hopkins Hospital was created, and was to become a model for modern medical centers, changing medical education from proprietary training to a scientifically-based, university enterprise. The medical school emphasized research for both students and faculty, and was the first to introduce the residency system in the United States.

I had a scholarship to Johns Hopkins, having been selected by U.S. Congressman George Fallon, in whose district we lived, and who was known personally by my father. In July, a month after graduation from high school in 1944, I began my studies at Hopkins which at the time was operating what was called an "accelerated program." This meant that students could sign up for any courses they wished to take, the only requirement being prior approval by one's advisor. My advisor was a well-known geneticist, Carl Swanson, who never vetoed any of my choices. Unknown to me at the time, I finished two years work in my first year. I thought at the time: "College is much harder than I had expected." The study habits that this crowded schedule required made my subsequent transition to medical school much easier.

Professor Carl Bruning, who taught freshman chemistry at Hopkins, surprised us by giving us a written examination on the first day of classes. I don't know what he had in mind. Calvert Hall had prepared me well for college. My teacher of German at Calvert Hall, Brother Gideon Francis, who was also the basketball coach, was such an effective teacher that the Professor teaching me German at Hopkins asked me why I was ashamed to admit that my parents spoke German at home. They didn't. My grandmother never spoke German, made any references to Germany, or told us why she had left Germany.

At Hopkins, I played varsity football and basketball, only because so many competent young boys were away in the military service. Our football team played only one game, with the opponent being members of the ASTP (Army Student Training Program) at Hopkins. The final score was 0–0. In basketball, we played several "big name" teams, such as Haverford (9 letters in the name) and Hampton Sydney (13 letters in the name).

After the war ended, I was subject to the draft. Because of my experience with sailing on the Chesapeake, I had developed a love of the water, and so took the competitive examination to enter the U.S. Coast Guard Academy. I was accepted and, at the age of 18, left by train from Pennsylvania Station in Baltimore for New London, Connecticut. I was sworn in as a cadet with William Brandfass and Nicholas Ivanosky. "Whitey" Brandfass subsequently became an orthopedic surgeon, specializing in sports injuries. Nick became a helicopter pilot after graduation and retired as an Admiral in the Coast Guard after 20 years service. One of the members of our class of 1949, the year our class graduated, rose to become Commandant of the Coast Guard. I enjoyed my time at the Academy, especially the cruises on the square rigger, the *Danmark*; the ice breaker, *Mackinac*; and the cutter, *Sebago*.

On August 6, 1945, we learned of the atomic bombing of Japan. The *Enola Gay*, a B-29 bomber named after the mother of the pilot, Paul Tibbets, who had taken off from the Pacific island of Tinian, headed for the city of Hiroshima. The bomb, named Little Boy, exploded over the city with a force equal to 20,000 tons of TNT. 100,000 persons died. The world would never be quite the same. On August 9, 1945, a second atomic bomb was dropped on Nagasaki, killing 80,000 more Japanese. On August 15, the Japanese surrendered. Kyoto had been considered as a possible target, but was taken off the list because of its cultural and historic importance.

Figure 20 Varsity basketball team at Calvert Hall in 1944. HNW on left in second row.

At the completion of my first year and second summer at the Academy, I decided to resign for several reasons: (1) Everyone wore his status on his sleeve, and we were expected to defer to those of higher rank, and look down on those of lower rank; (2) After three days of intensive psychological testing of all the cadets, in my interview with a consulting psychologist, I was told that the testing showed that I would make a superb gunnery officer because of my ability to visualize objects with an accurate three-dimensional perspective. (Perhaps this was a predictor of my subsequent activities in imaging in medicine.) I didn't look forward to the life of a gunnery officer; and (3) I still had a latent desire to become a doctor. So I returned to Johns Hopkins University, and joined the Air Force Reserve Officers Training Program (ROTC), graduating as a 2nd Lieutenant in the Air Force Reserve.

Among the Hopkins students in both undergraduate (males only at that time) and medical school were many veterans returning from World War II, studying under the GI Bill of Rights. They were serious students, eager to make up for lost time. The Beta Theta Pi fraternity became the focus of my social life. We went to dances at Hopkins and in various Baltimore hotels, including the Emerson with its Crystal Ballroom, the Belvedere with its Terrace, John Eager Howard Room, the Owl Bar, and two ballrooms on the top floor. Today, there is now a dancing school at the Belvedere with dinner dances every Wednesday night.

A sign of the times in the old days was when Captain Isaac Emerson, founder of the Bromo-Seltzer Company, was refused permission to take off his jacket on a hot day at the Belvedere Hotel. He got revenge by building the Emerson Hotel. Unescorted women could not be served in the bars of these Baltimore Hotels after 7:00 PM "lest chance acquaintances be formed." The striking Bromo-Seltzer Tower had a 37 foot replica of a blue Bromo-Seltzer bottle on its roof. We spent many evenings at the Peabody Book Store on Charles Street, which had tables in a back room and served beer and sandwiches. We "Betas" had a fraternity meeting serving beer on Thursday night. At times, the revery after the meetings would result in my missing some of my Friday classes. We had a party every Saturday night, and occasionally played golf at Bonnie View golf course. Our ambition was "to be able to live the way we were living."

3

First Taste of Research

Curt Richter, Professor of Physiological Psychiatry at Johns Hopkins School of Medicine, had the greatest influence on my scientific life, which began during my senior year in college at Johns Hopkins. I worked with him during the summers between my first three years as a medical student at Hopkins.

Dr. Richter, who was born in 1894, had been recruited in 1919 to Johns Hopkins from Harvard by Adolph Meyer, the first Professor of Psychiatry at Hopkins. He initially worked in the laboratory of Dr. John Watson and took over the lab when Dr. Watson left. Dr. Richter was famous for his work on self-selection of diets by rats, and the relationship between physiology and behavior.

I first learned of Dr. Richter's work when I worked as an attendant in the Henry Phipps Psychiatric Clinic at Hopkins during the summer between my third and fourth year of college. His labs were located on the third floor of Phipps, two floors above West 1, the floor that housed the most severely ill patients. He had limitless curiosity and enthusiasm, and respected the opinions of students working with him, making us all feel important. The metaphor for his professional life was walking along a shoreline, picking up interesting shells that he came across.

His most striking characteristic was his curiosity. He constantly asked questions, a habit persisting throughout his life. He never took a superior attitude in his relationships with colleagues, students, or technologists. We worked together carrying out experiments on rats, monkeys, and human beings. We developed instruments to measure motor activity in rats, the galvanic skin response in monkeys, and instruments to map the patterns of sweat gland activity in patients with a variety of diseases, including a congenital absence of sweat glands. He made it possible for us to go to scientific meetings, such as the Endocrine Society, present our scientific results and get comments on our work, even though we were medical students. There were no computers, only adding machines and machines to creat graphs of physiological data. We learned how to handle wild rats captured on the streets of Baltimore, and how to operate on the brains of wild and domestic rats, viewing the operative sites through a microscope. We became expert at removing the adrenals and other endocrine organs. Dr. Richter never characterized any of our ideas as "bad." Dr. Richter's model was Claude Bernard, the Father of Physiology, who became my model as well. His work continued to have an enormous influence on my scientific career.

At the 25th anniversary of the graduation of our Hopkins medical school class of 1952, the planning committee decided to invite to our celebration some of the faculty during our student days. At the first meeting of the committee, we decided immediately "O.K, we'll invite Curt Richter and Horsley Gantt. Now let's decide on some others." This unanimous opinion was remarkable, because neither Richter nor Gantt had ever taught us any formal courses. Their lasting impact had all been extra-curricular, which illustrates what it was like at Hopkins medical school during the 1950s.

Dr. Richter's enthusiasm and love for research made a great impact on me. He let me carry out my own experiments, published in my first three scientific papers. During my freshman year of medical school, I presented a paper at the meeting of the Endocrine Society in Atlantic City. The presentation was scheduled on a Friday, the day I was to take the final examination in Psychiatry. I went to Dr. John Whitehorn, Chairman of the Department, and said: "Dr. Whitehorn, I have a problem." He immediately offered me a chair and a cigarette, and said: "What is the problem?" I fell sure that he expected to hear of some psychiatric symptoms. When I told him my only difficulty was scheduling, he responded: "No problem. I'll just leave the examination on my desk, and you can take it before you leave." When I returned the next morning, the exam was nowhere to be found, but fortunately, Dr. Whitehorn arrived at his office, and helped me search his desk for the exam which he had left. We never found it, but again, he said: "No problem. I'll just give you an oral exam." I passed the exam.

Research was greatly encouraged at Hopkins, and the faculty was always willing to encourage students to participate. An episode that illustrates this point occurred during an oral examination that I was being given in the library by the Chairman of the Obstetrics and Gynecology Department. While Professor Eastman was questioning me about a problem patient, another student wandered into the library, and joined in our conversation. He answered some of the questions before I could. Neither the Professor nor I told him we were having an examination. After a short while, the student left the library, never realizing that he was interfering with an oral examination.

In those days before the National Institutes of Health had been founded, only 16 Hopkins medical students in the whole school carried out any research. Without any influence or participation of the faculty, we students met monthly in what we called the "Osler Society" to discuss our research results among ourselves. My chief interest (and Dr. Richter's) was in endocrinology and its influence on mental functioning and behavior. He had discovered self-selection of food intake in rats, and the effects of electric shock stress on the endocrine system of wild and domesticated rats. He had a contract with the City of Baltimore Health Department to develop an effective rat poison during the war. Among his discoveries was that rats could not taste the poison, alpha-naphthyl thiourea (ANTU), which produced acute pulmonary edema and death in rats. His findings led to its wide use as a rat poison. His observation that ANTU produced hyperplasia of the thyroid gland led to the idea and experiments that eventually resulted in the use of thiourea drugs to treat hyperthyroidism.

While working in Richter's lab, I was awarded a Henry Strong Denison scholarship because of my research activities. I had been the sole author of my first published scientific paper which appeared in the *Archives of Ophthalmology* in 1950, entitled: "Objective Testing of Vision with the use of the Galvanic Skin Response." Electrodes attached to the palms of human beings or monkeys measured the bursts of decreased electrical skin resistance that occurred with activation of the sympathetic nervous system. I administered an electric shock whenever a specific letter was projected on a screen before the subject (or monkey) but not with the projection of other letters. The galvanic skin response (the GSR, which today is part of lie detector testing) was the indication of whether the specific, conditioned letter had been seen. When the paper was submitted, I acknowledged the assistance of Anne Collins Barrett, even though she had not participated directly in the experiments. I had seen acknowledgements at the end of scientific papers that I read, and, after all, Anne had been helpful to me in general.

When we met, Anne was a student in the Bachelor of Science in medical technology program at Mount St. Agnes College. On March 11, 1948, Anne and I were attending an afternoon dance sponsored by the Newman Club, a club for Catholic students, in Levering Hall, the student activities building at Hopkins. Hopkins students often dated girls from Goucher College, Maryland College for Women, and Mount St. Agnes, all of which admitted only women in those days. Mt. St. Agnes was founded in 1867, and closed its doors in the late 1950s, transferring its academic programs to the Loyola College. The Mt. St. Agnes property was recently acquired by the Johns Hopkins University, after prior ownership by two insurance companies.

Anne, a lovely, cheerful brunette, had a smile that "looked as if all heaven had opened up." She has been responsible for all the pleasures of my life, our four wonderful children and nine grandchildren. Beyond question, Anne is the "best thing that ever happened to me." The first weekend after we met, we had dinner at the house of a fraternity brother who turned out to be Anne's third cousin. We went to see the opera *Carmen* at the Lyric Opera House, and finished the evening at a party at the Beta fraternity house. Her mother was annoyed when we arrived back at Anne's house at 5:00 AM after our first date. After 54 years of marriage, we still watch television in the living room of that same house. On our first date, as a colleague and I watched Anne descend the steps into the living room, she tripped and fell, to our uproarious laughter, she was unhurt, and we went off to an unforgettable evening.

We dated frequently throughout my senior year in college, and during my first three years at Hopkins Medical School. Upon graduation from college, Anne went to work with Dr. Paul Carliner, an internist who became famous for his discovery with Dr. Leslie Gay that the anti-histaminic drug, dramamine, was effective in preventing seasickness. This observation had extremely important military consequences. The first clinical trial of the drug to prevent seasickness was carried out on a troopship. Paul remained in private practice in his home office a quarter of a mile from our present house in Mt. Washington until his untimely death at age 46.

Anne and I occasionally met at St. Alphonsus Church in downtown Baltimore after her work and my classes, and spent a few hours in a local restaurant. During the obstetrical quarter in my third year at Hopkins, I told my father that Anne and I were going to be married the next weekend. He was shocked, convinced that getting married would keep me from becoming a doctor.

We were married in the Shrine of the Sacred Heart in Baltimore, on February 3, 1951. After a reception at the Johns Hopkins Club, our honeymoon consisted of a weekend at the New Yorker Hotel in New York City. Going up to our room, we were the only passengers in the elevator with the operator, a pleasant young man. Feeling somewhat awkward, I said to him: "It certainly is cold outside." He replied: "What do you care? You have your love to keep you warm." Only in New York! We lived for three months in Mount Washington with Anne's parents, and then moved to a second floor apartment in a row house owned by Johns Hopkins at the corner of Wolfe and Monument Streets, across the street from the Women's Clinic.

Once in our marriage, Anne left me and went home to her parents. Late one night, after I came home from the hospital, and was taking a shower, a large rat ran across the floor of our bedroom. The next morning, I called the hospital administrator, Colin Churchhill, who immediately sent over an exterminator. After a short period, the exterminator told us: "There is no problem. I have fixed it so he can't possibly get out." Anne immediately took the children to her partents' home for three weeks as we waited for the rat to die and disintegrate.

Ever since those days at Mt. St. Agnes, Anne still has cocktails and dinner once a month with the women who were with her at the dance where we met. I tell people that I married Anne and kept the other women as "controls." One of her classmates subsequently became a nun; most of the others are now widows.

At 1900 McElderry St., we were extremely happy, even though extremely poor. The office of the Dean of the School of Public Health, Dr. Alfred Sommer, is where our bedroom was in the front room of the row house that was torn down not many years after we had moved out.

While living on McElderry St. we had dinner every Saturday night with Anne's parents in Mt. Washington, and Sunday afternoon dinner with my parents. Every Saturday afternoon, Anne would do our laundry in her parents washing machine, while I was working as Teen Age Director at the Schenley Road Community Center in Roland Park. I worked from 12:00 to 4:00 PM and again from 8:00 PM to midnight. I was paid $15.00 for the 8 hours, which supplemented the $150.00/month borrowed from my father. Because I was working on Saturday afternoon, I missed all my pediatrics classes.

Hopkins rented our house on McElderry Street to Dr. Frank Williams, an Assistant Resident in Medicine and is wife. Frank was an assistant resident in medicine, and subsequently became the Director of the National Institute of Aging of the NIH. We paid Frank $32.00 per month for the second floor of the house. Once I was called by Colin Churchill, a hospital administrator, who asked why we should be allowed to sublet from Frank, rather than pay the rent directly to Hopkins. I countered his question by asking: "Who would be responsible for stoking the coal furnace in the basement?" He said no more.

We ate every evening in the Doctors' Dining Room. The doctors ate huge meals, which were free, while our wives and children had only drinks. Occasionally, some of us received letters from the Administration telling us that we were eating too much. Anne and our two children's clothes came from the "Grapevine," a consignment shop that Anne's mother ran on Roland Avenue. Most of the time, I wore the white jacket and pants of a house officer, first an intern and then assistant resident and chief resident. When we picked up the pants from the hospital laundry room, they were so heavily starched that it was impossible to get one's legs in them without first fighting your way with your fist from the waist to the ankles.

In the summer between my third and fourth year of medical school, I was not able to continue to work in Curt Richter's Laboratory because we were short of money, even though Anne still worked for Dr. Carliner. I learned that the pharmaceutical company, Charles Pfizer & Co., was going to hire medical students to act as detail men during the summer, visiting doctors to promote the use of the world's first broad spectrum antibiotic, Terramycin. Our group of 10 medical students from all over the country went to New York for orientation, before traveling over the country, speaking with physicians in their offices. It was a brilliant idea on the part of Pfizer, because nearly every physician was willing to see the "medical student in the waiting room." I still remember one lesson from our leader, the company representative: "If you were standing on a corner, handing out five dollar gold pieces, you would still have to persuade people to take them." He also taught us: "The sale begins when the customer says no." We were paid $400 per month plus expenses. Every week a bonus was awarded to the student who had managed to see the greatest number of doctors. We students decided to divide the bonus equally among ourselves, thereby eliminating the need to compete.

A major additional benefit of working for Pfizer was that, during my fourth year of medical school, I obtained a contract from Pfizer to carry out collaborative research to determine whether Terramycin was effective when administered by the intramuscular rather than the oral or intravenous route. I decided to determine its effectiveness in treating urinary tract infections. In those days, the only approval that was needed was that of the Chief Resident in Urology at Hopkins and that of Dr. George Mirick, Head of the Infectious Disease Division. There was no Institutional Review Board (IRB) in those days. This work not only provided financial aid, but led to publication of the results in the July 1952 issue of the *Bulletin of the Johns Hopkins Hospital* in an article entitled: "Experimental and Clinical Results with Intramuscular Terramycin."

I continued part time research work with Pfizer for the next three years, which was a financial God-send. One study was on the effectiveness of Terramycin in the treatment of experimental cholecystitis in dogs. The attitude of the faculty during my presentation of this work taught me that research was judged on its own merits, not on the academic position, or lack thereof, of the person performing the research. Increasingly, I became determined that I would go into academic medicine. In those days, internal medicine was the premier medical specialty, and most of the upper third of the graduating class entered this field. This is no longer the case, as medicine has become increasingly specialized.

Infectious diseases and endocrinology were the focus of attention in those days. I joined Dr. A.M. Fisher, who suffered from Huntington disease, and Richard Ross, Chief Resident in medicine, and subsequently, head of Cardiology and Dean of Johns Hopkins Medical School, to study of the clinical manifestations and effectiveness of antibiotic treatment of patients with staphylococcal endocarditis. This disease was prevalent then because of the occurrence of rheumatic heart disease. It is rarely seen now. HIV/AIDS did not exist, or at least was not recognized at that time.

During my last two years of medical school, we were saddened every day by news from the Korean War. There were great numbers of casualties during the advances and retreats of the United Nations forces, led by the 8th army of the United States. General Douglass MacArthur was a hero of the general public but distained by many politicians. When he returned home after being fired by President Truman in 1951, he was given a ticker tape parade in New York, and delivered a moving farewell address to the U.S. Congress. We only later learned that he had advocated the use of nuclear weapons during the war when the United Nations forces were in full retreat from North Korea. Paradoxically, it was the first demonstration of a nuclear weapon fired as an artillery shell in 1953 that convinced the North Koreans that they should seek a truce. By the time President Eisenhower went to Korea two weeks after he became president, an armistice was signed. Unfortunately, over 157,000 American casualties had been sustained by then, with 400,000 casualties among the other United Nations and South Korean forces. Although the exact number remains unknown, the number of North Korean casualties was in the hundreds of thousands.

Except for momentous news related to the Korean War, we on the Osler house staff had little contact with the world outside of the hospital. I was hospitalized for minor surgery during my second year of assistant residency and had the opportunity to watch the 1954 hearings in Congress in which Senator Joseph McCarthy defended a close friend of Roy Cohn of McCarthy's staff. Joseph Welch was the lawyer for the Army who, during an attack by the Senator on a young member of Welch's staff, named Fisher, made the famous remarks: "Let us not assassinate this lad further, Senator. Have you no sense of decency, sir, at long last? Have you left no sense of decency?" This hearing marked the end of the McCarthy era, marked by a national hysteria, the response to an unholy alliance between the American Communist Party and Soviet spies in the United States during the 1930s and 1940s, which ended in the early 1950s, after some of them had stolen the secrets of how to design an atomic bomb.

My second scientific publication was six months after my first. James W. Woods, a friend and co-worker who was a student working with Drs. Phillip Bard and Vernon Mountcastle in the Department of Physiology, and I published a paper in the *Archives of Neurology and Psychiatry*, entitled: "Interruption of Bulbocapnine Catalepsy in Rats by Environmental Stress."

Jim slept in the Physiology Department, setting an alarm clock to wake him up every hour during the night to adjust a heating lamp placed over experimental cats or monkeys who had had parts of their brain removed that interfered with their ability to maintain the proper body temperature. Jim was eventually replaced by a thermocouple, and had to resort to driving an oil tanker truck from Baltimore to Washington and back twice a night to get sufficient income to support his life as a graduate student.

The drug, bulbocapnine, was of great interest, because it could produce immobility that resembled patients with catatonic schizophrenia. We discovered that immobility induced by the drug could be interrupted by placing the rats in a pool of water where they would have to swim. Even when immobilized by bulbocapnine, they would begin to fight with each other in response to electrical shocks administered through bars in the bottom of their cages. This showed the difference between the effects of bulbocapnine and anesthetic drugs. We also found that the catalepsy could be interrupted by having the cataleptic rats breath high concentrations of carbon dioxide. The same effect was observed with the administration of scopolamine, a drug which blocks the parasympathetic nervous system, or cocaine. Audiogenic seizures, that could be produced in rats by loud noise, was not abolished by bulbocapnine. Thus, from the beginning, my research has been oriented toward the study of the biochemistry of mental functioning, an interest created by working with Curt Richter in Phipps Clinic. How wonderful it would have been in those early studies of wild and domesticated rats, if we had had techniques, such as positron emission tomography, to relate brain chemistry, mental activity and behavior. These new "tracer" techniques for "molecular imaging" would not be developed until decades later.

We mapped out the patterns of sweat gland activity in human beings heated in a heated enclosure. The third of my first three papers was the study of a mother and daughter with "hereditary ectodermal dysplasia of the anhidrotic type." These patients were intolerant of heat, because of the absence of sweat glands. We observed that the patients did manifest axillary and facial sweating, as well as on the palms and soles of their feet. I wanted to find out whether they had a congenital absence of epocrine, but not apocrine sweat glands, such as those in the axillae. After we showed that their axillae would indeed sweat, I wanted to get a biopsy specimen to confirm the hypothesis that this was due to functioning apocrine sweat glands. I went to the Chief Resident in surgery and asked if it would be possible to obtain a skin biopsy. Of course, we would obtain informed consent.

The Chief Resident asked: "How many biopsy specimens do you want?" Somewhat surprised at the question, I responded: "How many can I get?" He replied: "The knife is sharp." I said: "One will be enough." I published the results in the *Archives of Dermatology and Syphilology* in 1952, the year I graduated from medical school, and began my internship in internal medicine on the Osler Service at Hopkins Hospital.

Throughtout my entire professional life, I have been interested in the autonomic nervous system, which controls sweat glands as well as other autonomic functions. This interest was also stimulated subsequently by the study of three patients with orthostatic hypotension (their blood pressure fell dramatically whenever they stood up). We found that they could not regulate their blood pressure when given vasoactive drugs that either raised or lowered their blood pressure. They were able to excrete enormous amounts of urine when their extracellular fluid was expanded by administration of normal saline solution. Studies of these patients resulted in my receiving the Francis F. Schwentker award given to a member of the Johns Hopkins House staff for excellence in research.

Most diseases are characterized by a deficiency in one or more physiologic or biochemical processes. At other times, disease can be the result of normal processes being increased. For example, persons can exhibit deficient or excessive sweating. Hyper- and hypo-thyroidism are classic examples of an increased or decreased physiologic process, respectively. Patients with orthostatic hypotension have a generalized decrease in sympathetic nervous system activity affecting both their blood vessels and sweat glands. Persons with tumors of the adrenal glands manifest abnormally high secretion of nor-adrenaline. Hyperactivity of the sympathetic nervous system can result in coronary artery disease or essential hypertension. Diseases of the autonomic nervous system are now considered "molecular" as well as "physiological" diseases.

4

Medical School and House Staff Days

After my work with Curt Richter, starting even before I entered medical school, I decided to pursue a career in clinical research in an academic institution. After courses in anatomy, pathology, and microbiology, I had my first contact with patients in a course called "Physical Diagnosis," or officially, "Introduction to Clinical Medicine." For the first time we were able to speak to and examine living patients. We defined diagnosis as the identification of a disease by investigation of its signs and symptoms. Today I refer to these and other data about the patient as "manifestations" of disease. We took courses in clinical diagnosis, laboratory diagnosis, physical diagnosis, anatomical diagnosis, bacteriological diagnosis, x-ray diagnosis, and electrocardiographic diagnosis. All of these courses showed us how to characterize a disease. We used the term "identify," rather than "characterize" a disease, because we were then in an "ontological" period of medicine where diseases were thought of as "things" to be detected and, if possible, eliminated.

Bedside teaching and the "clinical pathological conference (CPC)" were the two principal teaching methods. The most admired physicians were those who could predict during the patient's life what would be found at autopsy. This was called "the case method of teaching medicine," a term introduced at Harvard by the distinguished physiologist, Walter B. Cannon. In those days, after daily "rounds" of patients with the "attending physician," we would go to the autopsy room to see what was found in the autopsy of patients who, before death, had been on the internal medicine wards which were divided into "white males (Osler 6), colored males (Osler 2), white women (Osler 3), and colored women (Osler 4)". Private patients were on Osler 5.

As second year students, we attended the CPCs every Wednesday at noon, and after we heard the discussion by Professor Harvey of the clinical and laboratory findings of a patient who had died and been autopsied, we awaited with bated breath, the report of the autopsy findings by Dr. Arnold Rich, Chairman of the Pathology Department. Professor Harvey based his diagnosis on the clinical findings prior to death. Because his diagnosis was usually correct, Professor Harvey was universally admired and respected by the medical students. He displayed great intelligence, had enormous experience and revealed a wry sense of humor. The Clinical Pathological Conferences were viewed as contests between Drs. Harvey and Rich, and were a high point of the grueling weeks during our second year.

Dr. Rich was the first Jew to become chairman of a department at Johns Hopkins Medical School. His daughter, Adrienne, one of the most renowned poets in the world, challenged her father during her first year at Harvard University: "Why haven't you told me that I am Jewish? Why do you never talk about being a Jew?" He answered: "You know that I have never denied that I am a Jew. But it's not important to me. I am a scientist . . . I have no use for organized religion. I choose to live in a world of many kinds of people. There are Jews I admire and others whom I despise. I am a person, not simply a Jew."

Adrienne was told that in the late 1940s, anti-Semitism was not in peoples' consciousness in Baltimore; racism made much more of an impression. Her associates would tell her: "I would almost have to think that blacks went to a different heaven than whites, because the bodies were kept in a separate morgue (this was true at Hopkins), and some white persons did not even want blood transfusions from black donors." (From *Rereading America*, by Gary Colombo et al., 1992, Library of Congress 90-71613.)

The autopsy no longer plays the important role that it played in medical education in my medical school days. There is an unwarranted belief that we nearly always know why a patient died even without an autopsy. With the invention and development of three-dimensional imaging procedures, CT, MRI, SPECT, and PET, "bedside" skills have also diminished. In 1998, it was reported that in 25 to 40% of the cases in which an autopsy is performed, it reveals an undiagnosed cause of disease. Yet the percentage of patients who were autopsied has fallen from 50% in the early days to less than 10% by 1991.

During my last two years of medical school and during house staff training, considerable attention was paid to mastering the art of "taking a medical history." In the course on "physical diagnosis," we learned how to listen to the patient's heart and lungs with a stethoscope, palpate the abdomen, look at the "eye grounds" with an ophthalmoscope, and then examine the blood and urine. Excellence at observation, palpation and auscultation were the coin of the realm. We were taught that the two most effective computers in making the diagnosis were the brains of the patient and examining physician. Doctor Phil Tumulty, a master at history taking and physical diagnosis, was one of our role models. He would often spend an hour in speaking to and examining each patient. He also taught us interpersonal and ethical qualities, as well as clinical skills. I remember his telling a patient that "we have a test that will not tell us what is wrong with you, but will tell us whether or not you are seriously ill." I waited with baited breadth to learn what this test was. After an appropriate pause, he continued: "The test is called the erythrocyte sedimentation rate." I had never before heard this test praised so highly.

Three times a week, we would "present" selected patients to Professor Harvey, Chairman of the Department of Medicine and the youngest physician (34 years old) ever to occupy the Chairmanship of a major Department of Medicine in a medical school in the United States ups to that time. He had carried out research even while serving with the Johns Hopkins Unit in the south Pacific during World War II. His picture appeared on the cover of *Time* magazine. His philosophy as Chairman was: "At Hopkins, we choose good people, and leave them alone." On rounds of the patient wards, by describing the illness of a specific patient in all of its details, he was able to cover all aspects of a specific disease for his residents and medical students.

Once, as Chief Resident on the Osler Medical Service, I was walking down a hospital corridor with Professor Harvey, and I woman stopped us, saying: "You won't remember me, Dr. Harvey, but I was nearly dead, and none of the doctors knew what was wrong with me until you came around with some medical students, and told them what was wrong with me. I began to get better right away." Professor Harvey asked: "What did I tell them was wrong with you?" She responded: "You told them I was moribund."

In our graduating class in 1952, there were 75 students. Two of the four women in the class subsequently went on to become the first women Chairmen of the Departments of Anatomy (Betty Hay) and Pediatrics (Mel Avery) at Harvard Medical School. We didn't use the term "Chairperson" in those days. One of the other women, Dr. Pat McIntyre, graduated first in our class in medical school, and subsequently became an extremely valuable colleague in nuclear medicine at Hopkins.

We were always conscious of the patient as a person, but were also scientifically oriented. Once, an attending physician told the intern on medical rounds that a specific patient with heart failure should be considered as a person, not just a patient. The intern responded: "I just want to know whether the patient is taking too much or not enough digitalis." Another anecdote concerned a young intern who was speaking soothing words at the bedside of a patient who was gasping for breath. The patient wrote something on a piece of paper as he lay breathless in his oxygen tent. After the patient died, the intern read what was written on the paper. The patient had written: "You are standing on my oxygen tube."

Today there is diminishing interest in the role of physical examination. The time spent with a specific patient has also decreased for economic reasons. The failure of insurance companies to appreciate the enormous value of personal contact with the patient has also played a role. We can only hope that the interest in autopsies will be revived. All will benefit, patients as well as doctors.

What we call progress is not always viewed as such by everyone. An example of the "old days" took place during my third year of medical school in the Obstetrics Quarter. Background music, called "Muzak," was played for the first time in the labor room. After the delivery, all of us gathered around the new mother's bed, and asked, expectantly: "What did you think of the music?" Her response was: "It didn't bother me none."

In July 1953, I started my internship in internal medicine at Hopkins. Each intern cared for 14 patients at a time, supervised by an Assistant Resident, and the Chief Resident who was responsible for all the patients on the entire medical service, which consisted of four wards, and rooms for private patients. A Senior Attending Physician visited the wards with the Chief Resident three mornings a week. The Chief Resident also visited each ward every evening after supper. Three mornings a week Professor Harvey joined the Senior Attending Physician, spending two hours on each ward in sequence. We all wore "whites"—the familiar short coat and trousers with a stethoscope in the side pocket of the white coat. Today it is customary to drape the stethoscope around the neck, probably because of the large numbers of women house officers.

We were responsible for the care of all of our patients (14 per intern; 28 per assistant resident; hundreds for the Chief Resident) all of the time. We had no scheduled time off. Today, house officers take care of many more patients and have scheduled time off. Efforts have been made to limit the working week to less than 80 hours per week, to prevent fatigue from interfering with patient care.

I carried out a survey, asking the interns, assistant residents, and chief resident independently who had the ultimate responsibility for the patients. Each of the three groups responded: "Me." During an assignment of house staff in the Emergency Room, one of their most important decisions was whether or not the patient should be admitted to the hospital. An assistant resident served as the "Admitting Officer." When the patients came into the Emergency Department, clerks would write down the patient's chief complaint, many of which began with the statement: "Patient's friend states . . . ," for example: "Patient's friend states that something is dripping from his penis." My internship, followed by two years as an Assistant Resident in internal medicine at Hopkins, was the greatest learning experience of my life. I learned how to recognize when a person was suffering from a serious disease, whether it was an infection, diabetic acidosis, heart failure, renal failure, or stroke. We were responsible for each patient from the time of admission until their discharge. This led to a great affection and strong sense of responsibility for each patient as an individual. One did not feel that "when 5 o'clock comes around, the patient will be someone else's responsibility."

In addition to the nightly visits to each ward by the Chief Resident in Medicine, the Chief Residents in Surgery also came to the medical wards to see if there were any patients who required surgery. An example of the extensive experience that these residents obtained on the surgical service at Hopkins is the fact that, Dr. Jerry Kay, during his year of Surgical Residency, corrected coarctations of the aorta in six patients. On one Christmas Day, Dr. Frank Spencer, Chief Resident in Surgery, carried out major surgery on four of my patients when I was an Assistant Resident.

Everyone on the house staff was in the same boat. There was great esprit de corps and group social activities for our wives and children. The house staff were in the Hospital from 14 to 16 hours per day. There was no apparent difference in lifestyle, regardless of economic status of the individual members of the house staff, because everyone was living in the same type of housing in the immediate vicinity of the hospital. The wife of Professor Harvey, Elizabeth Harvey, together with other wives of the Professors, established a nursery school in the living area for the children of the house staff. One of the medical student's wives, Jane Anne Norton, led in the establishment of a "Staff and Student Wives Association." Without financial backing, they organized a formal dance at the Belvedere Hotel, the finest in the city. Fortunately, the event was sold out, even oversold, because of the strong support of faculty wives, led by Elizabeth Harvey. Senior faculty, house staff, and students all had a wonderful time.

In those days, we had "no scheduled time off." We were expected to take care of our 14 patients from the time they were admitted until they were discharged (or died). We could "sign out" to co-workers if there was no acute problem to be solved with "our patients." This continuous responsibility of the same persons—intern, assistant resident and resident—around the clock was of great benefit to the patients.

We had two children during the three years when we lived at 1900 McElderry Street, across the street from the Women's Clinic of Hopkins Hospital. Nick, the older child, went to the Nursery School supported by the Women's Board of Hopkins. Gentleness, love, tolerance for everyone and the desire to get ahead were the principal values of the house staff and their families.

After two years as an assistant resident at Hopkins, I joined the National Institutes of Health in Bethesda, Maryland, as a Clinical Associate in the Laboratory of Kidney and Electrolyte Metabolism of the National Heart Institute. After two years at the NIH, I spent a year at Hammersmith Hospital in London.

After we moved from living near Hopkins Hospital to 3410 Guilford Terrace, during vacations we took automobile trips to the west coast and Canada in a Volkswagen Microbus. On these camping trips, which covered over seven thousand miles, we took along our weekly helper and best friend, Nellie Moody. Anne and I slept in one tent; Nellie and

Figure 21 Wagner family during the year at Hammersmith in 1957.

the children in a larger tent. Our children today tell us that these were the best vacations they ever had. They remember them in great detail.

Ten years after I joined the Hopkins faculty, we purchased 17 acres of land on the Chester River, and built a "Deck" house, where we went every weekend. The "house in the country" provided a focal point keeping all of us close together. Two of our children were married there. The "house in the country" was also a focal point for lab parties and informal meetings of leaders in nuclear medicine. The requirements for certification by the American Board of Nuclear Medicine were written at the dining room table of this house in 1970.

During my years as an Assistant Resident, because of my interest in infectious diseases, I established weekly infectious disease rounds, rotating among Hopkins, the University of Maryland Hospital, and Baltimore City Hospitals. I was very eager to learn from Dr. Ted Woodward, Chairman of the Department of Medicine at the University of Maryland. He was famous for his work in infectious diseases, including typhoid fever, Q fever, and dengue. He was Chief Medical Officer during the Fifth Army's movement up the boot of Italy during World War II, in the army led by General Mark Clark. Subsequently, I participated in several collaborative studies with people in Dr. Woodward's department at the University of Maryland, focusing on infectious diseases at the Maryland House of Correction, a maximum-security prison, where prisoners would volunteer in studies to test the effectiveness of vaccines. Such studies would never be permitted today.

During my assistant residency, I carried out a study indicating that hydrocortisone therapy did not have a beneficial effect in the penicillin treatment of pneumococcal pneumonia, a common disease at that time. Between October 1, 1954, and May 31, 1955,

125 patients were admitted with subsequently proven pneumococcal pneumonia to Johns Hopkins and Baltimore City Hospitals. We commented on the dramatic effect that penicillin had in the treatment of this formerly often fatal disease.

In George W. Gray's book, *The Advancing Front of Medicine*, he wrote:

"Come, physician, put my hand in poulticee,

Give me a drug in which some healing volt is,

Come, antibiotics, sulfanilamide!"

"Up, Doctor Jekyll, down with Mr. Hyde."

One of my classmates, Wilbur Mattison, whose first assignment on the Osler House Staff at Hopkins was in the Emergency Room, told me: "I never felt so alone in a crowd." Every patient was first asked for his or her "chief complaint," which was then recorded by a receptionist. In some cases, an accompanying friend would provide this information. Once when the chief complaint was: "Patient says she hasn't seen anything for two months," the receptionist referred the patient mistakenly to the ophthalmology department.

5

The National Institutes of Health

The intramural and extramural programs of the National Institutes of Health (NIH) have been the greatest force in biomedical research anywhere in the world. Many Nobel prizes have been won by scientists working both as part of the extramural program, as well as at the site of the intramural program in Bethesda, Maryland. The Clinical Center opened in 1952, and I had the wonderful opportunity to become a Clinical Associate on July 1, 1955, after finishing the second year of Assistant Residency in Internal Medicine at Hopkins. I worked under the direction of the late renal physiologist, Dr. Robert Berliner, who subsequently became Dean of Yale Medical School.

Those were heady days. We interacted every day with the most talented and accomplished biomedical scientists in the world, as they searched for the causes of human disease. Only about 10% of the principal investigators at the NIH were really happy. Their research was successful and they won the admiration of their peers. Many of them had no thought of the relationship of their work to human disease, although they expected that sooner or later something useful would come from their research. Their main motivation was to find out how living organisms work. The other 90% of the investigators at the NIH eventually decided that they would be better off working in medical schools, where they would have the satisfaction of teaching and taking care of patients, even if they were not particularly creative or successful at research.

Before coming to the NIH, Robert Berliner had worked with the first Director of the NIH, Dr. James Shannon, whom he had known since his days at New York University in New York. In 1949, Shannon was recruited to Bethesda to head the National Heart Institute. He brought his former collegues from New York University—Bernard Brodie, Robert Berliner, Thomas Kennedy and Sidney Udenfriend—with him. I had the opportunity to meet Dr. Shannon from time to time. I recall the two of us walking back to the Hadden Hall Hotel in Atlantic City from a restaurant during the meetings of the American Society for Clinical Investigation (ASCI), the Association of American Physicians (AAP), and the American Federation for Clinical Research (AFCR). I was elected to membership in the ASCI on May 14, 1962, the AAP in 1965, and later was elected President of the AFCR. The membership of the AFCR was limited to persons under 40 years of age. We were called the "young squirts," while the members of the ASCI were called the "young Turks," a term derived from political events taking place in Turkey at the time of the founding of the ASCI. The members of the original parent organization, the Association of American Physicians (AAP), were called the "old f..ts." In those days, the annual meetings were held in the Steel Pier Theatre, which had sloping, amphitheatre seats. If you sat in the back rows, the cigarette smoke was so thick, you couldn't see the speakers at the podium.

Shannon's greatest contribution to the NIH and American biomedical research was his creation a policy where investigators submitted proposals to be evaluated by "peer-review." The reviewers had to be investigators outside one's own institution. The proposals submitted to the NIH had to be signed by the investigator's institutional director. The "peer-review" policy with reviewers from outside the institutions gave enormous freedom to the individual investigator, who no longer had to depend on his superiors in his home institution for his salary and research funds. Indirect costs were given to the institution, which provided a strong incentive to the institution to support the programs, even though the relative power of the investigator was increased greatly relative to that of administrators. Winston Churchill said that democracy has many problems, but no other system is better. The same can be said about the peer review system instituted by Shannon at the NIH over half a century ago.

The procedure is as follows: when a researcher submits an application to the NIH for a research grant, it is referred to an NIH integrated review group, or IRG. The IRG assesses the scientific merit of the proposal, and then passes the application on to a study section. Each study section consists of 20 or more scientists from the applicant's area of research. Two or three review the application in detail, assess its scientific merit, and present written critiques to the study section. After discussion, each member of the study section assigns a priority score on a score sheet. The results are tabulated, and the highest rated applications are designated to be funded, depending on the priority of the application and the availability of NIH funds. Each year about 30% of the more than 30,000 applications are funded. By 2004, there were 45,000 research-grant recipients, fellows, and trainees at American Universities.

Under the peer-review system, individual young faculty members can bypass the power structures of medical schools. With steady, annual increases in the NIH budget, there was a great increase in the size of medical school faculties who submitted increasing numbers of research projects. This resulted in a steady increase in the number of creative basic and clinical scientists throughout the United States. The United States achieved the major role in biomedical research. Essential to the success of the peer review system was the honesty of the reviewers, the objectivity and confidentiality of the proposals, and the expert opinions of peers. Their reviews were returned to the investigators, and resubmissions were commonplace if the initial application failed to be funded. Scientists worked, not to gain the favor of their academic or political leaders, but to gain the respect and approval of their peers from throughout the country. Shannon was admired by the academic faculties of all American medical schools.

We young investigators soon learned the importance of having convincing evidence that our ideas would work by obtaining and submitting preliminary results. Otherwise the proposal could be rejected if one of the reviewers stated that the basic premise was unfounded. Usually a young person beginning to carry out research would join the team of an experienced, well-funded investigator. Only later would he or she strive to become a "Principle Investigator."

With increasing clinical demands for the faculty in medical schools today, it is difficult for the clinical faculty to have time to do research. They are now required to generate income for the institution by providing patient care and are often reduced to "spending a day in the lab." More and more grants are being awarded to those with both an MD and a PhD, or with a PhD only. Some clinicians, who are basically not effective in carrying out research, do so only to be able to climb the ladder of academic medicine. Unfortunately, after they become Professors, their research activities come to an end.

Johns Hopkins has always valued research as a principal function of the University and its medical school. No other environment could have better advanced my research career than my two years at the NIH. As Clinical Associates at the Heart Institute, we were encouraged to conduct research projects, both on our own and under the leadership of our superiors, while caring for the patients of other investigators. I was surprised to observe that some research scientists hesitated to have contact with patients, to the point of conducting "rounds" of patients without even entering the patient's room.

One of my first lessons at the NIH was that the opposite of "love" is not "hate," but "to ignore." When we presented our work to our peers and leaders, their critical comments were an indication that they took the work seriously. Once, after a colleague had presented some material, Dr. Berliner, who was Research Director of the Heart Institute at the time, made no comments. On the way out of the conference room, I asked: "Do you agree with that?" I was disappointed when he told me that he didn't agree with any of what had been said, but chose not to comment. I learned then that one's best friends are those who will tell you when you are wrong as well as when you are right. While on the house staff at Hopkins, I had always taken the words of our superiors as "gospel." I was disabused of this attitude after I arrived at the NIH. Even today, I still tend to believe that what people tell me is not true. Not that they want to deceive me, but that they "don't know what they don't know." This attitude is helpful to a professional scientist. Before they made scientific presentations, I taught my colleagues and students to ask: Is what I am saying clear? Is it true? Is it new? Is it significant?

While at the NIH, I carried out experiments involving the kidneys in trained, unanesthetized dogs. While we had technicians to help us, we had "hands on" involvement in all aspects of the studies, from injecting and examining the dogs to assaying samples of blood and urine. This training would serve me well throughout the rest of my professional life. Research depends on teamwork, with no first and second class citizens. Curt Richter had taught me that while practically everyone likes the "idea" of research, few like the hard work. My greatest incentive was an overwhelming desire to get the answer to the question that is being addressed by the experiments.

We showed that there was a balancing of the secretion of the pituitary anti-diuretic hormone and the amount of solute being excreted by the kidneys as the two determinants of the osmolality, that is, the concentration of solutes in the urine. This quantification of a physiological process taught me that diseases were not just "present" or "absent," but that there could be varying degrees of deficiencies. For example, deficiency of the antidiuretic hormone in the disease, diabetes insipidus, could be present in different degrees in different patients. This concept that there is a spectrum of deficiencies making up a given disease has been helpful to me throughout my professional life. It represnted a "physiological" approach to disease, different from the "ontological" approach which views disease is a thing that you either have or you don't. The ontological approach is expressed in the statement that you have "caught a cold." The germ theory of disease and the occurrence of genetic diseases were important factors in the acceptance of the ontological approach to disease, as distinct from the physiological approach.

Among the patients that I took care of while at the NIH were two older men and a teenager with a disease called "orthostatic hypotension," manifest by a dramatic fall in their blood pressure whenever they stood up. We were able to establish that the disease was a general, extensive defect in their sympathetic nervous system, which we called "Autonomic Insufficiency." I was eager to study these patients because of my familiarity with Claude Bernard, Walter Cannon, and Curt Richter.

The regulation of blood pressure is an example of one of the most fundamental principles of biology: homeostasis, a term first used by physiologist Walter Cannon to describe Claude Bernard's concept of the "constancy of the internal environment (milieu interieur)." All matter has a tendency to progress to total disorder, called an "increase in entropy." Life is maintained in a free and independent "steady state" in the face of this Second Law of Thermodynamics by the existence of mechanisms for monitoring body functions and balancing the rate of formation with the rate of breakdown of body constituents. Physiology is the characterization of these processes. Medicine today rests on a foundation of physiology. As Paul Bert wrote on February 12, 1878: "Claude Bernard is not merely a physiologist, he is physiology."

The physiological adjustments to maintain blood pressure when we stand illustrate Claude Bernard's principles. Gravitational forces have to be counteracted for primates to assume an upright posture. The activation of the sympathetic nervous system constricts blood vessels below the heart to maintain arterial pressure when we stand up, in order to maintain adequate blood flow to the heart and brain. Persons who have been in bed for a long period of time lose this ability, and become faint and lightheaded. During the Civil War and both World Wars, this condition of becoming giddy with blurring of vision with a fall in blood pressure was called "effort syndrome" or "neurocirculatory asthenia."

The three patients with autonomic insufficiency that I studied had difficulty in maintaining their blood pressure in a normal range in the face of anything that would either decrease or increase blood pressure. In normal people, administration of drugs that constrict blood vessels are counteracted by the ability of the autonomic nervous system to dilate other vessels and normalize blood pressure. This did not happen in these patients. Their blood pressure would rise when they were administered drugs, such as pitressin, that are not able to raise the blood pressure of normal persons. Removal of blood or vasodilatation by drugs in my patients resulted in a far greater fall in blood pressure than in normal persons. These patients also suffered from an absence of sweating, another process under the influence of the autonomic nervous system. The administration of drugs that constrict blood vessels or increase the volume of their extracellular fluid provided these patients great relief. I described these patients and their successful treatment was at the internal medicine meetings in Atlantic City and subsequently published the results.

Before one of these meetings, I was offered the job as Head of Nuclear Medicine at Marquette University in Milwaukee. I told the Chairman of Internal Medicine, who had made the offer, that I would give him my answer at the meeting in Atlantic City. We met for breakfast and I told him I had decided not to accept the offer. Unfortunately I told him this before we had settled the bill. He made me pay for own breakfast, teaching me the important lesson that if you are going to give a negative response at the meal, wait until after the bill has been paid.

Working in the laboratory of Curt Richter on homeostatic processes had shown me the excitement of doing research. The three clinical research projects during my 3 years on the Osler housestaff at Hopkins provided further experience in carrying out research that involved patients, even those seriously ill. The two years at the NIH provided me with a virtual PhD, improving my ability to carry out research. I came in close contact with many of the giants of biomedical research, such as Julius Axelrod, who were to win the Nobel prize. Beyond any doubt what I wanted was a career in academic medicine.

At the end of those happy days in Bethesda, I was not allowed to leave the NIH without taking the orientation course. So, every morning of my last week, I went down to the auditorium of the Clinical Center at NIH and signed in even though I never went to any of the sessions. We then headed off to England.

After two years at the NIH, Anne, then pregnant with our fourth child, and I, with our three children, left New York City on the Cunard Line ship, *Scythia*. It was the last voyage of the ship, so that first class tickets cost less than tourist on other ships. We took salt water baths, and the stewardess would bring a basin of fresh water to wash off the salt. Anne and I got little sleep during the 10 day crossing of the Atlantic, because we spent evenings with the older passengers and entertained our children during the day. One evening during dinner, a mental patient eluded his attendant and jumped overboard. A sailor threw him a ring life preserver, that fortunately the patient grabbed. A lifeboat was lowered, he was picked up, and we were underway after 20 minutes.

Our three children ate during the childrens' meal hours, with a steward or stewardess cutting their meat and otherwise taking care of them. Since Anne and I were 28 and 30 years old, we were part of the family crowd on the ship during the day and part of the young adult group in the evening. We would carry messages from one group to the other, informing people of their assigned playing times in the table tennis competition.

One evening, at a dance, everyone was to come dressed as the title of a song. I wore a dress with a pillow simulating pregnancy. Anne was obviously pregnant with our fourth child. The song I depicted was: "Anything you can do, I can do better." We won first prize, a bottle of champagne.

After a brief stop in Cork, Ireland, to let off a group of priests and nuns, we sailed on to Liverpool, where I suddenly found myself on the dock, with a wife and $3\frac{1}{2}$ children, our first time out of the United States, not counting the days when the U.S. Coast Guard had taken me and fellow cadets to several Caribbean islands. Traveling by train to London, row after row of chimney pots left no doubt that we were beginning a new adventure. When we arrived in London, we stayed for a few days at the Queen's Court Hotel but checked out when an elderly guest, waiting with us for an elevator (lift), growled: "Children: Bah!"

We moved to a single room in a boarding house, called J.D. Residential, in Earl's Court. Mr. John Derrick, the owner, suffered from bronchitis. His periodic clearing of this throat brought to our minds a typical Englishman. The cost for room and board for the five of us was $33.00 per week. I wanted to stay there for the rest of the year, an idea vetoed immediately by Anne. We went to a real estate agent in London. A week after I started to work at Hammersmith Hospital, we moved to a wonderful, old fashioned three-story house, called Broomfield, with a lovely garden in Woking, Surrey, a quaint town on the last stop on the train from London and an hour and 15 minute drive from Hammersmith Hospital. We inherited a maid, Mrs. Blondell, who came with the house. The owners of

Broomfield were the Berry's, who lived next door. Mr. Berry had been knighted for his service in India before his retirement. The previous tenants at Broomfield were five young fliers from the U.S. Navy. Mrs. Blondell told us of some of their great parties. She was great with our children, making it possible for Anne and me to take several trips to the continent. We had purchased a Borgward station wagon, which was two weeks late in arriving from Germany because of a slip-up in our ordering it in Bethesda before we left.

Our oldest child enrolled in a small private school in Woking, and in the first grade learned "reading and sums" as we were told he would when we registered him. One Saturday, I attended a parents meeting at the school, where every mother spoke eloquently on the topic: "Should games be compulsory on Saturday?" Our daughter attended kindergarten at a private school, called Halstead. Both schools were within walking distance of our house. When winter came, we wondered why people complained about their houses being cold. We told them we had no problems because we had electric room heaters. After six months, we got our electric bill, and it was so high, the electric company gave us industrial rates. Needless to say, we turned to covering our shoulders with a blanket when we went to the toilet, and put glass-topped beer bottles full of hot water under the covers before we got in bed. Once I got the flu and Anne wanted me to stay home. I told her that I was too sick to stay home, and left for the warmth of Hammersmith Hospital.

Dr. Roland Morgan, Chief Registrar in Radiation Therapy at Hammersmith Hospital, was a proper Englishman, who lived up the street from us. We alternated days driving to the hospital. One hot day, nine months after we had been enjoying these daily trips, Roland surprised me by asking: "Do you mind if I loosen my tie?" I replied: "Roland, I don't care if you take your shirt off."

In those days before the passage of controlling legislation, the fog and smog were at times so thick that one could drive only by following the red tail lights of the car in front. Occasionally, a driver would find a whole string of cars lined up behind him as he drove into his driveway. When we would drive by a field, we would hear disoriented motorists crying: "Anyone on the road, please call out." Having a left hand drive in our Borgward was a huge advantage because you could look out of the window for the next curb, as you crossed the intersection to the next block.

Anne had the interesting experience of having our fourth child in England at the Woking Maternity Hospital. We thought that Guy's Hospital in London was too far away. The Woking maternity hospital had been founded by Dr. Grantley Dick Reed, famous in those days because of his invention of "natural childbirth." The only medication was minimal use of the anesthetic, nitrous oxide. Anne and our new daughter, Anne Elizabeth, were kept in the hospital for a week, which was routine in those days. The total hospital bill for the week's stay was $18.00. After leaving the hospital, milk and fruit juice were delivered daily to our house, courtesy of the National Health Service.

On October 4, 1957, a 184-pound satellite, called Sputnik, was launched into orbit by the Soviet Union. My British colleagues could not resist the temptation to gloat over the triumph of Soviet science and technology over American. They also took the occasion to tell me that they weren't happy over our purchase of a German automobile. They thought the American invention of beer in cans instead of bottles was a step backward.

Physicist Edward Teller called Sputnik a "technological Pearl Harbor." Americans were undergoing a national crisis of confidence, made even worse one month later, when a second satellite, Sputnik II, six times larger than Sputnik I, carrying a small dog, Laika, had been launched into orbit. The Russians also announced that they would soon put a man into space. The failure of the U.S. Navy to successfully launch a 4-pound satellite, called Vanguard, shortly thereafter was mocked in British newspapers as "Kaputnik," "Flopnik," and "Stayputnik."

Although my work under the direction of Professor Russell Fraser concerned thyroid disease, I was fascinated by the work going on in the Medical Research Council facility at Hammersmith, using tracers made by a cyclotron built two years before I arrived.

One of my responsibilities at Hammersmith was to distill iodine-132 from its parent radionuclide, tellurium-132, which produced iodine-132 in the process of radioactive decay. One of my mentors, physicist John Mallard, told me that, when I returned to the United States, I should look into a new method of obtaining iodine-132. Stang and Richards at Brookhaven National Laboratory had used resin columns to separate "daughter" radionuclides from their "parents". In addition to the tellurium-132/iodine-132 system, they had also developed a "radionuclide generator," also called a "radioisotope cow," that made possible separation of technetium-99m from its parent radionuclide, molybdenum-99. The radionuclide technetium-99 was to become the foundation of nuclear medicine for decades beginning in the 1960s.

Technetium does not occur in nature. Both of its isotopes are radioactive and have therefore decayed since the formation of the earth. The element was discovered in 1937 by Segre and Seaborg in Berkeley, California. The technetium-99m generator was developed by Stang and Richards in 1960. In the case of technetium-99m with a six-hour half-life, the parent radionuclide has a half-life of 67 hours, long enough for a hospital to obtain a shipment of the generator once a week. The six-hour half life of the technetium-99m and its decay by the process of isomeric transition without emitting a beta particle made it ideal for nuclear medicine.

In 1961, the catalogue of commercially available radionuclides included technetium-99m and iodine-132 on its cover in letters 2 inches high. No one recognized the important characteristics of technetium-99m for radioisotope scanning until 1963, when Paul Harper, Catherine Lathrup, and Alex Gottshalk administered technetium-99m pertechnetate for scanning of the thyroid; technetium-99m sulfur colloid for the study of the reticulendothelial system in the liver, spleen, and bone marrow; and later pertechnetate for detection of brain tumors. Paul Harper also worked with indium-113m, and had a poster in his lab describing it as the "Isotope of the Week." Paul has written: "Henry Wagner's chemist came by for a visit (Manny Subramanian) and saw this poster on the wall, but he didn't say much about it. But it wasn't too long until Henry Wagner was working with indium-111 and it became his favorite isotope."

We began using technetium-99m pertechnetate for brain scanning at Hopkins in January 1964. Its attractive physical characteristics included the short physical half-life of 6 hours, the absence of beta emission, which greatly reduced the radiation dose to the patient, and the energy of the emitted photons was 140 Kev, ideal for their being detected and imaged with external scanners. By June 1964, we had scanned 137 patients for suspected brain tumors. The results with technetium-99m were far better than with the previous agents, iodine-131 albumin or Hg-203 chlormerodrin. The doses could be given either intravenously or by mouth, although the former was preferred.

Figure 22 Powell Richards, co-inventor of the technetium-99m generator in 1960, standing alongside Masahiro Iio, first Japanese traineed in nuclear medicine at Johns Hopkins.

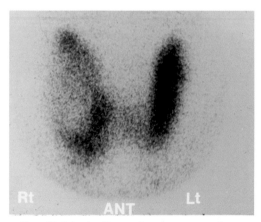

Figure 23 Technetium-99m pertechnetate gamma camera image.

By 1983 these were some of the questions being asked routinely at Hopkins with technetium-99 m radiopharmaceuticals:

1. Is a thyroid nodule functioning?
2. Is the thyroid overactive?
3. Does the patient have a brain tumor?
4. Does he or she have a cerebral infarction or subdural hematoma?
5. Has the patient had a pulomonary embolus?

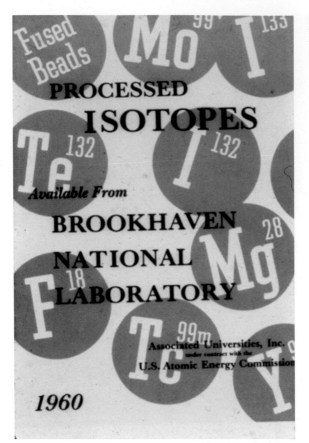

Figure 24 Cover of the Brookhaven National Laboratory catalogue, advertising technetium-99m generator three years before it was recognized as being ideal for nuclear medicine.

Figure 25 First PET scanner designed by Dr. Yamamoto of the Brookhaven National Laboratory in the 1970s.

Figure 26 (l to r) George Taplin, Benedict Cassen, HNW, and Paul Harper in Athens in 1964.

6. What is the blood flow distribution between the lungs or in different lung regions?
7. Is the patient able to withstand lung surgery?
8. What is the cause of the patient's chest pain?
9. How severe is the patient's heart disease?
10. Is the patient responding to treatment of heart failure?
11. Is the right ventricle failing from lung disease?
12. In a patient with shortness of breath, is the problem the right or left ventricle or the lungs or both?
13. Does the patient have acute cholecystitis?

Figure 27 Dr. Benedict Cassen, inventor of the first rectilinear scanner.

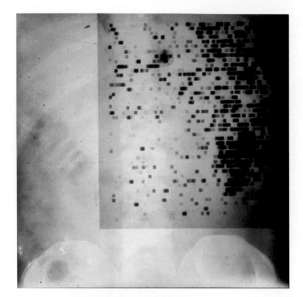

Figure 28 Imaging of lesions in the spleen with Hg-203 BMHP, a technique that never caught on.

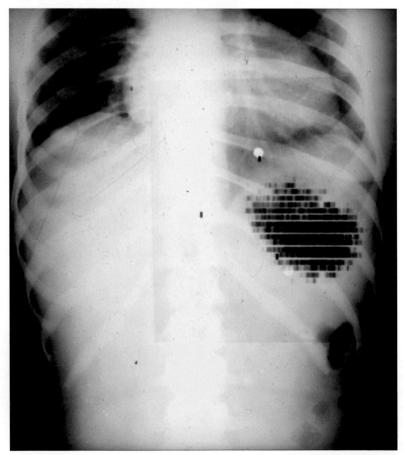

Figure 29 Imaging of the normal spleen superimposed on an abdominal radiograph, an early "fused" image of structure and function.

Figure 30 Early iodine-131 albumin rectilinear scan of a patient with a brain tumor superimposed over a radiograph of the skull. This is an another example of a "fused" image performed routinely at Johns Hopkins beginning in 1960.

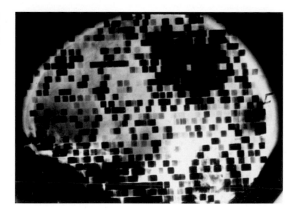

14. Does the patient have skeletal lesions?
15. Is there osteomyelitis?
16. Is a renal transplant working well?

By 1983, we had carried out and followed for one year 243 consecutive patients in whom technetium-99 m pertechnetate was used in the diagnosis of hyperthyroidism.

During my year at Hammersmith in 1957, I had read the classic book on nuclear medicine written by Norman Veall of Guy's Hospital and another by William Beierwaltes at the University of Michigan. These, plus the work with cyclotron-produced radionuclides, were an eye-opener. I concluded that the use of radioactive tracers could help solve the medical problems that I had been learning about since I started medical school.

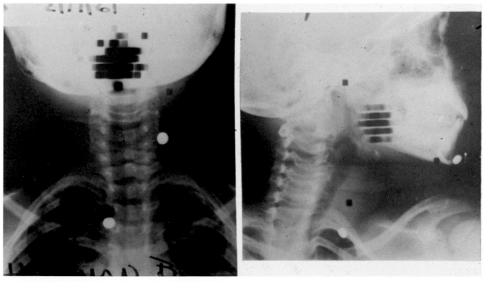

Figure 31 Image of radioiodine accumulation in thyroidal tissue at the base of the tongue. Its location was established by superimposition of the rectilinear scan on to a radiograph of the neck. This was routine from the beginning in nuclear medicine at Johns Hopkins and represents an early version of "fused" images. The identification of the nature of the lesion prevented surgical removal.

I tried to imagine what Claude Bernard could have done if he had had radioactive tracers and the means to detect and quantify their distribution within the body. Here were tools to directly measure the "dynamic state of body constituents" within the bodies of the living animals.

Bernard wanted very much to relate chemical processes in the laboratory to those occurring within the complexity of the living body. In this famous book *Principles of Experimental Medicine*, he wrote of the "great error" being made by those who thought that the physicochemical phenomena within organisms were identical to those which took place outside of them. "It is that error into which certain chemists have fallen, who reason from the laboratory to the organism, whereas it is necessary to reason from the organism to the laboratory." This same error is often occurring in genetic medicine today.

Advances in science often occur when experimental techniques developed for one purpose are brought in close and effective relationship with different and independent scientific disciplines. A classical example is when Einstein related energy and matter. A whole new science was created. The sciences of chemistry and physiology were brought together in 1843, when Carl Lehmann became Professor of physiological chemistry at Leipzig. His *Lehrbuch der physiologischen Chemie* in 1842 provided an organized view of this new science. Its goal was "to discover precisely and in their causal connections the course of the chemical phenomena that accompany vital processes," and to "derive them from known physical and chemical laws." What remained was to determine the "topography" of these processes within the fluids, tissues, and organs of the body. A hundred years later, we are witnessing the evolution of "molecular imaging"! We must remind ourselves always what Bernard often stated: "the physiological point of view must dominate the chemical point of view."

After a year at Hammersmith, I returned to Hopkins as Chief Resident in Internal Medicine on July 1, 1957.

6

A New Medical Specialty

"The history of the living world can be summarized as the elaboration of ever more perfect eyes within a cosmos in which there is always something more to be seen." Nuclear medicine provides new ways to look at the function and biochemistry of all parts of the living human body. People try to pinpoint the "birth of nuclear medicine" to a specific date or event, but it is difficult to do so. Nor did nuclear medicine develop along a continuous growth curve, but rather was the result of "revolutionary" discoveries, some of which had difficulty in being accepted.

The story begins one night in 1895, when the German physicist Conrad Roentgen noticed that certain crystals, inadvertently left near a highly evacuated electric discharge tube, became luminescent even when the tube had been enclosed in a light proof box. He found that the invisible radiations, which he dubbed x-rays, ionized the air through which they traveled, and, moreover, exposed x-ray film. The x-ray "shadowgraphs" could reveal the internal structures of objects opaque to light. He realized the potential medical applications immediately, and took the famous x-ray picture of his wife's hand.

Two months later, the French physicist, Henri Becquerel, was exploring the consequences of Roentgen's discovery, and placed a phosphorescent uranium salt in a light-proof envelope containing an unexposed photographic plate and set it in the sunlight. He reasoned that if the plate showed black spots where the salt was, this would be due to x-rays excited by the sun. He was gratified to find the black spots. Then there was a period of cloudy weather. By force of habit, he developed some photographic plates that had been lying in a dark drawer with bits of uranium salt on them without any exposure to the sun. To his astonishment, the darkened areas were again observed where the uranium salt had been. As has often been said, discovery depends on a "prepared mind." Becquerel concluded that the uranium salts were giving off penetrating x-rays, even when in their natural, unstimulated state. Radioactivity had been discovered.

Becquerel was a professor of physics at the Ecole Polytechnique. One of his most promising students was Marie Sklodowska, bride of Pierre Curie, a good friend of Becquerel. Marie had emigrated from Poland to Paris, and eagerly accepted the suggestion of Becquerel that she try to find out what in uranium was responsible for the radiation that he had discovered.

Using a radiation detector invented by her husband, Pierre, Marie (who coined the term "radioactivity") set out to isolate the active fraction of pitchblende, and that was so interesting that Pierre dropped his own research and joined Marie in her efforts. Their first discovery was that there was another source of radioactivity in

Figure 32 Winners of four Nobel prizes. Pierre and Marie Curie with their daughter, Irene. Marie won two; Pierre and Irene, one each.

Figure 33 Marie Curie with her daughter Irene in 1908.

Figure 34 Frederick Joliot and Irene Curie, winners of the Nobel prize for the discovery of artificial radioactivity.

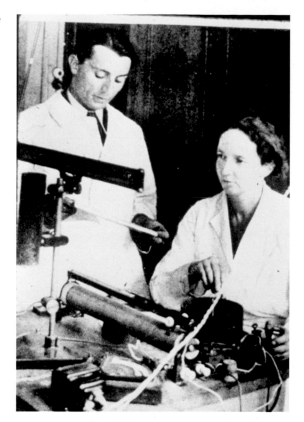

Figure 35 Irene Curie shortly before her death from leukemia.

pitchblende in addition to uranium. It turned out to be another element, thorium. Even more intriguing was the finding that the radiation did not emit low energy x-rays, but far more energetic and penetrating rays.

As they isolated fractions of pitchblende, they found some that contained more radioactivity than pure uranium. They concluded correctly that there must be other radioactive sources present besides uranium. By a series of remarkable and laborious chemical separations, involving tons of pitchblende, in July of 1896, they discovered an element with a radioactivity 400 times greater than an equal amount of uranium. They named the newly discovered element polonium, after Marie's native Poland. Later that same year, they discovered another new element, radium.

Soon thereafter, they discovered that the radiation from radium produced biological effects. Among the first observations was that made by Becquerel himself in 1901. Traveling to London from Paris to deliver a lecture, he carried a sample of radium in his vest pocket. After returning to Paris, he noticed a reddened area on the skin beneath the pocket. This led to great interest in the biological effects of radioactivity. In 1903 the American inventor Alexander Graham Bell suggested that placing vials of radium near cancerous skin lesions might cure them.

In 1913, the most important principle of nuclear medicine—the tracer principle—was invented in Vienna by Hevesy and Paneth. They monitored the movement of radioactive lead from the soil into plants and then back into the soil when the plants died. In this way, they could monitor dynamic processes that characterize life. Early instruments for measuring radioactivity included the piezoelectric device used by the Curies, photographic plates used by Rutherford, the cloud chamber, invented in 1895 by the Scottish physicist Wilson, the gold leaf electroscope, and the Geiger counter, perfected in 1928. Hevesy and Paneth relied on the gold leaf electroscope for their experiments. This instrument was so sensitive that radioactive lead could be given in such small amounts that there were no toxic effects of the lead on the plants. This is another important characteristic of the "tracer" principle: radioactive tracers in small amounts have no biological effects on biochemical processes within the organism being studied. These early experiments were forerunners of what lay ahead—a new branch of medicine based on the use of radioactive "indicators," as they were called by Hevesy, to reveal the biochemical pathways of molecules labeled with radioactivity as they moved through the biological system. If larger amounts of the tracer atoms or molecules were given, the biological process could be suppressed, which provided a basis for treatment.

An amusing historical anecdote occurred when Hevesy began to suspect that his landlady was returning food to the dinner table that was left over from previous meals. So he spiked his uneaten food with radioactive lead. At the next meal, he quietly pocketed a sample of the food, which he put under his gold leaf electroscope. Sure enough, the specimen was radioactive! He confronted his landlady with the evidence. Shorter in temper than in scientific vision, she promptly evicted him.

In these early days, the studies were limited to the use of the few naturally radioactive materials that were available. These were uranium, thorium, radium, polonium, bismuth, and lead, elements that had little biological significance. Blumgart and his colleagues conducted the first clinical studies with radioactive tracers in Boston in the late 1920s. They injected solutions of radium C (bismuth-214) into an arm vein, and measured the time it took for the tracer to go through the heart and lungs and be detected in the opposite arm. This took about 18 seconds in people who did not suffer from heart disease. They called their measurement the "velocity of the circulation," which was abnormally slow in patients with heart failure.

In the 1940s, the first reports of the use of radioactive iodine to treat diseases of the thyroid, and radioactive phosphorus to treat leukemia were greeted with great enthusiasm by the medical profession and the public. These classic findings were immediately recognized as forerunners of a whole series of similar uses of radioactive "magic bullets" to seek out and destroy diseased tissues within the human body. They predicted what has proved to be true: radioactive "tracers" have helped in the diagnosis or treatment of millions of patients with overactive or cancerous thyroids, and many other types of diseases. It is estimated that today one of three patients admitted to a hospital has a radioactive tracer procedure as part of the diagnostic process or treatment.

Radioiodine accumulation is blocked by the administration of triiodothyronine (T3) in normal thyroidal tissue but not in autonomously functioning nodules. Nuclear medicine was recognized as a medical specialty in 1971 by the American Medical Association. Today, over 5,000 hospitals in the United States provide nuclear medicine services to their inpatients and outpatients.

The specialty is based on the creation of images of the spatial and temporal distribution of radiolabelled molecules to reveal the chemistry and function of the organs of the living human body in health and disease. The specialty provides the unique ability to examine both the location and site of these process to reveal regions that are abnormal.

Nuclear medicine was the 22nd medical specialty to be recognized by the American Medical Association, being incorporated on July 28, 1971 when nuclear medicine was recognized as a medical specialty. The American Board of Radiology (ABR) recommended that radiology residents obtain 6 months of training in nuclear medicine. The

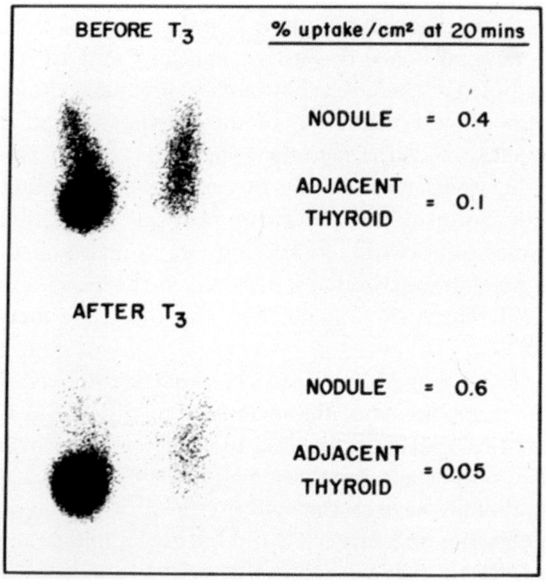

Figure 36 Suppression of radioiodine accumulation by administration of tri-iodothyronine (T3) in normal thyroidal tissue but not in an autonomous nodule.

American Board of Internal Medicine (ABIM) accepted one year of training in nuclear medicine as part of training in internal medicine. The American Board of Nuclear Medicine (ABNM) defined nuclear medicine as "the scientific and clinical discipline concerned with the diagnostic, therapeutic and investigative use of radionuclides."

Dr. Jack D. Myers, Chairman of the Department of Medicine at the University of Pittsburgh, and head of the ABIM played a major role in obtaining recognition of the ABNM by the American Medical Association. He proposed a new type of Board, a "conjoint" Board that would bring together as "allergists," those physicians originally trained in Pediatrics and Internal Medicine. Nuclear Medicine was proposed to be a Conjoint Board of the Boards of Radiology, Internal Medicine, and Pathology.

In 1968 the ABIM had not been willing to support the founding of an independent or affiliate board in nuclear medicine, because they thought that nuclear medicine should be a sub-specialty of internal medicine. Only the great efforts of Dr. Merrill A. Bender, President of the Society of Nuclear Medicine in 1968, the radiologist Dr. David Kuhl, and Jack Myers of the ABIM, brought about the recognition of a Board of Nuclear Medicine, affiliated with the boards in medicine, pathology and radiology. Representatives of the ABR who supported the new "conjoint" Board of Nuclear Medicine were Drs. E. Richard King, William T. Moss, Donald S. Childs, and James W.J. Carpenter.

On April 26, 1968, Dr. Myers had written to Dr. Bender: "We remain deeply concerned about nuclear medicine because of the high proportion of internists in the field . . . For those specialists, basic certification in internal medicine is highly desirable, with further certification in the subspecialty (of nuclear medicine)."

The initial agreement by the ABR to support the formation of a Conjoint Board of Nuclear Medicine was subsequently rescinded, but their opposition came too late, and the Conjoint Board of Nuclear Medicine was recognized in 1971, during the year of my Presidency of the Society of Nuclear Medicine.

On March 25, 1972, more than 1,000 physicians gathered in 10 cities throughout the United States to take the examination by the American Board of Nuclear Medicine (ABNM). Several proposed boards have died during the birth process, for example, an American Board of Gastroenterology and an American Board of Tuberculosis. Many radiologists believed that nuclear medicine should be a subspecialty of Radiology, and opposed the formation of the ABNM, despite the fact that backgrounds of the first nuclear medicine specialists were 43% radiologists, 35% internists, 15% pathologists, 8% other specialties. During the process of obtaining approval by the Council on Medical Education of the American Medical Association, the withdrawal of support of the American College of Radiology was a major obstacle. This opposition was expressed in a resolution, eventually defeated in the House of Delegates of the AMA in November 1971. The ACR Council then supported the formation of the ABNM by a vote of 52 to 39.

Nuclear medicine requires expertise derived from almost every medical specialty. Even opponents of the formation of the ABNM were intrigued by the idea of existing medical specialties pooling their efforts and working together to establish a new board. Not only would the three major related specialists of radiology, internal medicine, and pathology have an opportunity to work together, but also others, such as physicians in pediatrics and surgery, could become qualified in the use of radioactive tracers in their practice if they wished to do so. Opponents of the recognition of the Board of Nuclear

Medicine feared that the ABNM might result in restrictive practices, excluding some physicians from participating in the practice of nuclear medicine.

Dr. R.S. Gambino, from Columbia-Presbyterian Medical Center, feared that diplomats of the ABNM planned to restrict the use of radioactive tracers in medicine. We founding members of the ABNM stated our position that we would oppose any attempt to prevent a person trained in the use of radioactive material from using them. We defined specialists in nuclear medicine as those who chose to spend 100% of their professional efforts in this new specialty of medicine. We wanted to prevent the use of radioactive materials in medicine from being limited to an exclusive clique. This was a justifiable concern.

In the early 1950s a resolution had been passed by the House of Delegates of the AMA limiting the use of radioactive tracers to those physicians certified by the American Board of Radiology. As the result of the efforts of dedicated nuclear pioneers, such as Dr. William G. Myers of Ohio State University, this resolution was subsequently repealed.

The immediate consequence of establishment of the ABNM was to permit hundreds of internists, as well as radiologists and others, to be examined and certified that they were competent to use radioactive tracers in the practice of medicine. If internists had been excluded, many subsequent advances would not have been made. Dr. Robert Newell, Chairman of Radiology at Stanford University, and who coined the term "nuclear medicine," said "certification by a specialty board is an ornament, not a license." Today, most radiologists have embraced the value of what is now called "molecular medicine," and have incorporated extensive training in nuclear medicine into their residency programs. This was not the case in the early days.

William G. Myers said that he had had a life full of "twinkling" atoms and "scintillating" people. He also said: "Laughter lubricates life." The World Federation of Nuclear Medicine and Biology (WFNMB) was founded on October 26, 1970, in San Jeronimo Lidice in Mexico, and it was decided to have the first World Congress in Japan in 1974, with the late Hideo Ueda as President, the late Masahiro Iio as Secretary General, and

Figure 37 Bill Myers (right) with Bill Beierwaltes, head of nuclear medicine at the University of Michigan, whose pioneering textbook was the first book about nuclear medicine that I read at Hammersmith Hospital in London in 1957.

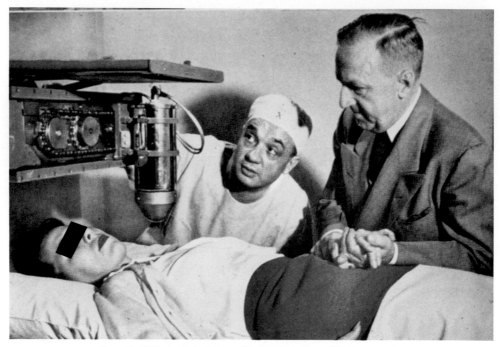

Figure 38 Bill Myers with the Chairman of Medicine, Charles Doan, of Ohio State University.

Figure 39 Bill Myers with Henry Wagner and Sadatake Kato, Treasurer of the First World Congress of Nuclear Medicine and President of Nihon-Medi-physics Company.

Figure 40 (l to r) Sadatake Kato, Masahiro Iio, Anne Wagner, Mrs. Kazuko Kato, and Mrs. Hiroko Iio.

Figure 41 Munho Lee, one of the founders of nuclear medicine in South Korea.

Figure 42 Henry Wagner during an IAEA meeting in Greece.

Figure 43 Constantine Constantanides, one of the founders of nuclear medicine in Greece, a frequent visitor to Johns Hopkins.

Figure 44 (l to r) HNW, Marco Salvatore, Pietro Muto; leaders of Italian nuclear medicine.

Figure 45 Professor N. Bekhtereva, a Russian neuroscientist, who introduced positron emission tomography in the USSR.

the late Sadataki Kato as Treasurer. The first World Congress was held at the Pacific Hotel in Tokyo with a satellite meeting in Kyoto. The Japanese had taken a great risk financially in holding the First Congress. Fortunately, the meeting was a huge success, but had to be subsidized financially to a great extent. Dr. Iio was a great leader. Once he convinced a leading Japanese department store to have an exhibit explaining nuclear medicine to the public on a whole floor of the store in the center of downtown Tokyo.

Dr. Masahiro Iio was the first research fellow from Japan to come to Johns Hopkins to study nuclear medicine in 1961. In his first research project, he showed that human serum albumin aggregates labeled with iodine-131 or chromium-51 could be administered safely to human beings in amounts sufficient to be able to measure their clearance from the blood and accumulation in the reticuloendothelial system.

In September 1963, shortly before Dr. Iio returned to Japan to become head of nuclear medicine at Tokyo University, he injected me with iodine-131 macroaggregated albumin and produced the first lung scan ever performed in a human being. The first patient was a 75-year-old woman with a mid-thigh amputation who came to the Hopkins emergency room in shock. She had huge perfusion defects in her lungs, and was operated on by Dr. David Sabiston in October 1963. In 1980, Current Contents reported that over 245 citations had been made to our article on the development and first use of lung scanning in the *New England Journal of Medicine* in January, 1964. Since 1961, over 70 Japanese physicians and scientists have studied in nuclear medicine at Hopkins. Many of them have gone on to become leaders of nuclear medicine and radiology in Japan. Among the many contributions of Japanese nuclear medicine physicians was their holding the First World Congress of Nuclear Medicine and Biology, in 1974. The 2nd World Congress was held in Washington, D.C. in 1978.

On the morning of September 17, 1978, the day of the meeting, Anne and I called room service at the Washington Hilton Hotel. If there was an answer, I knew that an impending strike of hotel employees had been averted. Staying at the Hilton (where President Reagan would be shot 3 years later) were political leaders, including Senator Ted Kennedy, Menachen Begin, and Egyptian and Israeli political figures attending the Peace Conference at Camp David, where Egypt and Israeli settled the issue of the Sinai desert, the "Camp David Accords." The site of these political leaders in the lobby of the Hilton Hotel in the evenings enhanced the international flavor of the Second World Congress of Nuclear Medicine and Biology.

At the opening ceremony at 5:00 PM on September 17, 1978, greetings from President Jimmy Carter were read, followed by addresses by Donald Frederickson, Head of the NIH, and Nobel Laureate Roslyn Yalow, followed by a wine and cheese reception. On September 18, the Congress sponsored a concert by the Preservation Hall Jazz Band from New Orleans at the Kennedy Center followed by a rooftop reception. On September 19, there was a banquet at the Hilton Hotel, and the following day, a private opening of the National Gallery of Art including a concert by the National Gallery Orchestra, playing the Beethoven Overture to Prometeus and Mozart's Symphony #3. There were tours of Washington for $11.00; a Presidential Tour including Mt. Vernon, Arlington Cemetery and the Jefferson and Lincoln Memorials for $18.00, a tour of the Kennedy Center for $13.00; and a tour of the Smithsonian for $6.00.

The organizational meeting at which it had been decided to have the 2nd World Congress of Nuclear Medicine and Biology in Washington was held in Philadelphia on June 17, 1975, at the annual meeting of the Society of Nuclear Medicine. We tried to have the 2nd World Congress held jointly with the 1978 annual meeting of the SNM, but an agreement could not be reached. The organizers of the Washington World Congress agreed to pay $25,000 for the SNM to be in charge of the commercial exhibits, under the direction of the Executive Director of SNM, Judy Glos. Jim MacIntyre, President of SNM, agree to this arrangement.

Ed Matson of Abbott Laboratories and John Ryan of 3M chaired the industrial affiliates committee, whose members contributed $100–200 each. The affiliates agreed to pre-pay the charges for exhibit space beginning immediately. On July 2, 1975, a group of women in a company called Courtesy Associates agreed to provide administrative services immediately and delay charging us until the year of the meeting 3 years later. A contract signed on March 24, 1976 agreed to pay them $65,000 to $72,000 depending on how much time their efforts took. The registration fee for the meeting was $100 for attendees and $50 for spouses.

The theme of the 2nd World Congress was the worldwide promotion of nuclear medicine. Another goal was to "glamorize Washington and the USA for overseas visitors," as well as to ensure the survival of the WFNMB, which was to hold Congresses every four years with a Congress in South Korea in 2006 and South Africa in 2010.

The minutes of the organizing committee for the 1978 Washington meeting stated that the "major thrust is to help nuclear medicine achieve a clearer identity vis a vis the public, government, colleagues in other specialties, and people within the specialty itself."

The 3rd World Congress, held in Paris in 1982 was also very successful with a budget of $700,000. The 4th meeting was held in Buenos Aires in 1984. Geriatric medicine was emphasized because it was becoming an important focus of nuclear medicine. In 1983, I joined Dr. Iio in writing a book on Geriatric Nuclear Medicine. We proposed that the ideal radiotracer for studying the aging brain should:

1. be labeled with the positron emitting tracers, carbon-11 or fluorine-18
2. cross the blood–brain barrier to permit intravenous administration
3. have a high affinity and selectivity for neuroreceptors, including receptors for neurotransmitters, glucose, and amino acids
4. slow dissociation from targets to facilitate imaging
5. little or slow metabolism within the brain
6. rapid clearance from the blood.

Before coming to Hopkins, Dr. Iio worked in the Department of Internal Medicine at Tokyo University, headed by a cardiologist, Dr. Hideo Ueda. When Dr. Julius Krevans visited Professor Ueda, they agreed that Dr. Iio should spend some time at Hopkins in cardiology. When this was not possible, Dr. Iio was accepted to join our new Nuclear Medicine Division at Hopkins. He became enamored about the field of nuclear medicine and dedicated the rest of his life to it, even though he subsequently became Chairman of the Department of Radiology at Tokyo University. Without any special training in Radiology, he studied hard on his own, and passed the examination required for certification in Japan as a radiologist.

Nuclear medicine has long been a major focus of Japanese medicine. In the early 1950s in Japanese universities, such as Keio University, a program in nuclear medicine was developed under the direction of Professors H. Yamashita, I. Kuramitsu, and F. Kinoshita. In 1951, Professor Yamashita was sent to the United States by the Japanese government to study the use of radioisotopes in medicine. He became friends with many prominent nuclear medicine physicians and scientists in the United States, including Drs. Charles Dunham, R.T. Overman, Marshall Brucer, W. Chamberlain, and Hal Anger.

The early studies of Dr. Yamashita's group involved cobalt-60, phosphorus-32, and iodine-131. Radionuclide therapy as well as diagnostic uses of radiotracers was emphasized.

Another institution emphasizing nuclear medicine in Japan was Nagoya University, a program under the direction of Dr. T. Nagai, who worked at the National Institute of Radiological Sciences from 1959 to 1974. He also spent some time at the Oak Ridge Institute of Nuclear Studies in Oak Ridge. From 1968 to 1973, he was a Senior Scientific Officer at the International Atomic Energy Agency (IAEA). In 1974, he was appointed as a Chairman of the Radiology Department in Gumma University

Structure, Function and Chemistry

Computed tomography (CT) and magnetic resonance imaging (MRI) are oriented toward structure while nuclear medicine defines the molecular manifestations of disease. For example, in 2004 in the United States there were 2,000 sites producing 1,000,000 patient examinations with PET imaging using a single radioactive tracer, fluorine-18 fluorodeoxyglucose (FDG).

More and more, radiologists are becoming involved in positron emission tomography (PET) and single photon emission computed tomography (SPECT) often combined with CT or MRI. Less than a third of the physicians certified by the ABNM end up in the full-time practice of nuclear medicine. In many hospitals, a "nuclear radiologist" interprets nuclear medicine images as well as radiographs. There is increasing integration of nuclear medicine within radiology to provide efficient, high quality and effective patient care and to interpret PET/CT, SPECT/CT and other images that fuse anatomy, biochemistry and physiology.

Computed tomography or magnetic resonance images fused with PET or SPECT images must be of diagnostic quality. Training in CT is now required for all nuclear medicine physicians, as is now required for radiology residents. Training to achieve expertise in PET/CT and SPECT/CT are required in radiology residencies. Both nuclear physicians and radiologists must be able to interpret PET/CT and SPECT/CT images.

Particularly because of the increasing growth of radionuclide therapy, full-time nuclear physicians have increasing involvement in total patient care. They are involved in vitro imaging of gene and other arrays of data, interpreted together with in vivo molecular imaging. The increasing complexity of nuclear medicine procedures warrants the full time efforts of full time nuclear physicians. Integration of imaging with other specialties should be at an institutional level, not restricted to a single department.

The exciting results by John Lawrence and colleagues with phosporus-32 to treat more than 100 patients in the 1930's led William Donner to provide funds for the construction of the "Donner Laboratory" at the University of California in Berkeley, dedicated to the "application of physics, chemistry and the natural sciences to biology and medicine." In 1948, John Lawrence became Associate Director of the Donner Lab. Over the ensuing years, he and his colleagues pioneered the use of radioiron to label the hemoglobin of red blood cells. His colleagues carried out pioneering studies of erythropoietin, the hormone controlling the production of red blood cells.

At the end of the war, in October 1945, President Truman appointed a commission to control the production and use of atomic power in the United States. He proposed that a civilian "Atomic Energy Commission (AEC)" should direct all atomic energy programs, extending from bomb production to nuclear power to medical and research uses of radioactive materials. He stated that "atomic power can become a powerful and forceful influence towards the maintenance of world peace," and called for talks leading to international cooperation to expand the peaceful uses and control of the military uses of atomic energy. After prolonged debate about the relative merits of military and civilian control of atomic energy, five civilians were appointed to head the newly created AEC. The Atomic Energy Act of 1946 gave the AEC total control of atomic energy. All research, development, and the production of atomic bombs was to go through the Commission, as well as every peaceful use of atomic energy. The AEC was to be under the jurisdiction of the Joint Congressional Committee on Atomic Energy (JCAE), aided by a civilian nine-member General Advisory Committee.

The mission of the AEC was as follows:

"... it is hereby declared to be the policy of the people of the United States, that, subject at all times to the paramount objective of assuring the common defense and security, the development and utilization of atomic energy shall, as far as practicable, be directed toward improving the public welfare, increasing the standard of living, strengthening free competition in private enterprise, and promoting world peace."

On Friday, June 14, 1946, in the journal Science, an announcement made by the Manhattan Project Headquarter, Washington, D.C. said:

"Production of tracer and therapeutic radioisotopes has been heralded as one of the great peacetime contributions of the uranium-chain-reacting pile. This use of the pile will unquestionably be rich in scientific, medical and technological applications."

On August 2, 1946, the first shipment of radioisotopes was sent out from Oak Ridge, Tennessee to the Bernard Free Skin and Cancer Hospital in St. Louis. It was a compound labeled with carbon-14. By 1962, the Oak Ridge National Laboratory alone had made over half a million shipments totaling 1,600,000 Curies of radioactivity. By July 31, 1964, there were 1,201 institutions licensed by the AEC to receive and use radioisotopes.

Not everyone greeted the creation of the AEC as a boom for the U.S. and the world. Some, such as Ralph Nader, believe: "The history of the AEC and the JCAE can soberly be described as outrageous." He and his followers accused the AEC, in their zeal to put atomic energy under civilian control, of being "a totalitarian monster" that would be out of control in the promotion of nuclear power.

A major debate at that time was whether, as the military wished, atomic energy would be kept secret and totally held within the government. Others believed that secrecy was a hopeless task, so that the path should be "control" after worldwide release of nuclear energy and other "peaceful uses of atomic energy." The explosion of a Soviet atomic bomb on September 23, 1949, supported the belief that atomic energy could not be kept secret. On December 8, 1953, President Eisenhower delivered his famous "Atoms For Peace" speech before the United Nations General Assembly in New York.

The year 1953 also marked the signing of the armistice ending the Korean War. The structure of DNA was published by Watson and Crick, Edmund Hilary and Tensing reached the summit of Mount Everest, and the Society of Nuclear Medicine (SNM) in the United States was conceived in the Davenport Hotel in Seattle by twelve persons who visualized the potential value of radioactive tracers in medicine and biomedical research. Five were radiologist, two internists, a cardiologist, a physicist, an engineer, and a nuclear medicine physician. Thomas Carlyle was the first president; Asa Seeds was secretary. Joseph P. Nealen, S.J. was a priest, chosen in case "divine intervention should become necessary." Jeff Holter, who later would invent the "Holter" monitor for detection of cardiac irregularities, had conceived of the idea of a Montana Society of Nuclear Medicine. They decided that every 4th president be a scientist, not an M.D. The first annual scientific meeting of the SNM was held on January 19, 1954.

At the time of President Eisenhower's speech, the world was a terrifying place. The Soviet Union possessed hydrogen bombs, with 1,000 times more devastating potential effects than occurred at Hiroshima and Nagasaki. In his speech, Eisenhower stated three goals were: (1) To work with the Soviet Union to transform military to peaceful uses of

Figure 46 Paul Aebersold and John Lawrence. Paul was the first successful promoter of radioisotopes in biomedicine by the Atomic Energy Commission, and John was the first to use ^{32}P in treatment of leukemia.

atomic energy; (2) To negotiate non-proliferation agreements with the Soviet Union; and (3) To involve nations throughout the world, large and small, in peaceful efforts to develop atomic energy for peaceful, rather than military, uses.

The President said: "It is not enough to take this weapon out of the hands of soldiers. It must be put in the hands of those who know how...to adapt it to the arts of peace ... This greatest of destructive forces can be developed into a great boon for the benefit of all mankind ... if the entire body of the world's scientists and engineers had adequate amounts of fissionable material with which to test and develop their ideas, this capability would be rapidly transformed into universal, efficient and economic usage."

His proposals resulted in the establishment of the International Atomic Energy Agency (IAEA), which came into being in 1959.

In 1946 a resolution was passed unanimously at the Untied Nations to establish a United Nations Atomic Energy Commission "to make proposals for exchanging basic scientific information, confining atomic energy to peaceful purposes, eliminating atomic and other weapons of mass destruction from national armaments and effectively safeguarding complying states." From the time of the first atomic bombing in Japan, American scientists and others believed that the "positive uses" of atomic energy for electricity generation and biomedical uses would play an important role in the process of eliminating or limiting military uses. They believed that the development of atomic energy for peaceful and military purposes were intimately related and interdependent. They believed that development of peaceful uses would attract "the best people of good will, imagination and ingenuity." This was all at a time when the United States had a monopoly on nuclear weapons.

Even before the atomic bombings, many nuclear physicists who had worked on the development of the atomic bomb believed that other great powers would never permit the United States to retain its monopoly on the uses of atomic energy. They feared that the inevitable race for atomic weapons would turn the world into a "flaming inferno." Their goal was to try to insure that no nation would ever be able to produce fissionable material that could be suitable for atomic weapons. They believed that atomic weapons were a threat to civilization. Others, led by Ernest Lawrence believed that, by strenuous efforts, the United States would be able to retain leadership in the field of atomic energy and nuclear weapons, that nuclear materials should be stockpiled for both military and industrial use, and the door opened for widespread industrial and medical applications. On the other side, Leo Szilard and others were convinced that most people did not grasp the true threat of nuclear weapons. Robert Oppenheimer took the position that by fostering uses of atomic energy under international control, the United States could emphasize the peaceful uses.

A Committee on Political and Social Problems, chaired by physicist James Franck with other physicists who worked on the development of the atomic bomb, including Leo Szilard and Glenn Seaborg, made two proposals: (1) that secrecy or attempts to corner the supply of raw nuclear materials would be futile; (2) that the only hope was international control; and (3) that atomic bombs should not be used on the Japanese without warning them. They proposed a demonstration of the power of the atomic bomb in an uninhabited area, rather than using the bombs on the Japanese. Others believed that this would not convince the Japanese, and would result in a huge loss of life in the invasion of Japan that was imminent. The decision was made that, even viewing the Franck report, the bombs should be dropped on military sites in Japanese cities without warning. Hiroshima had military installations surrounded by homes and other buildings that would be likely to be damaged.

A report, submitted by other scientists, proposed that the U.S. government support a huge program to make possible fundamental scientific research, in addition to military, industrial, and medical research. Secrecy was to be minimized, and there should be maximum collaboration with other world powers. Others believed that people in the military would prejudice attempts to control the further development of atomic weapons.

On March 28, 1946, Acheson and Lilienthal released a report that supported international control of nuclear energy including nuclear weapons. Both the *Chicago Tribune* and the *Washington Times Herald* newspapers denounced the report as a scheme to give away the secrets of the atomic bomb to the Russians. Most of the public, however, favored the recommendations of the report. Alfred Friendly of the *Washington Post* saw the report as offering the best hope for lifting the "great fear" that had descended over the world when they heard of the atomic bombings. On June 14, 1946, Bernard Baruch proposed that the United Nations recommend the establishment of an international development authority and presented an "American Plan for International Control." The United Nations through this authority would be entrusted with all phases of the development and use of atomic energy. The authority would exercise managerial control and ownership of all activities potentially dangerous to world security. It would have power to control, inspect, and license all others. It would foster beneficial uses and assure research and development of these uses.

These attempts to achieve international control foundered on the question of whether punishment for violations would be subject to veto by the members of the Security Council. Baruch proposed that "there shall be no legal right, by veto or otherwise whereby a willful violator of the terms of the treaty or convention shall be protected from the consequences of violation of its terms." Andre Gromyko, representing the Soviet Union, charged that the American plan defied the principles of the United Nations Charter by insisting on the veto of punishment allocations. The Soviets criticized the United States for interfering with acceptance of a plan that had overwhelming international support. Some questioned the United States' sincerity in its proposal for international control.

It took until 1954, the year after President Eisenhower's Atoms for Peace speech, for the Atomic Energy Act to be amended to make possible private ownership of nuclear reactors and other uses of fissionable materials, rights previously limited to the Atomic Energy Commission. In collaboration with industry, the AEC initiated a five year Reactor Development Program. By 1956, this program had received $15 billion in federal funds.

There are four types of people in this world: those who make things happen; those to whom things happen; those who watch things happen; and those who don't even know things are happening. Paul Aebersold made things happen. He was a major contributor to the development of nuclear medicine, long before it was recognized as a medical specialty. He was a prize student of Ernest O. Lawrence, and learned much from his mentor, the first organizer of "Big Science" in the United States. He brought together a team of researchers to obtain the funds from private sources and government to work on huge projects.

As a graduate student in the Radiation Laboratory at the University of California at Berkeley, Paul wrote an article, entitled The Cyclotron: A Nuclear Transformer (*Radiology*, vol 39, 513–540, November 1942). In his clear writing and speaking style, that was to be a great asset in his crusade, he told us that: "While the solar system is held together by gravitational forces between masses, the atomic system is maintained by the electrical forces between charges."

"While others were seeking methods of attaining very high voltages, Professor E.O. Lawrence conceived the idea of speeding a particle up to very high energies by giving it successive accelerations. Instead of trying to attain a potential at some point in an apparatus great enough to give a particle all of its energy in one push, a much lower voltage would be used to give the particle a series of pushes."

Paul Aebersold described the history of development and characteristics of all the cyclotrons built in the Radiation Laboratory, a must reading. He concluded his article: ". . . man is investigating his universe in one instance by building a huge observatory and instrument to look out into the vast realms of space at things millions and millions of times larger than himself, whereas in another he constructs an apparatus for the purpose of looking into the most infinitesimal realms, millions and millions of times smaller than he." He was clearly alluding to the Rockefeller Grant to California Institute of Technology in 1928 to build the 200-inch telescope at the Palomar Observatory.

Paul met Lawrence in 1932, when he went to Berkeley to try out for the U.S. Olympic Team. In the words of Nobel Laureate, Glenn T. Seaborg, "He didn't make the Olympic track team, but he did qualify for the physics team." Walking through the Le Conte Hall on a Saturday afternoon, he saw a man standing near a strange device. Paul asked if Professor Lawrence was around. The man said: "I'm Professor Lawrence and this is a cyclotron."

In the words of Glenn T. Seaborg in the Paul Aebersold Memorial Lecture on December 2, 1969, at the American Nuclear Society: "I have many memories of Paul as a warm and vital person ... pre-war members of the Radiation Laboratory share many memories of the festive dinners at the DiBiasi Restaurant ... held when there was some scientific or other achievement to celebrate, such as Professor Lawrence's winning the Nobel Prize in 1939 ... these were always organized by Paul and staged under his leadership as master of ceremonies ... In 1946 Paul was placed in charge of radioisotope work at Oak Ridge, Tennessee, where he directed the first peaceful distribution and use of nuclear products." The first announcement of the availability of reactor-produced radioisotopes for public distribution was published in the June 14, 1946 issue of *Science*.

Eleven years later he was transferred to Atomic Energy Commission (AEC) Headquarters in Washington where he presided over the phenomenal growth in the production and use of radioisotopes. "He was responsible, more than any other individual for the acceptance and application of radioisotopes in this country, and contributed a great deal to their use abroad ... He really became "Mr. Radioisotope" as the result of his energetic, influential, and successful efforts to introduce the use of radioisotopes into a broad spectrum of scientific and practical applications. He retired as Director of Isotopes Development in the AEC in 1965.

On June 14, 1973, the first presentation of the newly created Paul C. Aebersold Award of the Society of Nuclear Medicine was presented to Bill Myers. In his presentation remarks, President Monte Blau said: "He (Bill Myers) holds the distinction of having introduced into medicine more nuclides than any other single person: Cobalt-60, gold-198, chromium-51, iodine-125, iodine-123, strontium-87m, carbon-11, and others." Mrs. Aebersold presented the award. In her remarks, Mrs. Abersold said: "He (Paul) was a man with a mission, and a sense of humor. He taught our parakeet to say "R-A-D-I-O-I-S-O-T-O-P-E-S.""

Bill Myers pointed out that Paul was behind the formation of the Society of Nuclear Medicine from the start, and was its first Honorary Member. He referred to the classic article The Development of Nuclear Medicine, by Paul in the *American Journal of Roentgenology, Radium Therapy and Nuclear Medicine*, Vol. LXXV, No. 6, June 1956. In this paper, Paul concluded: "Great advances have been made in nuclear medicine in the last two decades. The work in the first decade with accelerator-produced radioisotopes laid a firm groundwork for development in the last decade with reactor-produced radioisotopes ... On the basis of past developments and the present state of nuclear medicine, we can expect a continuous and steady growth in techniques and uses. It is impossible to anticipate all the future benefits mankind will derive from this new field, nuclear medicine." He wrote this half a century ago.

Figure 47 Paul Aebersold with Neils Bohr.

In January 1957, a new agency, the Energy Research and Development Agency (ERDA) was established to incorporate and expand the research activities sponsored by the Atomic Energy Commission. No other federal agency or private industry had the needed combination of isotope production facilities, specialized laboratories, including Brookhaven, Oak Ridge, and Los Alamos National Laboratories, experienced research teams and experience with technology transfer from basic physics and engineering research. In a Feberuary 1979 Task Force report from ERDA at the time, it was stated: "Instrumentation development may well hold the most promise for providing, in the near future, a more effective and efficient biomedical research program and improved diagnosis of disease."

President Eisenhower first proposed the creation of the International Atomic Energy Agency (IAEA), a part of the United Nations in his "Atoms for Peace" speech before the UN in New York in 1953. It was founded in 1957. More than 2,000 persons from more than 80 Member States of the UN work at the IAEA headquarters in Vienna, Austria. Over the past 40 years, the IAEA has played an important role in introducing nuclear medicine in 39 countries. A major accomplishment has been the establishment of 350 radioimmunoassay (RIA) laboratories, particularly in developing countries. Since 1991, the IAEA has also donated 48 gamma cameras in these countries. In countries all over the world, professionals in the field of nuclear medicine have been encouraged, educated and supported by the IAEA. Over the past decade, more than 600 specialists in nuclear medicine have been trained under IAEA-sponsored programs.

Among IAEA sponsored physicians who came to Hopkins was Professor Fevzi Renda. In November 1945, he came to the U.S. Marine Hospital in Baltimore in internal medicine, where he worked for 15 months under Dr. Luther Terry, who was subsequently to become head of the U.S. Public Health Service, and famous for his successful campaign to try to get people to stop smoking.

Fevzi then went to Washington University, where he worked with Professor Barry Wood, a famous infectious disease specialist, for 5 years, returning to Ankara, Turkey, in 1951. In 1959 he was selected to start a nuclear medicine division by Professor Mahmet Ali Tanman, head of the Radiobiological Institute in Ankara. He took a 5 week nuclear medicine course at Sheffield and Leeds in the United Kingdom.

In 1965 he came to Johns Hopkins Nuclear Medicine under IAEA sponsorship, and then returned to Ankara University, where Nuclear Medicine became an independent specialty in 1974. He also worked for the U.S. Embassy in Ankara for 49 years, examining all Turks emigrating to the U.S. In 1988 he retired from nuclear medicine.

Another source of funds for trainees in nuclear medicine was the United Nations Enviroment Programme, which was concerned with environmental medicine, as well as nuclear medicine. Dr. Medat Mohammed Osman from Egypt was one of the trainees who subsequently joined Dr. Cahid Civelek in nuclear medicine, first at Hopkins and now at St. Louis University.

The IAEA has been the focus of controversy since its creation, perhaps none greater than today. Helping prevent proliferation of nuclear weapons in countries throughout the world has been its greatest challenge. Another goal of the IAEA is to promote the peaceful uses of nuclear energy, including nuclear power. Nuclear energy has become a major source of the world's energy.

In the United States, the Department of Energy (DOE) has played a major role, its office of Biology and Energy Research (BER) dedicated to advancing nuclear medicine and biology. In a lecture at Oak Ridge, Tennessee, on September 19, 1983, I stated:

Figure 48 Dr. C. Civelek, a trainee in nuclear medicine at Johns Hopkins who stayed to become an Associate Professor at Johns Hopkins, in charge of nuclear cardiology.

Figure 49 Professor Ludwig Feinendegen, a pioneer in nuclear medicine from Germany who often visited nuclear medicine at Johns Hopkins.

Figure 50 Michel Bourgignon, a trainee in nuclear medicine at Hopkins from Orsay, France.

"The field of nuclear medicine will continue to benefit from the efforts of the DOE. No force in the country or in the world has done more to develop nuclear medicine than the DOE." Enormous number of radioactive tracers, radionuclides, and instruments, such as the rectilinear scanner, the Anger camera, computers, and the human genome project have all been developed originally in the National Laboratories and by the extramural research supported by the DOE. The DOE, as did its predecessor, the Atomic Energy Commission (AEC), continues to play an important role nuclear medicine today. Unfortunately, there have recently been severe cuts in the budget for nuclear medicine research in the Department of Energy (DOE).

On November 18, 2004, I presented the 13th Annual Edward Teller Lecture at the annual meeting of the Citizens for Nuclear Technology Awareness. During a visit to the Savannah River National Laboratory, I saw the efforts being made to try to transfer the technology developed over the past half century to the advancement of biomedical research. For example, they have discovered bacteria that can divide and grow at radiation levels 4,000 times those that would be fatal to human beings. The genomics of these organisms is of great interest.

Compared to the other National Laboratories of the DOE that have contributed directly to the birth, development and growth of nuclear medicine, the Savannah River Site (SRS) has not in the past been involved in biomedical research, but plans to do so in the future. It has recently been designated as a National Laboratory. Originally, the SRS was not part of the Manhattan Project, but was created in 1950 after the world learned of the explosion of a hydrogen bomb by the Soviet Union in 1949. The government designated the E.I. duPont de Nemours Company as contractor for the design, construction, and operation of the site near Aiken, South Carolina, which was chosen because of its weather, water supply, available labor, suitable space (over 300 square miles), and little public objection. During the construction, it was necessary to relocate six small towns and their inhabitants, greatly taxing services such as schools, fire and police protection. The largest town was Ellenton with a population of 600 and Dunbarton with a population of 230. The site, which contained 300 miles of roads and 60 miles of railroad tracks, was to produce most of the plutonium and tritium needed for the construction of H-bombs. The current budget, 80% of which is devoted to waste removal by the Defense Waste Process Facility (DWPF), is $1.6 billion/year, with an employment of over 10,000 persons. The high level radioactive waste, now in 48 tanks each containing 1 million gallons of high level radioactive waste, was begun in 1989 to be incorporated into borosilicate glass and stored in steel canisters $^3/_4$ of an inch thick.

One of the most important contributions of the biological programs of the DOE has been the Human Genome Project, which became a joint effort of the DOE and NIH in 1990. It goal was "... to characterize all the human genetic material—the genome—by improving existing human genetic maps, constructing physical maps of entire chromosomes, and ultimately determining the complete sequence ... to discover all of the more than 50,000 human genes and render them accessible for further biological study." The initial DOE work focused on the assessment of radiation effects on genetic material, but was subsequently expanded. The original plan was modified in 1993 to "to develop technology for high-throughput sequencing, considering the process as integral from template preparation to data analysis," and "to develop methods for identifying and locating genes."

Perception of Risks

Perception of the risks from radiation has always been a problem. In 1957, Representative Chet Holifield, a member of the Joint Committee on Atomic Energy (JCAE) from California, said: ". . . the AEC approach to the hazards from bomb fallout seems to add up to the party line—play it down." Neither the AEC nor the nuclear industry was successful in educating the public about what is and what is not risky in developing the peaceful uses of atomic energy.

The same day that Holifield spoke, 100 miles east of the Bikini Atoll in the South Pacific, Japanese fishermen aboard the *Lucky Dragon* were hauling in their nets. They had not heard the radio broadcast warning from the Japanese Maritime Board that there was going to be an atomic bomb test in the vicinity of the Marshall Islands.

On March 10, the AEC had reported that 236 Marshall islanders had been evacuated from their homes before the test as a precautionary measure. On March 16, 1957, all over the world the news was broadcast that twenty three fisherman had suffered severe radiation exposure.

As early as 1946, Operation Crossroads was carried out on the Bikini Atoll after 162 natives had been moved to another atoll. In July 1947, the AEC established a proving ground for testing nuclear weapons on Eniwetok Atoll in the Marshall Islands. Tests were carried out in 1948, the same year that the Soviet Union exploded their first atomic bomb in Siberia. On January 27, 1951, a second testing operation, Operation Ranger, was developed by the United States near Las Vegas, Nevada.

When Shinzo Suzuki first saw a whitish-yellow glow on the horizon, he thought it was the sunrise, but then realized that it came from the west, not the east. When the glow changed to a flaming orange color, he shouted: "It's a pika-don," a term used for the atomic bombing of Hiroshima and Nagasaki eight years before.

What they were seeing was the result of the explosion of the world's first hydrogen bomb. Soon they felt an enormous shock wave, and saw a giant mushroom cloud extending into the sky. As their boat speeded away, ash rained down from the sky, which by then had become completely clouded over. Hours later, several men felt nauseous, and saw their skin darkening. When they reached shore at their home port of Yaizu two weeks later, 21 of the men were hospitalized. Their thyroids contained large amounts of radioactive iodine; the skins, especially their scalps and chest, were burnt. Their white blood cell counts continued to decrease for a month. In 6 weeks, all but one had recovered. The radio operator, who became jaundiced after 3 months died six months later, his death being attributed to blood transfusions which he had received.

Japanese physicians and scientists were unprepared for this catastrophe because they had been forbidden by the government, under the order of General Douglass MacAuthur, to conduct research or become involved in nuclear science. They were forbidden to obtain radioisotopes even though carbon-14 and other radionuclides had been made available to other qualified scientists and physicians throughout the world for scientific and medical purposes. Indeed, the U.S. Army dumped into Tokyo Bay two cyclotrons that had been been built under the direction of Yoshio Nishina, the foremost nuclear physicist in Japan. Nishina directed the building of the first cyclotron in Japan, the second in the world, the first being Lawrence's in Berkeley. The second cyclotron built by Nishina was as large as Lawrence's.

Figure 51 Yoshio Nishina, who introduced cyclotrons into Japan.

When the news of this destruction of the two cyclotrons in Japan reached the United States, there were great protests from the community of physicists. They used the event to support the concept that nuclear energy in the United States should be under civilian rather than military control.

Professor Nishina had studied with Niels Bohr, the creator of the model for the structure of the atom, in Copenhagen before the war. Returning from Denmark, he led a group of Japanese physicists in assessing the possibility of the Japanese developing an atomic bomb, concluded it was impossible, and dropped the project. The Institute for Physical and Chemical Research where they worked was destroyed in April 1945 by an incendiary bomb, dropped by B-29s during the massive bombing of Japanese cities.

Nishina's goal had been to make Japan a center of international science and prevent what he feared was going to be the destruction of the human species. After the destruction of his cyclotrons, which for him symbolized the destiny of Japan to reconstruct itself after the devastation of the war, his personality changed completely, and, with bitterness, he devoted the rest of his professional life to the re-industrialization of Japan.

By the summer of 1954, at a joint Japanese/American conference in Tokyo, two American physicists, J. Harley and M. Eisenbud, presented to the Japanese their first scintillation detector for measuring radiation. The first use of this instrument was called Operation Troll. Aboard a U.S. Coast Guard cutter, Japanese scientists collected sea water and seafood samples to assess whether fallout had resulted in contamination. They concluded that the levels measured would not present any health hazards. They also obtained important information about currents throughout large parts of the Pacific Ocean.

Momentous events in the 1960s affected the field of nuclear medicine. Everyone remembers what he or she was doing the day President Kennedy was shot. I was on my way over from the hospital to the "dog lab," in order to carry out experiments evaluating different tracer methods to measure brain blood flow. Lyndon Johnson became President, the next 7 years were dominated by the Vietnam War.

I remember the night of the Democratic Convention in Chicago in 1968. Our whole family was on a week-long cruise of the Chesapeake Bay in a rented Rainbow class 21-foot sailboat. We had a portable radio and were listening to the description of riots outside the Hilton Hotel. I had previously met Hubert Humphrey, Vice President under President Johnson, and the nominee for the Presidency, in 1968 at a dinner honoring Senator Lister Hill, who had done so much for American medicine. Vice President Humphrey was the Principal speaker. He began his talk by reminding Senator Hill that: "flattery is OK, as long as you don't inhale it." An afternoon symposium and the dinner in the evening was a great success. To honor Lister Hill, Senator Joseph Tydings from Maryland sponsored both events, which were held in the Senate Office Building and Washington Hilton Hotel. Thirty-five members of Congress, as well as Vice-President Humphrey, attended.

In 1968, in Chicago, Humphrey described the problems facing America. Bobby Kennedy and Martin Luther King has been shot; there was violence and riots in cities all over the United States; the Cold War was at its height with revolts against the communist government in Czechoslovakia; and year long conflicts after the Tet offensive in Vietnam in January 1968. Some wanted the United States forces to leave Vietnam immediately. Outside the convention hall there were jeeps in Grant Park opposite the Hilton Hotel. All the social conflicts of the Sixties were on display on the streets of Chicago.

For the Democrat party, Chicago '68 doomed the candidacy of Hubert Humphrey and set off shock waves of reform, but intensified the revolutionary fervor that would spawn street violence and bombings. Fears of nuclear devastation by nuclear weapons became more widespread, and fear of radiation increased.

Ten years later, we in nuclear medicine in the medical school and radiological sciences in the School of Public Health realized that we needed to become more involved in the issue of radiation risks. Throughout the development and growth of nuclear medicine, we had always keep in mind the benefits versus the risk, both real and perceived. The effects of low levels of radiation are too small to be measured, and for decades an assumption had been made that any exposure to radiation carried some risk of harm, either to the person receiving the radiation or to future generations. This is called the "zero threshold hypothesis." Regulatory agencies assume that there is no safe dose of radiation. Public concern about all aspects of radiation—nuclear weapons, nuclear power, and medical radiation—were greatly increased by the accidents at Three Mile Island and Chernobyl, and led to a striking decrease in the numbers of persons applying for entrance into radiation related programs, both medical and non-medical, as well as a hesitancy of institutional review boards to approve research protocols that involve

Figure 52 Vice President Humphrey honoring Senator Lister Hill.

Figure 53 Vice President Humphrey, Robert Williams (Chief of Medicine at the University of Washington), and Senator Joseph Tydings from Maryland.

human exposure to radiation. There was also a hesitancy to develop facilities for the safe disposal of high and low level radioactive waste.

After the accident at the nuclear reactor at Three Mile Island in 1979, I was appointed Chairman of the Maryland Governor's Committee on Three Mile Island. Our charge was to insure that the interests of Maryland were served during the clean-up process after the accident. In 1983, we created the Center for the Advancement in Radiation Education and Research (CARER) in the Johns Hopkins School of Public Health to provide public education about radiation. For years, CARER sponsored a series of educational symposia, covering nuclear energy in Maryland, radon in homes, and radiation and women. The center also sponsored lectures in public schools.

We also developed procedures to better inform patients having procedures involving exposure to radiation. The following advice was given:

1. Ask your doctor why the recommended x-ray or nuclear medicine procedure is needed.
2. Ask how the procedure will influence treatment decisions or outcome.
3. Don't insist on or suggest an x-ray or nuclear medicine procedure either directly or by inference.
4. Don't refuse a procedure involving radiation just because you are worried about the harmful effects of ionizing radiation. Refusing to have the procedure may be harmful.
5. Tell your physician if you are or might be pregnant.

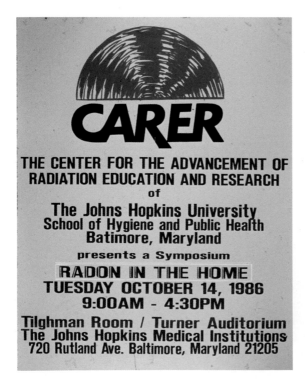

Figures 54–56 Announcements of symposia sponsored by CARER.

RADIATION AND WOMEN

A Symposium
Presented by

THE CENTER FOR THE
ADVANCEMENT OF RADIATION
EDUCATION AND RESEARCH

of

The Johns Hopkins University
School of Hygiene and Public Health
Baltimore, Maryland

Wednesday, April 2, 1986
8:15 a.m. – 5:00 p.m.

AFTER CHERNOBYL

Briefing on the Chernobyl
Nuclear Reactor Accident

Presented by
The Johns Hopkins University
CENTER FOR THE ADVANCEMENT OF
RADIATION EDUCATION AND RESEARCH

Wednesday May 21, 1986
1:00 – 4:00 PM
East Wing Auditorium
School of Hygiene & Public Health
615 North Wolfe Street
Baltimore, Maryland

PRESS INVITED   **OPEN TO ALL**

6. Ask for lead shielding of your reproductive organs during the procedure.
7. Keep a record of all procedures that you have that involve radiation.
8. Exaggerated perception of risk, grounded on misinformation or distrust of authority figures can be harmful.

CARER's profile was raised by the Soviet accident in which there was a reported meltdown of a nuclear reactor at Chernobyl. We sponsored a news conference on April 29, 1986, where four experts assessed the damage caused by the accident. We cautioned that the information from the Soviet Union was too sketchy to predict long-term effects. I had been in telephone contact with Swedish and Danish colleagues. According to Bengt Langstrom in Uppsala, Sweden, rain contained radiation levels 10–20 times the usual background measurements. The radiation was coming chiefly from cesium-137. All concluded that "there is no evidence that the American population is in any way threatened . . . The U.S. is in no danger . . . If you're in western Russia, you're in real trouble."

I had back-to-back appearances on the MacNeil/Lehrer Newshour, and fielded many telephone calls from reporters in Los Angeles, Baltimore, and other cities throughout the country. The speed with which CARER responded to the TMI and Chernobyl accidents justified its creation in 1983. Subsequently, I testified to the Senate Foreign Relations Committee on the identities of the radionuclides being detected in Scandinavia.

The accident at Chernobyl resulted in as many deaths as all prior nuclear accidents in the world between 1944 and 1986. Despite the accident, the Russians (USSR at that time) stated they still intended to increase the amount of electricity generated by nuclear power from 170 billion kilowatt-hours to 360 billion kilowatt hours by 1990. The number of nuclear power plants was expected to increase from 15 to 31 by the year 2000.

The Asian Rare Earth Company (ARE)

CARER was later involved in an intensive debate about the siting of a low level radioactive waste disposal site in Ipoh, Malaysia, that would eventually shut down the Asian Rare Earth (ARE) plant. A trial in the high court of Ipoh involved experts and anti-radiation activists from all over the world. The ARE started operation in May 1982 as a joint venture between Malaysians and Japanese. Rare earths were shipped to Japan to yield yttrium and other elements used in making television sets, computers, steel, and superconducting magnets.

To mine rare earth elements, tin and other ores are extracted from sand. The residue, called monozite, contains the radioactive elements, uranium and thorium. Before the operation of the plant, the monozite was not handled as radioactive material, even though it contained 6% thorium. After processing in the plant, the thorium was concentrated to 12%, and then handled as radioactive waste in an enclosed radioactive waste storage area. One of the problems was that the storage site was close to a Chinese cemetery, aggravating tension between two often-opposing groups in Malaysia, the native Malays and the Chinese-Malaysians.

After the plant had been operating for $2\frac{1}{2}$ years, villagers from Bukit Merah, near the storage site, obtained an injunction restraining the company from producing, storing or keeping radioactive materials in the plant. In July 1984, the residents formed an Anti-Radiation Waste Committee, and were joined by the Environmental Protection Society of Malaysia, the Papan Action Committee, and the Friends of the Earth. The trial began in 1987, and was held intermittently over a period of two years. I had accepted the invitation to serve as an expert witness for the defense on February 10, 1987.

At the end of the trial, which extended over two years, Judge Peh of the High Court denied the claim by the plaintiffs that abortions and other medical conditions were the result of the operation of the plant, but ruled that the plant should be closed nevertheless.

The Company appealed the decision to the Supreme Court. On December 23, 1993, the Supreme Court stated that the evidence of the ARE's experts had been accepted, and there was no legal basis for the injunction to close the plant. Even though the decision made by the original trial judge ordering the plant to be closed in 1990 had been reversed by the Supreme Court, the ARE owners decided to close the plant for good. The lawyer

Figure 57 Entrance to the courthouse in Bukit Merah, Malayasia, at the trial of the Asian Rare Earth Company.

Figure 58 Courtroom scene in Ipoh, Mayalsia, at the trial of the Asian Rare Earth Company.

for the defense, Datoh Gill, was subsequently installed as a Judge on the Malaysian High Court on April 27, 1992.

The judgment by the Supreme Court stated that "the bulk of the evidence consisted of scientific evidence of experts. The Plaintiff's expert conducted tests and took measurements for a very short period of time, ranging from one day to 3 months, whereas the Defendant's experts took measurements over a period of 3 years. The Learned Trial Judge found the evidence of the experts neutralized each other but he preferred the evidence of the defendants. The Learned Trial Judge erred in law with regard to the factual and scientific evidence."

An important issue at the trial was the genetic effects of radiation. The "Zero-threshold" hypothesis for genetic effects was developed because there was no clear-cut scientific evidence to prove the existence or absence of a threshold. At that time there was no evidence of the existence of naturally-occurring mechanisms for the effective repair of damage to genetic DNA. Such evidence now exists, and the concept of a "practical threshold" for harmful effects of low level radiation has been proposed. The threshold is defined as the dose level below which there will be no harmful effects in view of

Figure 59 Lawyers for the defense in the trial of the Asian Rare Earth Company. Dato Gill on the left; R. Cheliah on the right.

Figure 60 Radiation monitoring at the Asia Rare Earth Company.

Figure 61 Radiation control site in Ipoh, Malaysia.

the life expectancy of the exposed person. For example, from the study of radium-dial painters, it had been concluded that if the radiation dose to bone had been less than 39 millirem/day, the time of appearance of bone cancer would exceed the lifespan of the exposed person.

We receive external radiation from outer space in the form of cosmic rays, and radio-active materials that occur naturally in the earth's crust. In Colorado, for example, radiation levels are 140 millirems/year because of the high altitude, compared to 38 millirems in Florida at sea level. The larger amount of air at the lower altitude absorbs cosmic radiation. Internal radiation from naturally occurring radioactive potassium within our body amounts to 20 millirems/year in men and 15 millirems in women. Of every 10,000 potassium atoms in the body, one is radioactive. Thus, exposure to radiation can never be completely eliminated. The question is what is a safe dose of radiation from human activities?

Radon coming from the soil causes a constant exposure to radiation, and has caused great public concern in the past, affecting the choices of house sites and the building industry. Radiation from medical x-rays accounts for at least 20% of the exposure of an average person in the United States. Today the greatest fear of radiation remains fear of a nuclear weapon exploded in a major city by international terrorists. The IAEA, which has played an important role in limiting nuclear weapons to a club of six nations, is facing immense challenges. An important challenge remains to define a safe dose of radiation, particularly if it ever becomes necessary to clean up the effects of "dirty bombs," i.e. those which deliver large amount of radioactivity over large areas.

A safe dose is that from which no deleterious effects would occur. Identifying safe doses is not simple because the effects can often not be measured. One in five persons in the United States dies of cancer from all causes, so it is impossible to detect an increased incidence from radiation from human activities. No more than 2% of cancer is caused by natural or man-made radiation.

Many view all aspects of the "atomic" age as a curse, not a blessing. They view the nuclear age as a nightmare that will some day result in a nuclear catastrophe affecting the whole world. Eight countries—the United States, Russia, the United Kingdom, France, China, Israel, India, and Pakistan—still possess nuclear weapons.

When the cold war ended in 1991, the fear of a nuclear war was lessened, but on September 11, 2001, Leon Wieseltier wrote: "When those planes flew into those buildings, the luck of America ran out." Today's nuclear nightmare is that a terrorist group or irrational national leader might acquire nuclear weapons in order to shock or blackmail the world to achieve their nefarious goals. The willingness of terrorists to commit suicide has changed the mindset of the entire civilized world. The ability of terrorists to steal or produce nuclear weapons is the curse that could some day destroy the human race and render the planet uninhabitable. Einstein had observed: "Each step appears as the inevitable consequence of the one before. At the end, looming ever clearer, lies general annihilation." The accidents at the Three Mile Island nuclear power plant in 1979 and at Chernobyl in 1986 stoked the embers of fear. The attack on the World Trade Center on September 11, 2001, proved that terrorists or ruthless dictators could threaten the world with "weapons of mass destruction (WMD): nuclear weapons, dispersed infectious biological agents, and "dirty bombs", (radiation dispersal devices" [RDD]).) The word "radiation" generates more anxiety than any word in the English language. Today, the possibility that a nuclear weapon might fall into terrorists' hands represents the greatest threat to the civilized world. Non-proliferation efforts have been, at best, a partial success. Nuclear proliferation has increased greatly as a result of the activities of Abdul Qadeer Khan, the flamboyant, German-trained metallurgist widely regarded as the revered father of Pakistan's nuclear bomb. The A.Q. Khan Research Laboratory of Pakistan advertised for sale the essential components for the building of nuclear weapons developed in Pakistan's three decade-long project of building a stockpile of enriched uranium. Such material is usable for making atomic bombs, as well as fueling nuclear reactors to produce energy. Pakistan's atomic bomb developed in this laboratory was tested in 1998. On January 30, 2004, Khan was dismissed from his position as a science adviser to the prime minister. The Nuclear Command Authority, comprised of senior military and civilian officials of Pakistan, informed Gen. Pervez Musharraf, Pakistan's president, that Khan had sold nuclear secrets to Iran, North Korea, and Libya. Admitting his role in sending abroad the materials needed to produce nuclear weapons, Khan confessed, but was pardoned by President Musharraf, and confined to his home.

What seems clear is that from the late 1980s Khan and his associates throughout the world developed a network that shipped parts and designs for the equipment to permit the enrichment of uranium to bomb-grade material. This material was provided to Iran, and to Libya, the latter country also receiving a blueprint for making a nuclear warhead. North Korea was provided with expertise and materials in exchange for ballistic missile technology vital to Pakistan for developing a military deterrent against India. Libya has now discontinued its program for production of nuclear weapons.

Mohamed Elbaradei, director general of the International Atomic Energy Agency, the agency responsible for monitoring nuclear energy all over the world for half a century, stated that in 2003, 35 to 40 nations had the knowledge needed to build an atomic weapon. He has called obsolete the nuclear non-proliferation treaty of 1970 devised to control which countries could possess or pursue nuclear arms. He said: "Unless you are able to control the actual acquisition of weapon-usable material, you are not able to control proliferation."

Large amounts of weapons-grade, highly enriched uranium—enough to make 1,000 nuclear weapons—still exist in 43 countries, most in Western Europe or other friendly nations. This uranium was distributed by the United States over past decades. Bomb-grade material was loaned, leased, or sold to these countries starting in the 1950s under the Atoms for Peace program of 1953. The material was to be used for the production of nuclear power, and biomedical radioactive materials or for educational purposes in universities. The distribution continued until 1988. Over the past 50 years, only 2,600 kilograms have been recovered by the United States out of the 17,500 kilograms that were distributed. Thus, 15,000 kilograms are still in foreign hands. The efforts to return the material to the United States continue under the mandate of the U.S. Department of Energy.

More stringent, intrusive inspections have now been added to the Nuclear Non-Proliferation Treaty, but only half of the treaty's participants have agreed to them, and less than a fifth have ratified the new version. Today, most Americans feel more vulnerable and fearful of nuclear weapons than at any time since the cold war.

Almost 80% of Ameicans believe they are at more risk today than 30 years ago. There is also concern about the risks of genetic modification of food as well as about food irradiation. There is more fear of risks imposed unwillingly upon us than those that we accept ourselves. Fortunately, nuclear medicine procedures fall in the latter category. I personally have never had a patient who refused a nuclear medicine procedure because of fear of radiation.

My First IAEA Meeting

The first meeting sponsored by the IAEA that I attended was in Vienna in 1964, where I presented our results in lung scanning of pulmonary embolism with chromium-51 and iodine-131 macroaggregated albumin.

After presenting our results in normal persons and patients with pulmonary embolism, I described studies in which we had injected macroaggregated particles into pilots while they were flying fighter planes in a parabolic path that resulted in zero gravity. Our hypothesis was that the hypo-perfusion of the tops of the lungs that we observed in normal persons when they stood up would not be seen if the injections were made at zero gravity. We also carried out injections to examine the distribution of pulmonary blood flow while the subjects were exposed to four Gs in a human centrifuge.

We wanted to see the difference in pulmonary arterial blood flow between the apices and bases of the lung. Since the radioactive aggregates remained trapped in the lungs, the injections could be made during parabolic flight, the planes could land, and the pilots could come to Hopkins for the scanning procedure.

Once, while one of the pilots was walking down the corridor of Hopkins Hospital in his flying suit with his helmet under his arm, an elderly woman passing him said: "Oh, you poor dear. Did your plane crash"?

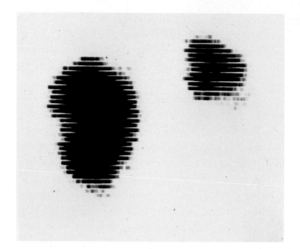

Figure 62 First perfusion lung scan ever performed in a patient with massive pulmonary embolism (1963).

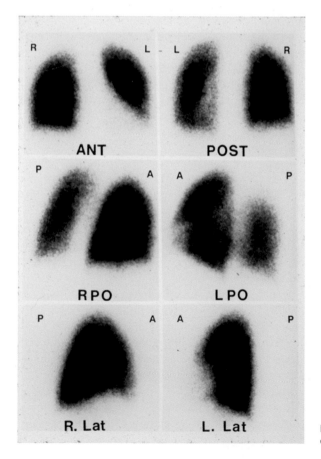

Figure 63 Lung scans from different projections obtained in 1975 with a scintillation camera.

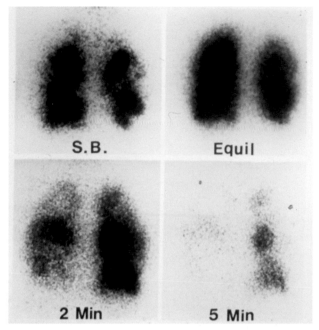

Figure 64 Scintillation camera images in a patient with multiple pulmonary embolism.

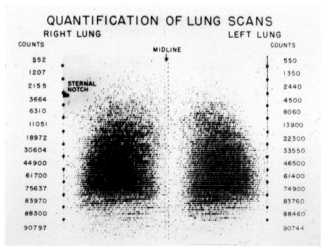

Figure 65 First quantification of perfusion lung scans by accumulating counts from each lung separately.

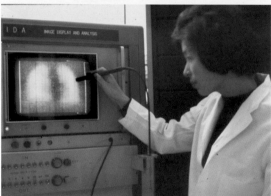

Figures 66 and 67 First use of a mini-computer (IDA) to quantify regional perfusion of the lungs in 1968.

We also measured regional ciliary activity in the lung by measuring the clearance of the radioactive aggregates from the tracheobronchial tree in patients with tuberculosis. We often saw regions of decreased ciliary activity thought to be due to regions where cilia had been destroyed. Such areas are often associated with metaplasia of ciliated columnar epithelial cells. In studies of the rate of clearance of particles from the nose by ciliary activity, we measured the rate of clearance of particles to be 6 mm per minute. No difference was observed between smokers and non-smokers.

On the way to an IAEA meeting in Vienna, David Kuhl and I spent 16 days in Australia on a trip jointly sponsored by the U.S. Atomic Energy Commission and the Australian Atomic Energy Commission. The purpose of our trip was to promote nuclear medicine in Australia. The Australian government operated a nuclear reactor and radio-pharmaceutical manufacturing program that was supplying the country.

Between the two of us, Dave and I delivered 48 lectures in five cities: together in Sydney, Melbourne, and Canberra; I went to Brisbane and Tasmania alone, while Dave went to Perth and Adelaide. One of the principal topics of our lectures was "Cameras vs.

Figures 68 and 69 IDA (Image Display and Analysis) images of the cardiac ventricles and a background region.

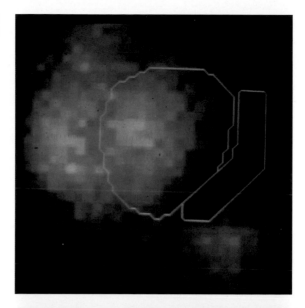

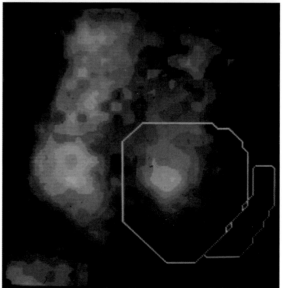

Scanners." The Anger camera had recently been introduced by Nuclear Chicago, and would eventually replace the rectilinear scanner.

Shortly before we left Sydney, our hosts asked if there was anything we wanted to see. We suggested that it would be a shame to come to Australia and not see a duck-billed platypus. No problem, they said. They provided a black governmental limousine that dropped us off outside the Sydney zoo. Soon after we entered, we heard over the loud-speaker system, broadcast all over the zoo: "Would the two American atomic scientists please report to the duck-billed platypus cage?" We headed for the cage, and arrived

Figure 70 The IDA system in 1968, showing the large size of "mini-computers" in the early days.

Figure 71 A picture of the entire IDA system.

together with hundreds of others who came to see the American atomic scientists. As we stood outside the cage, there were no platypi to be seen, but the zoo-keeper banged on a wooden shoot, lay down on the grounds, reached up the shoot, grabbed a duck-billed platypus by a leg and brought it over to us to see. We showed our appreciation by snapping a few pictures. He then asked: "Do you know why there are so few duck-billed platypuses in Australia?" We gave up and he told us that it was "because the aboriginals eat them."

In the 1970s, our first use of ECG-synchronized imaging of the heart was to decrease the effect of cardiac motion on the images, using potassium-43 and subsequently thallium-201 to image regional myocardial blood flow.

Two images were created, one with the data during obtained end systole and the other during end-diastole. It soon occurred to us that, after the injection of Tc-99 m albumen, by viewing each of the two images alternating in rapid sequence, one could see the motion of the ventricles. Subsequently, investigators under the leadership of Mike Green at the NIH used the computer to divide the cardiac cycle into 16 time periods. One could then visualize ventricular wall motion, and measure the ejection of blood from each cardiac chamber. These "gated blood pool images," together with Tl-201 myocardial blood flow images, led to the birth of nuclear cardiology. Between 1975 and 1982, over

5,000 thallium-201 studies had been performed at Hopkins. Today over 10 million nuclear cardiology procedures are carried out in the United States.

The computer was the greatest advance in nuclear cardiology since the Anger scintillation camera. I was 19 years old when the first computer, ENIAC, was built in the Moore School of the University of Pennsylvania in 1946. It took up 15,000 square feet, required 18,000 vacuum tubes, and cost $400,000. Today the same amount of computing can be done by a quarter-inch microchip that sells for $500. By 1973, 36,500 minicomputer units

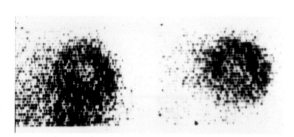

Figure 72 First myocardial perfusion images ever performed on a patient with potassium-42 at rest and during exercise. Liver blood flow decreased with exercise, and a perfusion defect is seen in the inferior myocardium.

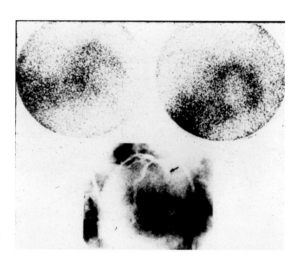

Figure 73 Early myocardial perfusion images obtained with the scintillation camera and thallium-201 in a patient with an occluded left anterior descending coronary artery.

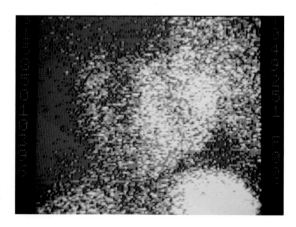

Figure 74 Myocardial perfusion images with thallium-201 in a patient with right ventricular hypertrophy. The images are displayed on the computer screen.

had been installed in the U.S, bringing the world total to 92,000 units, with 39% supplied by Digital Equipment Corporation.

The greatest impact of computers in nuclear medicine was that they made possible the transition from static to dynamic imaging, particularly necessary in the study of the heart, although the derivation of time/activity curves was useful for the study of other organs, such as the thyroid, brain, gall bladder, and kidneys.

An international company, Nuclear Data, Inc., was a pioneer in computing for nuclear

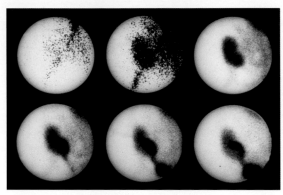

Figure 75   Serial images of the passage of technetium-99m DTPA through a kidney transplant in the groin.

medicine, providing an MED 50/50 computer-based analytical system for interfacing with gamma cameras in 1969. Nuclear Chicago developed the CLINCOM SYSTEM for processing and distant viewing of gamma camera images. The French/International company INFORMATEK, with Charles Zadje, founder and President, developed the SIMIS "Interactive Data Systems" in 1970, and in the early 1980s provided 430 systems in France.

In 1977, a Department of Computer Systems was established at the American Medical Association (AMA), charged with "undertaking specific activities designed to support the physician's pursuit of computer-based mechanisms to assist in the delivery of patient care." The "Current Procedural Terminology (CPT)" codes were developed for "easier, faster, more precise reporting."

In April 1975, the Institute for Electrical and Electronic Engineers surveyed computer experts who predicted that by 1981 computers would be the basis for medical diagnosis. The pioneers in the use of computers in nuclear medicine were the first to recognize that huge benefits could be obtained by the interaction between the physician and computer, which is now routine. Computers play a major role in nuclear imaging, not only for processing and display, but for data storage, instant retrieval and transmission. It was also recognized that a computer would never replace the physician/patient relationship, i.e. the ability to talk to and examine the patient.

In the 1950s and 1960s, some people predicted that the computer would replace the physician. It hasn't happened, and never will. In nuclear medicine today, the widest use of computers is to give the physician access to the knowledge base of the world, and to provide quantitative answers to questions about regional body functions and chemistry. Nuclear medicine led the way to the use of computers in bedside medicine. All over the world, computers play an essential role in nuclear imaging, and are leading a whole new approach medical diagnosis.

Another major development affecting nuclear medicine was the revolution in genetics. In 1953 the same year as the founding of the Society of Nuclear Medicine in the United States, Watson and Crick elucidated the structure of DNA. Over the next decade, Crick and other scientists began to map how DNA's sequence dictates the synthesis of proteins. In 1968, Hamilton Smith and Daniel Nathans established genetic engineering by discovering that "restriction enzymes" can slice DNA into snippets. In 1972, Paul Berg and others showed how these fragments of DNA could be spliced together. This process of "recombinant DNA" was used to develop the first genetically engineered recombinant DNA drug, human insulin.

In 1981, there was the first "transgenic" insertion of foreign genes into the DNA of mice and fruit flies. In 1983, Huntington's disease was linked to a mutation on chromosome 4. In 1990, the Human Genome Project of the Department of Energy and the National Institutes of Health (NIH) began with the goal to decode the structure of DNA. At about the same time, the first gene therapy was carried out, the patient being a 4-year-old girl with inherited Acquired Immune Deficiency Syndrome (AIDS). By the year 2000, the NIH and Celera Genomics, a Maryland biotechnology firm, had announced drafts of the entire human genome. The final human genome was published in the spring of 2003.

In 1958, when I returned to Hopkins from Hammersmith, I made up my mind that nuclear medicine was the career that I wanted. John G. McAfee, a Canadian-American graduate of the University of Toronto, was Chief Resident in Radiology at Hopkins at the same time that I was Chief Resident in Internal Medicine. The first scanner at Hopkins had been obtained in 1957 by a neurosurgeon, Donald McQueen, who used the positron-emitting radionuclide, copper-64, to detect brain tumors, but soon lost interest in its use. Simultaneously, John, in collaboration with a physicist, James Mozley, had just finished building a rectilinear scanner with a three inch sodium iodide crystal and a photorecorder of the type designed first by David Kuhl at the University of Pennsylvania and Merrill Bender at the Roswell Park Memorial Hospital of the University of Buffalo. The funds to build the scanner came from the Ruth Estrin Goldberg Memorial For Cancer Research on May 27, 1957, and were acknowledged by a brass plaque attached to the scanner throughout its period of service to the patients of Hopkins Hospital.

The biggest advance in nuclear oncology would come two decades later—the use of a glucose analogue, F-18 fluorodeoxyglucose, to detect cancer. Working with Martin Reivich in the Cerebrovascular Research PET Center at the University of Pennsylvania, David Kuhl, Al Wolf, and his colleagues at the Brookhaven National Laboratory were the first to synthesize F-18 fluorodeoxyglucose (FDG) for the quantitative determination of regional cerebral glucose utilization in human beings.

In 1958, my first study in oncology at Hopkins was scanning the liver of over 150 patients with the radiotracer iodine-131 Rose Bengal, a dye excreted by the liver. We found that the procedure was of enormous help in the care of patients with primary and

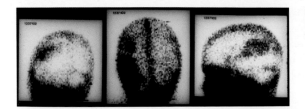

Figure 76 Early rectilinear scan in a patient with a stroke involving the left middle cerebral artery. The tracer was technetium-99m pertechnetate.

Figure 77 James Langan, first technologist in nuclear medicine at Johns Hopkins beginning in 1958 and continuing as Chief Technologist for decades.

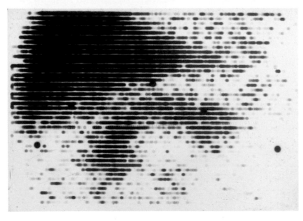

Figure 78 Rectilinear scan of the pancreas with S-75 selenomethionine, a procedure that never caught on.

secondary liver cancer. I reported the results at the meeting in Atlantic City of the Association of American Physicians. We also had an exhibit of radioisotope scanning of other organs at this meeting.

We summarized the clinical uses of liver scanning:

Figure 79 Alfred Wolf, a major contributor to cyclotrons in medicine and co-developer of F-18 FDG at Brookhaven National Laboratory (BNL). Many important chemists in the field of nuclear medicine worked under his direction at BNL.

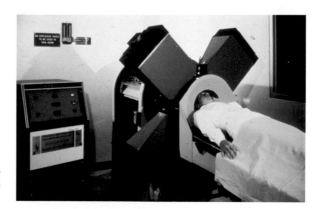

Figure 80 First tomographic scanner invented by David Kuhl at the University of Pennsylvania in the 1960s.

1. Differential diagnosis of right upper quadrant abdominal pain
2. Demonstration of space-occupying lesions of the liver
3. Detection of malposition of the liver, both congenital and acquired (subphrenic abscess, eventration of the diaphragm, medial dislocation
4. Differential diagnosis of abdominal masses
5. Detection of parenchymal disease of the liver
6. Detection of obstructive liver disease
7. Preoperative assessment of the extent of malignancies
8. Quantitative assessment of liver size
9. Localization of liver lesions prior to needle biopsies
10. Objective evidence of response of lesions to irradiation or chemotherapy

In the discussion of our paper, Dr. George Cotzias, who subsequently would become famous for his introduction of L-Dopa treatment of Parkinson's disease, commented:

"The extensive usefulness which was demonstrated this morning is amenable to even further development, in our opinion, since it is possible with the same apparatus to obtain biochemical information about the liver ... Our own work has shown that manganese accumulates in the mitochondria, so it seems to me that the same beautiful demonstration that we have witnessed today might actually, if properly applied in a different sense, give easy and ready biochemical information, including the happiness or unhappiness of liver mitochondria."

In March of 1961, in the *Archives of Internal Medicine*, we concluded:

"The photoscans of the liver of 150 patients were particularly helpful in the differential diagnosis of right upper quadrant abdominal pain, indicating whether the patient had a subphrenic abscess or space-occupying intrahepatic lesion, in the differential diagnosis of abdominal masses, and enabling an accurate follow up of therapy in intrahepatic abscesses. Major surgery was avoided in many patients when the hepatic photoscan revealed space-occupying lesions that were biopsied by needle aspiration." These were the days of "exploratory laparotomy," which are no longer needed to make a diagnosis.

My first publication in nuclear medicine, entitled "Medical Radioisotope Scanning," was published on September 10, 1960, with John McAfee and physicist, Jim Mozley, in the *Journal of the American Medical Association*. In those early days, we focussed on trying to "see" organs and lesions such as those in the liver and spleen that could not be seen on x-rays. We defined "radioisotope scanning" as the "visualization of an internal organ by determining the spatial distribution of a radioisotope within the body." We showed the scans of a patient with a substernal goiter imaged with iodine-131; a liver abscess imaged with iodine-131 Rose Bengal; vascular occlusion of the kidney imaged with mercury-203 chlormerodrin; and a pericardial effusion imaged with iodine-131 human serum albumin.

By 1964, we had developed at Johns Hopkins radioisotope imaging procedures based on the following methods of localization of radioactive tracers:

1. Active transport, for imaging the thyroid, kidneys and liver
2. Phagocytosis for imaging the liver, spleen, and bone marrow
3. Cell sequestration for imaging the spleen
4. Capillary blockade for imaging pulmonary arterial blood flow and other circulatory beds, such as muscle
5. Simple or exchange diffusion for bone, brain, and pancreas imaging
6. Compartmental localization for imaging pericardial effusion, mediastinal aneurysms, and placentas (the latter soon became obsolete because of radiation exposure to the fetus).

We did not focus solely on imaging, but were also interested in in vitro nuclear medicine. An example was the "T3 red cell uptake test," which measures the degree of saturation of the serum thyroxine-binding protein (TBG) in normal persons and patients with thyroid disease. We found in 176 persons with thyrotoxicosis, the uptake of the iodine-131 triiodothyronine (T3) tracer by the patient's red blood cells was increased, believed to be due to saturation of the patient's serum thyroxine-binding globulin (TBG). Today this test is of historic interest only, but represents one of many in vitro procedures that take advantage of the specificity and ease of measuring radioactivity.

Another in vitro technique in nuclear medicine in the 1960s was neutron activation analysis (NAA), a sensitive analytical technique useful for performing multi-element analysis of trace elements in biological samples. The element to be quantified is made radioactive by neutron bombardment of the sample in a nuclear reactor. After the element has been made radioactive, it can be identified by its characteristic gamma ray emissions, and rate of decay. Although activation analysis is used for detecting environmental contaminants, it is not widely used in nuclear medicine today, despite the fact that about 70% of the elements have properties suitable for measurement by NAA. Samples containing unknown amounts of elements to be measured are most often bombarded in nuclear reactors with their high fluxes of neutrons from uranium fission. An example is the measurement of Selenium-75, having a specific activity of 1,000 Ci/g, which was used in the discovery of dependent enzymes and other biologically important proteins. Trace-element and mineral nutrition are important aspects of human and animal health. The basic nutritional requirement at the cellular level can be studied using NAA and radiotracer techniques. NAA is one of two methods that can be used to study nutritional bioavailability and absorption of essential trace elements in the human using enriched stable isotopes. Samples such as hair, nails, blood, urine, and various tissues can be analyzed by NAA for both essential and toxic trace elements.

In 1961, I was studying nuclear medicine in a 6-week course at the Naval Medical Center in Bethesda, Maryland, under the direction of Dr. E. Richard King and Thomas Mitchell. The program included the use of an extremely low power nuclear reactor used for educational purposes. We used this reactor in neutron activation analysis to determine the specific activity of iodine-131 (I 131/I 127) in the urine of 34 normal persons. We found that the thyroid accumulates a constant fraction of the extrathyroidal iodide pool, that is, the uptake is a first order reaction. After I returned to Hopkins from the Navy course, a graduate student, Ed Smith, used activation analysis to measure protein-bound iodine (PBI) in human plasma.

Figure 81 Henry Wagner receiving certificate of attendance at the 6-week course in nuclear medicine at the U.S. Naval Medical Center in Bethesda, Maryland, in 1959.

One of the teaching exercises in our program in Bethesda had us gather at various points in a field around the building that housed the nuclear reactor. We each had a "Cutie Pie" radiation monitor. The reactor was then activated, and we all took readings of the radiation levels at the places where we were standing in order to map out the radiation levels throughout the field outside the reactor building. Then, we all gathered with our instructor to plot out the radiation levels that we had recorded when the reactor had been activated. It turned out that the radiation levels were greatest at the point where the instructor was recording the results in his notebook that rested on a 55-gallon drum that had been lying in that field for years. Unknown to us, someone had decided to use the drum to store long-lived radioactive material.

Another technique of nuclear medicine that we developed that did not involve imaging was the use of carbon-14 lactose to detect a deficiency in the absorption of milk by persons with a genetic deficiency of the enzyme lactase. A simple collection/detection apparatus was used to quantify carbon-14 dioxide in the breath as an indication of the absorption of lactose. Studies were carried out at Hopkins and subsequently in Japan by Dr. Yasuhiro Sasaki, the third Japanese physician to study nuclear medicine at Hopkins. He subsequently became head of the Japanese Department of Radiological Sciences in Chiba, Japan. Thirty-five Japanese and 18 Americans were examined. We examined the normally milk-tolerant Japanese and compared the results to milk-tolerant Americans.

Dr. Sasaki and colleagues also studied patients with steatorrhea (fat in the feces) and diarrhea. In 1971, they gave carbon-14 glycine-1-cholic acid orally and measured the carbon-14 dioxide in the breath. When there was increased metabolism of intestinal bacteria, this was increased. The procedure was also useful in assessing the response to treatment. These studies provided insight into the importance of the type and amount of intestinal bacteria in human beings.

Two persons arrived at almost the same time as the first trainees in nuclear medicine at Johns Hopkins. Dr. Mohammed A. Razzak, an Egyptian internist, spent two years in training at Hopkins beginning in 1960. He now heads a nuclear medicine department at

Figure 82 Yasuhiro Sasaki, the third Japanese trainee in nuclear medicine at Johns Hopkins and a leader in Japanese nuclear medicine.

the University of Cairo. Dr. Wil B. Nelp, a research trainee in nuclear medicine who subsequently became head of nuclear medicine at the University of Washington in Seattle, met Mohammed at the airport when he arrived in Baltimore. For 7 years after starting nuclear medicine in 1958, we did not have any residents in nuclear medicine at Hopkins, although we had many research trainees from the United States and abroad who helped perform clinical as well as research studies.

A well-known feature of our training program was the morning conference, which began at 8:00 AM, and was attended by all faculty, technologists when not involved in performing studies, residents, and students. In the beginning, every patient who had a nuclear medicine procedure was presented to the faculty without a priori information by the resident or fellow who had seen the patient before and during the procedure. An interpretation of the nuclear medicine results was given by the professor, followed by the presentation of the clinical problem; all other data were incorporated into the final interpretation. Two interpretations were always returned to the referring physician, one based on the independent interpretation of the study without other data, and a final interpretation based on all data. These conferences, which often lasted for several hours, were an important stimulus for research projects. They also helped define the value of nuclear medicine procedures in the diagnostic process.

For decades in the nuclear medicine department at Johns Hopkins, we had a sign on the wall of the morning conference room that read "Everybody here knows everything about everything, so if I were you, I would keep my mouth shut." One new foreign trainee, not recognizing that this was a joke, did not say anything for two weeks after his arrival.

Figure 83 Professor Hironobu Ochi, a trainee in nuclear medicine at Johns Hopkins who subsequently became head of nuclear medicine at Osaka University.

Figure 84 Hinohara Tomaki, a leader of the second generation of nuclear medicine physicians in Japan, specializing in nuclear cardiology. He trained with Dr. William Strauss, after Dr. Strauss had left Hopkins.

Figure 85 Dr. Junji Konishi, head of nuclear medicine at Kyoto University.

Figure 86 Wil B. Nelp, first trainee in nuclear medicine at Johns Hopkins who subsequently became head of nuclear medicine at the University of Washington and President of the Society of Nuclear Medicine.

Figure 87 James Conway and Sue Weiss, founders of the Oenology Society, which meets annually at the Society of Nuclear Medicine meeting.

For 17 years, beginning in 1970, every summer we conducted a one-week course in nuclear medicine at Colby College in Waterville, Maine. Our course was part of a large program that Colby used to keep its dormitories open and staff working during the summer vacation. We provided a faculty of about 6–7, who brought their families and stayed in the college dormitories. It was very congenial, and our younger children loved the experience of "being in college." In those days, all of nuclear medicine could be covered in one week. (In those days, when I was invited to deliver a guest lecture at other institutions, the title of the lecture would often be "Nuclear Medicine," and I would cover the entire field.)

Most of the faculty of the Colby course was from Hopkins, although we invited some from other institutions, including David Kuhl, Craig Harris, Alex Gottshalk, and Jim Potchen. At times, we would bring a sailboat and sail on Belgrade Lake. We always had a lobster-bake, where I learned that children usually don't like lobster, and their mothers don't like adults who talk their children out of their lobsters.

The cost for attendees which included room and board for the week was $350, including Colby's Down East lobster and clam bake, a reception, and use of the tennis courts.

Figure 88 (l to r) James Langan, Eileen Nickoloff, William Straus, Irv Goodof, HNW, and Steve Larson, faculty for many of the post-graduate summer courses at Colby College, Waterville, Maine.

soon after injection of the technetium-99m or indium-113m colloidal particles. An occasional problem is that an enlarged left lobe of the liver can obscure part of the spleen.

In 1963, describing the current status of nuclear medicine, I wrote: "The variety and extent of the application of radioisotope techniques to clinical and investigational medicine raise the question of whether nuclear medicine may not be in the infant stage of becoming a medical specialty."

Among our early problems was that we had to have our research fellows help take care of patients referred to nuclear medicine, because we had no residents.

On October 3, 1974, I wrote to Bob Heyssel, Director of Hopkins Hospital:

"Over the past 16 years we have provided services to patients, but have had no residency positions in nuclear medicine. There are no residents in radiology who take care of patients in nuclear medicine. We have an approved residency program in nuclear medicine." The American Board of Nuclear Medicine had been established in 1971.

He replied on October 14: "To take money for subspecialty training from the ordinary sources of patient care is difficult to justify." We were, however, able to get two nuclear medicine residency positions in 1974–1975, thanks to the efforts of Associate Dean Julius Krevans.

Four years after the founding of the Division of Nuclear Medicine at Hopkins, we obtained in 1962 funding for a program/project from the NIH, entitled: "Nuclear Instrumentation and Chemistry in Medicine" (GM 10548). In a description of the background for the proposed research, we wrote: "In biochemistry, the application of beta-emitting radioisotopes as tracers of the biologically important elements such as hydrogen, carbon, phosphorus, and sulfur has resulted in tremendous progress . . . In comparison, application of radioisotopes in the diagnosis and therapy of disease has been relatively slow and its impact on clinical medicine has been rather insignificant . . . Following the development of the scintillation detector and gamma ray spectrometer, it became possible for the first time to detect very small amounts of gamma-emitting materials in vivo. Tracer studies in patients with insignificant doses of radiation became possible . . . It has been hoped by workers in nuclear medicine that commercial instrument manufacturers would provide a liaison between the nuclear engineer and the physician, but they have failed to exploit new technological developments for medical applications . . . To overcome the present weaknesses in the field of nuclear medicine, it will be necessary for university centers and other research institutes to establish laboratories in the fields of nuclear instrumentation and radiochemistry—the basic sciences of nuclear medicine . . . Our program is long term in nature."

When Professor Harvey asked me what field of internal medicine I was interested in, I told him "nuclear medicine." He tried to talk me out of it, suggesting that I consider "working with Larry Shulman in arthritis." Years later I realized that investigators, such as Dewitt Stetten and colleagues at that time were working in the arthritis institute at the NIH, making fundamental advances in instrumentation and research in molecular biology and genetics. Professor Harvey apparently had the work of these investigators in mind in making his suggestion, but I declined, and began to work full time with John McAfee to establish nuclear medicine at Johns Hopkins in 1959.

The first Division of Nuclear Medicine at Hopkins was in both Internal Medicine and Radiology. The former connection eventually disappeared, and the latter blossomed. I

Figure 92 John McAfee.

have often asked myself why internal medicine did not seize the opportunities afforded by nuclear medicine. I concluded that the scientists in internal medicine prefer to study simpler systems than the whole human body. They select and try to solve problems at a more basic, easier to manipulate, level, such as cells or body fluids. Furthermore, internists are interested in the long term care of patients with chronic illness.

We opened the Division of Nuclear Medicine in September 1959 with a budget of $30,000 from the endowment of the Johns Hopkins University. A Radiological Science program was to be started during 1963 in the new wing of the School of Hygiene and Public Health of the Johns Hopkins University. In this building would be housed a radiochemistry laboratory and machine shop. Studies of human beings were to be conducted in 1,300 sq. feet on the third floor of the Clinical Science Building of the Hopkins Hospital. On the 11th floor next to animal facilities there was a laboratory for carrying out biodistribution studies in animals. The amount requested from the NIH for the period from October 1, 1962, through September 30, 1963, was $290,633. Included in the application was a request for $5,000/year for the addition of a radiochemical technician, and a physical chemist for $8,000/year. A small amount of research support (approximately $2,000/year) had been provided since July 1961 from the American Cancer Society. By then we had achieved the first visualization of the spleen and kidneys, liver scanning with colloids and brain tumor imaging with several investigational radiotracers, iodide imaging of myocardial infarcts, and detection of pericardial effusion in human beings.

The objectives described in the Program/Project grant application were:

1. The design and construction of new nuclear instruments for medical research and diagnosis.
2. Establishment of a program of radiochemical synthesis, with personnel trained in organic chemistry and pharmacology, for the development of new radioactive labeled compounds for medical research, diagnosis and therapy.
3. Application of neutron activation analysis to medical problems.
4. Study of the temporal and spatial distribution of radioactive materials in animals and man.

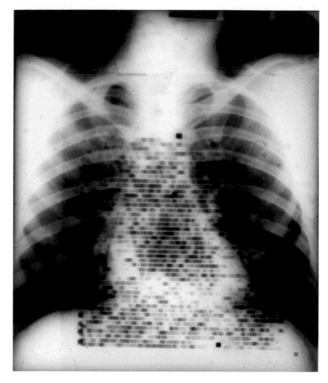

Figure 93 Iodine-131 albumin scan of a patient with a pericardial effusion. Superimposition of the scan on the chest radiograph with fiducial markers revealed the discrepancy between the ventricular blood pool and the cardiac shadow on the radiograph. To avoid magnification, the chest x-ray was obtained with two exposures, each over the centers of the two clavicles.

We proposed to build another scintillation scanner (with an 8-inch diameter sodium iodide crystal) before embarking on the design of a scintillation camera. The latter was to consist of 1,500 cesium iodide crystals covering a five inch diameter. We tried to get financial support from the hospital to help construct the 8-inch scanner. This was denied, and on June 27, 1974, I wrote to Bob Heyssel, Director of Hopkins Hospital:

"I was sorry to hear about the problem of financing the Ohio Nuclear scanner and data processor that we have had now for 7 months at no cost, and which they will let us keep for six more months without cost, provided we have some way to pay for it eventually. We have done over 900 studies on this machine in the past 6 months, so sending it back would put a big dent in our clinical services." We were eventually able to get the funds, the scanner was built, and helped solve many clinical as well as research problems.

Most of the funds for the development of nuclear medicine at Hopkins came from National Institutes of Health (NIH) grants, one of which was renewed every five years for over 40 years. Without NIH support, nuclear medicine at Johns Hopkins would not have developed as it did. The NIH not only provided the funds, but also made it possible for the Principal Investigator of the grants to function to a large degree independent of the administration, although the Department Chairman and Dean had to sign off on the grant proposals. The indirect costs provided by the NIH for the administration of the grant were usually sufficient to get their approval of projects, even those that they themselves would not have given a high priority. Principal Investigators of large NIH grants had three essential ingredients for success: money, authority, and accountability.

An example of how the study sections who review NIH grants function is the review of our brain research grant on June 10–12, 1979. The statement concluded: "The proposed reseach is directed toward the in vivo localization of dopamine and opiate receptors utilizing C-11 spiroperidol, a dopamine receptor binding agent, and C-11 deprinophine, an opiate receptor binding agent … Approval for 5 years is recommended … The purchase of the PET scanner will not be made until the third year … A price of $300,000 is budgeted." The funding for the first year was approved for $748,085, the amount we requested.

We in nuclear medicine were given strong support and considerable freedom by Dr. Russell Morgan, Chairman of the Department of Radiology. His greatest contribution to nuclear medicine was the establishment in 1961 of the Department of Radiological Sciences in the Johns Hopkins School of Public Health, across Wolfe Street from Hopkins Hospital. This provided space for research laboratories that were not available in the Hospital or medical school. We were able to carry out studies of radioligand binding to receptors, in vivo microdialysis, fluorometric and electrochemical analysis, high efficiency annihilation detection, and computerized animal activity monitoring.

Having a Division in the School of Public Health also made it possible for us to accept students in both Masters and PhD programs, not available in clinical departments in the medical school. Graduates of our PhD program played key roles in our research over the next 40 years. Among them were Bob Dannals, Jonathon Links, Tomas Guilarte, Kwamina Baidoo, James Frost, and Dean Wong, all of whom subsequently joined our faculty. Dr. Morgan followed the precept of many Department Chairmen at Hopkins: "Select good people and leave them alone."

Dr. Tomas Guilarte is an excellent example of how people born outside the United States made major contributions to our Nuclear Medicine program. Born in 1953 in Oriente, Cuba, Tom received his PhD in Radiological Sciences at Hopkins in 1980, and became an American citizen. He is now a full Professor at Hopkins and heads a major neurosciences program in the School of Public Health. Among his first projects at Hopkins was the development and application of a new method for detecting the amounts of vitamin B12 in food. The method puts a sample of the food to be tested in a vial with bacteria, a liquid growth medium, and a compound containing radioactive carbon-14. The bacteria metabolize the compound only if the vitamin is present. As metabolism takes place, carbon dioxide containing radioactive carbon-14 is produced and measured in an instrument called BACTEC, developed originally in our division to detect bacteria in blood cultures.

It is not possible to cite all the wonderful people from all over the world who trained in nuclear medicine and radiological sciences at Johns Hopkins. Among them was Dr. Nese Ilgin, from Turkey who developed tracers for PET imaging in Parkinson's disease, and has continued her excellent research and teaching after her return to Hacettepe University in Ankara, Turkey.

Other outstanding trainees from overseas included Dr. Saeeda Asghar and, later, her student, Dr. Misbah Masood Rasool, two nuclear medicine physicians from Pakistan who conducted excellent research during their years as research fellows at Hopkins. Dr. Mohammed Mohamadiyeh, a Jordanian physician working in Saudi Arabia, was one of many trainees from the Middle East.

Many outstanding trainees came from India. In 1972 I was honored by receiving the first Vikram Surabei Medal of the Indian Society of Nuclear Medicine.

Figure 94   Michael Maisey; G. Hounsfield, Nobel laureate and inventor of computed tomography; and HNW.

In 1990, Dr. Kwamena Baidoo from Ghana was promoted to Assistant Professor, and subsequently carried out pioneering research in the development of technetium-99m and rhenium-186 radioligands that bind to recognition sites, including neuoreceptors and hormones, including somatostatin. He had been an Instructor since receiving his PhD degree in our program in 1988.

Dr. Julie Price joined our group as a Research Assistant in September 1987 and received her PhD in 1992, working with the development of tracers for imaging opiate receptors. She subsequently has carried out important work at the University of Pittsburg in developing tracers for detection of the characteristic plaques that define Alzheimer's disease.

Many medical students spent elective time in nuclear medicine and radiological sciences. Among them was Stanley Kim, a Korean/American, who worked with us during his senior year of college at Hopkins in 1992, and subsequently in 1993 and 1994. As I wrote in a letter supporting his application to medical school, "Stanley was one of the most conscientious, intelligent, careful and creative young persons with whom I have ever worked ..." He was subsequently accepted into medical school, graduated with honor, and entered the field of radiology.

He worked with a high efficiency annihilation detection system that we developed, and patented on December 15, 1987. (Patent # 4,712,561) The abstract of this patent states: "The effectiveness of drugs and substances that affect brain chemistry can be efficiently and relatively inexpensively monitored. A radioactive tracer comprising a ligand that binds to presynaptic or postsynaptic neuroreceptors is administered to a patient ... The number of neuroreceptors and the degree of occupancy is calculated utilizing a mathematical model ... Dopamine, serotonin, opiate and other receptors all may be monitored, and the procedures are particularly applicable to the treatment of Parkinson's disease, schizophrenia, and drug addictions."

The patent included the description of a simple, inexpensive dual-detector coincidence system for measurement of positron-emitting receptor binding drugs in the human brain. Measurement of the binding of C-11 carfenanil to opiate receptors in the human brain could be made after administration of drugs that bind to opiate receptors. Stanley's research project, which was supported by a pharmaceutical company, ANAQUEST, showed that the new drug, nalmefin, was able to block opiate receptors in the brain of human volunteers for a period of 12 hours, significantly longer than the 2–3 hours blockage with naloxone, a previously approved drug used to arouse patients operated on under opiate anesthesia. His work led to the approval of nalmefin by the FDA.

The success of nuclear medicine at Johns Hopkins for over thirty years would not have been possible without the contributions of Ms Mary Lou Mehring who began work as an Administrative Assistance in nuclear medicine in 1960, and held this position until her retirement in 1993. Foreign physicians and students came to her for advice and comfort soon after their arrival, and continued to rely on her during their entire stay. I am indebted to Mary Lou for meeting all my executive administrative needs for over three decades. Her intelligence, devotion to her job, and personal qualities were outstanding, and my colleagues and I will be forever grateful to her. Also worthy of lasting praise and gratitude are James Langan who was Technologist-on-charge of Nuclear Medicine for decades after its inception in 1958; Kate Stolz, who began as Clinical Nurse in Nuclear Medicine, and then Executive Director of the Center for Radiation Education and Research (CARER) in the School of Public Health; and Judy Buchanan, who started as a technologist in 1959 and subsequently became a Research Associate, her career continuing to this day (2005).

Beginning in 1979, after the accident at the Three Mile Island nuclear power plant, Kate Stolz devoted her full time to the activities of the CARER center, which had been founded to address the health effects of radiation in human activities, particularly those related to medicine, nuclear energy, and environmental radiation. Among the symposia sponsored at Hopkins by CARER were "Perception of the Risk of Radiation" on April 7, 1982; "Nuclear Energy: The Maryland Experience" on May 8, 1985; "Disposal of low level radioactive waste: Options for Maryland" on December 4, 1985.

Kate had been the nurse working full time in nuclear medicine at Hopkins before taking over the responsibilities of CARER. When an effort was made on June 1, 1982, to delete the nuclear medicine nursing position at Johns Hopkins, Sharon O'Keefe, Director of Surgical Nursing wrote to the administration: "Patients often have emergency scans done and often go to nuclear medicine from the intensive care unit. They need attentive nursing care; intravenous mediations are being administered, and vital signs need monitoring. We have a fair percentage of patients who are diabetic or have other complications that make them a particularly difficult management problem." The position of a nurse in nuclear medicine was restored.

More than any other medical specialty, nuclear medicine is concerned with space and time. We define health and disease in terms of regional function and biochemical processes, and try to answer four basic questions that define the practice of medicine: What is wrong with the patient? What is going to happen? What can be done about it? Is the treatment effective? We do so by examining the molecular processes involved in energy production, bodily movement, and communication among the molecules, cells, tissues and organs of the living human body. Nuclear medicine makes possible quantitative measurements of the dynamic state of body constituents. Steven Hawking has pointed out that life is a continual struggle against the second law of thermodynamics to maintain an ordered state of life. He wrote: "Disorder increases with time because we measure time in the direction in which disorder increases." Human life is a small corner of order in an ever-expanding, disordered universe. As long as the DNA of our cells remains active, it can direct our ribosomes to synthesize new proteins involved in molecular processes, networks and patterns that resist the second law of thermodynamics. As errors in molecular processes accumulate during our lives, our cells age and die. When enough of the cells die, we die. Centuries ago, Homer wrote: "We fight because we know we are going to die."

We use the tracer principle, which had its greatest stimulus to growth after World War II. In 1942, Paul Aebersold, a graduate student of Ernest Lawrence, was responsible for the worldwide distribution of radionuclides by the Atomic Energy Commission. He published an article: The Cyclotron; A Nuclear Transformer. (*Radiology* 39, 513–540). In June 1946, an announcement was made in the journal *Science* that radionuclides would be available for scientific research to qualified persons throughout the world. The first shipment was to the Bernard Skin and Cancer Hospital in St. Louis.

Figure 95 Bill Myers receiving the Nuclear Pioneer award of the Society of Nuclear Medicine.

Another pioneer in the early days of nuclear medicine was William G. Myers. In February 1941, Bill received his PhD at Ohio State University with a thesis entitled: "Applications of the Cyclotron and Its Products in Biomedicine." Despite his best efforts, he was not able to persuade the Chairman of Radiology to install a cyclotron at Ohio State. At the 14th annual meeting of the Society of Nuclear Medicine in Seattle in 1967, Myers reported that: "Intravenous injection of C-11 insoluble carbonates ... provide particles for scanning lungs . . ." He pointed out that C-11 should achieve major significance in nuclear medicine because 18% of the human body is carbon.

His vision could not be easily converted to practical production of positron-emitting tracers, such as carbon-11; the production of radionuclides was too complex using the cyclotrons of the 1930s and 1940s. The cyclotron was put on a back burner as a source of radionuclides after the exciting news of the invention of the nuclear reactor by Fermi and colleagues at the University of Chicago in the 1940's. The reactor made possible production of great quantities of carbon-14, tritium, phosphorus-32 and other radionuclides, which eventually led to the field of biochemistry as we know it today.

Because it is so prevalent in living cells, carbon-11 is likely to someday become the number one nuclide in nuclear medicine, although technetium-99 m and fluorine-18 hold that position today. Nobel laureate, Glenn Seaborg, discoverer of 9 elements, including technetium-99m, wrote: "If we imagine a carbon atom enlarged to the size of a football stadium, the (four) electrons (in the outer shell) would be like flies flying around the outside of the stadium, and the nucleus would be the size of a golf ball in the center . . ."

Carbon became important in the evolution in living organisms because it can donate four "extra" electrons, like a metal, and accept four electrons, like a halogen. Melvin Calvin, another Nobel laureate in chemistry, has written: "Carbon is a constituent of 500,000 known chemical compounds." The length of the molecular chains of carbon and hydrogen atoms in nature extend from a few atoms to long chains, including polymers, such as DNA. Many are enormously reactive because they contain double bonds. What we need to do now is develop a system of "molecular classification of disease," as creative and useful as the periodic table for classification of the elements.

The longer half-lives of single photon emitting radionuclides, such as iodine-123 and technetium-99m, greatly simplify their availability, use and cost, compared to cyclotron-produced radionuclides. Carbon-14 and tritium emit only beta particles with a very short range in tissue, so that they cannot be used to examine regional biochemistry in living human beings or experiments animals. Decades were to elapse before physicians' and scientists' emphasis on regional biochemistry led to a rebirth of cyclotrons in biomedical research.

The physical characteristics of technetium-99m—its half life, type of decay and energy of its photon emissions—make it ideal for radiation detection from outside the body. The invention of generators make short-lived radionuclides available away from the site of their production. Examples of positron-emitting radionuclides available from generators include Fe-52/Mn-52m for production of manganese and the Xe-122/I-122 generator.

In 1953, Hal Gray in the Medical Research Council facility at Hammersmith Hospital in London proposed the use of high energy radiation to treat cancer. He and his colleagues were greatly helped by Michel TerPogossian, a physicist at Washington University in St. Louis. On January 28, 1955, the first internal beam was accelerated to 1.25 MeV. The cyclotron was inaugurated by Queen Elizabeth. On June 29, 1956, the first beam was extracted, and studies were begun with oxygen-15. I was fortunate to be a research fellow at Hammersmith two years later.

Among my duties at Hammersmith Hospital in London in 1957 was to perform radio-iodine uptake measurements in patients with iodine-132, that I obtained by distillation from tellurium-132. Physicist John Mallard told me to look into the newly developed

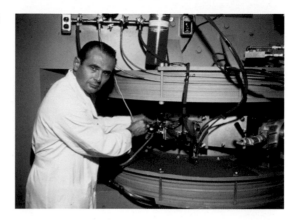

Figure 96 Michel TerPogossian with his cyclotron at Washington University.

Figure 97 Henry Wagner having a PET scan at Washington University in 1975.

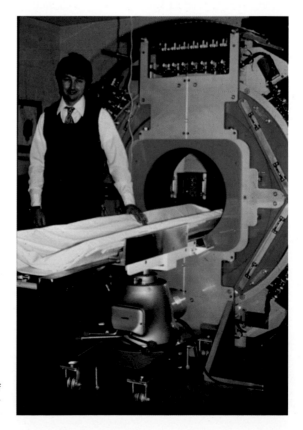

Figure 98 Michael Phelps, a graduate student of TerPogossian, and a leader in the invention and development of positron emission tomography.

Figure 99 Anne and Henry during a visit to London in 1961.

"radionuclide generators" being produced by Stang and Richards at Brookhaven National Laboratory after I returned to Hopkins as Chief Medical Resident in June 1958.

In 1965, we in the Nuclear Medicine Division at Johns Hopkins submitted a new research proposal to the NIH entitled: "Short Lived Radionuclides in Biomedical Research." The plan was to reactivate and redesign the 60-inch cyclotron built in 1939 by Merle Tuve at the Carnegie Institute of Washington, built as a copy of the 60-inch cyclotron of Lawrence. During the period 1944–1947, the cyclotron had operated 8 hours per day producing beams of 7.5 Mev protons, 15 Mev deuterons, and 30 Mev alpha particles. The proposal was to move the cyclotron from Washington to the Applied Physics Laboratory of Hopkins. The scientific proposal to the NIH was approved, but the funds for a building could not be found, and the project was dropped.

John Totter, head of the Biomedical Program at the Atomic Energy Department, and Alfred Wolf, in charge of the cyclotron at the Brookhaven National Laboratory, were among the site visitors reviewing the Hopkins proposal in 1965, and both were convinced of the important role cyclotrons could play in nuclear medicine.

(In 1985, Al Wolf, Director of the Cyclotron facility at Brookhaven National Laboratory, and I were members of a group that presented lectures on nuclear medicine in cities all over India, including Bombay, Dehli, Agra, and Madras.)

As a member of the Visiting Committee for the Brookhaven National Laboratory (BNL) where Al worked, I wrote: "Brookhaven National Laboratory is a unique and important research facility in this country. A strong Medical Department is necessary in order to carry out appropriate biologic and medical experiments utilizing the unique resources at this facility." Among the many contributions of BNL was the introduction of the treatment of Parkinson disease with L-Dopa, work carried out by George Cotzias and his associates.

On December 16, 1981, I joined the advisory committee of the Brookhaven Linac Isotope Producer (BLIP) program to develop and provide radionuclides to the biomedical community in the United States.

Soon after the 1965 site visit to Hopkins, the AEC funded the first three "in hospital" cyclotrons in the United States, to be run by physicists: Michel TerPogossian at Washington University in St. Louis in 1965; Gordon Brownell at the Massachusetts General Hospital in Boston in 1967; and John Laughlin at Memorial Hospital Sloan Kettering in New York in 1967.

Figure 100 Anne and Henry in Agra, India.

Figure 101 Vijay Varma and G. Kristnamurthy, two founders of the Indian/American Society of Nuclear Medicine, which meets annually at the annual meeting of the Society of Nuclear Medicine and sponsors periodic meeting to promote nuclear medicine in India.

In an article in *Nucleonics* in 1966, TerPogossian and I wrote: "The most important radioactive tracer in biological research is reactor-produced Carbon-14, but it has never been widely used in nuclear medicine ... The use of these nuclides (carbon-11, nitrogen-13, oxygen-15 and fluorine-18) in biomedicine justifies the additional effort to prepare them locally at the laboratory or hospital that plans to use them." By 1979 we had obtained a cyclotron at Johns Hopkins, and by 1991, the cover of the *Journal of Nuclear Medicine* and my editorial proclaimed: "Clinical PET: Its Time Has Come."

In a letter to me on November 20, 1992, Gerd Muehllehner, President of UGM Medical Systems and a pioneer engineer in the development of SPECT and PET, wrote: "Those of us who thought of PET as a potential clinical tool, found often that on the one hand the researchers were not interested in developing clinically applicable protocols and on the other hand, clinicians viewed PET with suspicion and as a threat rather than an opportunity. Thus, when FDG cardiac imaging was demonstrated to show viable cardiac

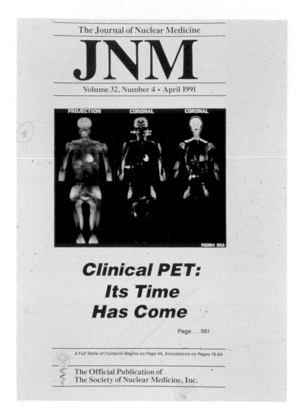

The Journal of Nuclear Medicine

JNM

Volume 32, Number 4 · April 1991

PROJECTION   CORONAL   CORONAL

P02504 UCLA

**Clinical PET:
Its Time
Has Come**

Page. . .561

A Full Table of Contents Begins on Page 4A, Annotations on Pages 7A-8A

The Official Publication of
The Society of Nuclear Medicine, Inc.

Figure 102 Cover of the *Journal of Nuclear Medicine* in April 1991, referring to an editorial by HNW declaring that PET was ready for clinical use as well as research.

tissue, the response was not that the typical clinically oriented Nuclear Medicine physician wanted to jump on the PET bandwagon, but instead thallium imaging was improved so that it could be claimed that thallium was just about as good as FDG ... the most important task now is to get the Nuclear Medicine physician to view PET not as hostile competition, but rather as just another tool in our armamentarium ... I am particularly excited about the clinical usefulness of whole body PET tumor surveys, which seem to get a warm and enthusiastic reception from Nuclear Medicine physicians and oncologists."

In a lecture that I delivered at UCLA in 1971 as President of the Society of Nuclear Medicine, on the occasion of the dedication of their new cyclotron, I stated that cyclotron-produced radionuclides would become widespread in nuclear medicine when a procedure was developed that required a cyclotron. Five years later, the procedure turned out to be tumor-imaging with F-18 FDG.

Since the earliest days of nuclear medicine, the U.S. Veterans Administration has been a major force in the advance of nuclear medicine. For example, the VA Hospital in West Haven, Connecticut, opened a PET Center with a $2.5 million PET scanner under the direction of Dr. Robert Soufer in 1991. The same year, Dr. George Ojemann, a neurosurgeon at the University of Washington in Seattle, reported that every individual appears to have a unique pattern of organization of language function as unique as fingerprints. The activated regions could be examined with PET. The University of British Columbia installed the first Canadian brain PET scanner in 1991 under the direction of Dr. Thomas Ruth. St. John's Mercy Hospital, a community hospital in St. Louis, began collaboration in 1991 with Barnes Hospital and St. Louis University Medical Center to develop an air ambulance service in order to share the PET Center facilities. Between 1966 and 1978, the number of cyclotrons dedicated to biomedical research increased from two to 66, five in the United States.

On July 14, 1978, we submitted an NIH proposal entitled "Multi-User Application for Specialized Research Equipment," in order to obtain funds to purchase a cyclotron (Cyclotron Corporation Mode CP-16) for our nuclear medicine division. All proposed projects were directed toward in vivo imaging of neuroreceptors, starting with C-11 spiroperidol and C-11 diprenorphine. We would concentrate on Parkinson's disease, Huntington's disease, tardive dyskinesia, schizophrenia, and chronic pain. "We believe that the ability to study non-invasively such neuroreceptors in man will lead to fundamental new information of unique importance . . . Although the concept that mental function is basically chemical is not new, recent technical developments have made it possible for the first time to attempt to track the miniscule amounts of chemicals within the brain and map out where and how chemical reactions are occurring in the living brain."

Receptors are plentiful on the cell membranes facing the extracellular fluid, and are involved in the transfer of information from one cell to another. Although all cells communicate with each other, the most highly specialized system for information transfer within the body is the brain and nervous system. Information is carried by "molecular messengers," called neurotransmitters, which include amines, amino acids, peptides, and proteins. In addition to membranes on cell surfaces, there are intracellular receptors, for example, estrogen receptors.

Neurotransmitters secreted at the terminals of presynaptic neurons interact with receptors on postsynaptic neurons, and information is transferred by changes in the state of ion channels or by the formation of "second" messenger molecules. In the brain, neurotransmitter receptors are present in picomolar quantities, comprising about one millionth of the total weight of the brain. More than 40 neurotransmitters have been found in the mammalian brain. The existence of receptors was proposed by pioneers, such as Paul Ehrlich and John Langley, based on their finding that many drugs, such as nicotine and curare, were active even when given in miniscule amounts. They postulated that the drugs must be acting as communicators of information to systems that would carry out their instructions and bring about the effects of the drugs. The small amounts of the drugs did not contain sufficient energy to bring about their effects themselves, but rather served as "molecular messengers." Our goal was to examine these neurotransmitter systems in living human beings by positron emission tomography.

Accompanying our application to the NIH was a letter from Bob Heyssel, Director of Johns Hopkins Hospital: "I want to assure you of my enthusiastic support for your grant proposal entitled 'Biodistribution and Kinetics of Receptors in Human Brain.' The Johns Hopkins Hospital stands ready to make space available in Room B1-189 for installation of the cyclotron." On May 4, 1978, the hospital administration approved our proposal to install a 16 Mev cyclotron in room B1-189 in the nuclear medicine department. This approval was needed prior to our submitting an NIH application to examine brain chemistry.

After looking into various cyclotrons that were commercially available, we decided to buy a Swedish cyclotron from the Scandatronix Company. While it was being delivered to the Nuclear Medicine division in the basement of Hopkins, it fell off the two rails along which it was being taken. Fortunately, it was not damaged and was safely installed in space that had previously served as a storage area in nuclear medicine, space that we controlled so that we didn't need to be given new space, which might be unavailable.

On December 6, 1979, attorneys for the Cyclotron Corporation who had bid on our purchase of a cyclotron under Public Health Service Grant # NS 15080-1 protested our choice of the Swedish Nucleotronics Cyclotron on October 26, 1979. In response, Hopkins attorneys summarized the circumstances of our purchase:

"On June 1, 1978, a grant application NS-10548 Study of Neurorecepter Binding in Man, was submitted by Dr. Henry N. Wagner, Jr, requesting 5 years of support ... the application indicated that three cyclotrons were under consideration, namely, those manufactured by Cyclotron Corporation, Japanese Steel and Nucleotronix. On August 2, 1979 Dr. Wagner was informed that the grant application had been approved ... The decision to purchase a Nucleotronix cyclotron was the result of a careful weighing of twenty-one factors, in addition to price." The 12 page letter concluded: "The undersigned feels constrained to express this University's consternation at the unfounded nature of Cyclotron Corporation's protest by which a disappointed vendor can hope to achieve no more than to hamper and delay some of the most significant experimental work in medical science being conducted in this country today." The NIH examined the protest and found that the complaint of Cyclotron Corporation had no merit.

The cost of the cyclotron was $1,300,000 plus $1,500,000 for the PET scanner. The facility modification cost $200,000. Cost financing over 8 years was $546,276/year. The prospective break even point was 55 studies/week at $500.00 per study.

During the year 1983, the new radiopharmaceuticals that were developed using our new cyclotron included C-11 methylspiperone and C-11 carfentanil. In 1984 we developed C-11 suriclone, C-11 methylketanserin, C-11 methyl bromo LSD, and C-11 dexetamide. In 1985, C-11 levetamide, C-11 SCH 23390, C-11 methionine, and F-18 fluorodeooxyglucose (FDG). In 1986, C-11 iodobenzamine, C-11 diprenorphine, C-11 pyramine, C-11 doxepin, and C-11 deprenyl were added to the list of new radiotracers.

Figure 103 Visit to the Fu Wai Hospital in Beijing in 1983.

On Nov 9, 1979, Dr. Heyssel had written to me:

"The growth in the volume of ancillary services is not an objective of the Institution. In fact, if anything, our objective is the opposite, which is to contain the growth and even where appropriate shrink those services . . ."

This letter was in response to our proposal to establish PET, based on the use of cyclotron-produced tracers into clinical oncology, cardiology, and neurosciences at Hopkins. Then, and now, the enormous increase in the cost of health care is the result of new, beneficial technology, including computed tomography (CT), magnetic resonance imaging (MRI) and positron-emission tomography (PET). In 1978, recognizing this problem the Hopkins administration created a "Diagnostic Testing Review Committee" that would review and approve or disapprove new tests performed by Diagnostic Radiology, Nuclear Medicine, University Laboratories, and other Hospital Laboratories that provide diagnostic tests. "The need (for prior approval of new tests) is only in part financial . . . Increased patient hazard and/or discomfort can result from ill-chosen or duplicate tests which may not provide added information of significant value." Paradoxically, the committee was disbanded after a few years because its activities called the attention of the house staff and their supervising faculty to the new tests, which increased rather than decreased their use.

The clinical use of PET was slow to achieve recognition. Moreover, reimbursement by Medicare and other third party payers was slow. Initially there was regulatory and reimbursement approval on an indication-by-indication approval process rather than general approval of radiotracers, such as F-18 fluorodeoxyglucose (FDG), in the care of patients with cancer.

The road from the development of the atomic bomb to molecular medicine had many obstacles, at Hopkins as well as in other institutions. For a long time, the cyclotron was thought to be little more than an exotic tool for physicists. On August 17, 1981, we requested from the Department of Radiology $450,000, which was 40% of the total cost of a PET scanner. We provided $300,000 from our neurological research program. The total price was $1,085,000.

I wrote to Martin Donner, Chairman of Radiology on April 24, 1981: "The reason this equipment was not in our five year plan for the hospital budget was that we did not believe that the PET scanner could be justified clinically. Now it can be. We have a real chance to be #1 in PET studies, if we can keep up our program."

The same week, I wrote to Richard Gaintner, Deputy Director for Administration of Hopkins Hospital: "I hope you enjoyed Dave Kuhl's lecture and got a better idea of what we are trying to do. We have the opportunity to go even beyond Dave's excellent work, by extending it to include dopamine and opiate receptors. At your convenience, I would like to give you a personal progress report."

We proposed to begin clinical applications of PET in 1982, based on the results of research supported by our existing NIH program/project grant. We also proposed to transfer this grant from the medical school to the School of Public Health, in order to use research space in the Department of Radiological Sciences.

In May 1981, we began to transfer faculty, including chemists, Donald Burns, Al Kramer, Tim Duelfer, and John Waud, physicists Thomas Mitchell, Jonathan Links, Ed Nickoloff, and Ken Douglass, physicians Edwaldo Camargo, Patricia McIntyre, Min-Fu Tsan, Pablo Dibos, Lawrence Holder, David Moses, Barry Friedman, George Sachariah, and others

Figure 104 David Kuhl, inventor of tomographic imaging in nuclear medicine and a major contributor to the study of Alzheimer disease. He was head of nuclear medicine at the University of Pennsylvania, the University of California in Los Angeles, and the University of Michigan, and a founding member of the American Board of Nuclear Medicine.

from the Radiology Department to the Department of Environmental Science in the School of Public Health. We paid tuition to the School of Public Health for Drs. David Chen, Alan Maurer (who was to become President of the Society of Nuclear Medicine), and Dean Wong, who was to become a Professor at Hopkins. The tuition was paid by our NIH training grant in nuclear medicine.

The U.S. Public Health grant Research Training in the Use of Tracer Principles in Oncology was funded from September 1, 1979, to May 5, 1994. This grant responded to

Figure 105 Edwaldo Camargo, a member of the faculty of nuclear medicine at Johns Hopkins and a founder of nuclear medicine in Brazil.

Figure 106 Don Burns and Susan Lever, two faculty members of the Division of Radiation Health Sciences of the Johns Hopkins School of Public Health who made major contributions to the development of new radiopharmaceuticals. Dr. Burns subsequently became head of Merck's nuclear medicine activities in drug design and development. He was a co-founder of the Society of Nuclear Imaging in Drug Design and Development (SNIDD), whose first meeting was in Baltimore.

the enormous need for more physicians trained in nuclear medicine. In 1968 a survey conducted by the Stanford Research Institute and the National Center for Radiological Health of the U.S. Public Health Service indicated that nearly 3 million patient procedures based on radioactive tracers had been performed during 1966. The rate of growth was 15% per year. A study by William Beierwaltes, head of nuclear medicine at the University of Michigan, revealed that by 1962, 79% of 7,160 hospitals in the United States with more than 100 beds had a "radioisotope unit." He predicted that there would be a need for 6,000 nuclear medicine physicians in the U.S. by 1980.

In 1959, we had started a research training program in the Division of Nuclear Medicine. In 1960, the programs of the nuclear medicine division in the medical school were combined with the program in the Department of Radiological Sciences in the School of Public Health. This was a great advantage because we were then able to award a Master's or PhD degree to appropriate trainees, which was not possible in a clinical department of the medical school. By 1985 we had trained 180 MDs and 47 non-MD trainees. These do not include the 200 MD's and 15 non-MD post-doctoral trainees who spent less than one year in our program. Nor do they include 68 persons studying for a Master's degree, and 17 special students not in degree programs.

In 1981, we in nuclear medicine at Hopkins negotiated with the Division of Laboratory Medicine to form a conjoint laboratory for the performance of in vitro procedures that involved radioactive tracers. Radioimmunoassay procedures had been performed in the nuclear medicine division since 1958. Nuclear medicine had been active in both research and clinical applications of in vitro procedures, including the development of an automated microbiology, called BacTec which was subsequently developed commercially. The joint effort with Laboratory Medicine lasted for a decade but eventually Laboratory Medicine took it over its functions completely.

On October 10, 1983, I wrote to Bob Gayler, Deputy Director of the Radiology Department: "As a result of recent publicity concerning the PET scanner, the Maryland Health Services Administration sent an inquiry to the Hopkins Hospital Regulatory Specialist wanting to know how we got this unit without a Certificate of Need. We responded that the equipment was purchased in April, 1983, largely through NIH research grant money, and all patients studied were research patients, and no charges were made for the studies." We promised to get a certificate of need when the studies were no longer considered experimental. That year (1983), 230,000 examinations of all types were carried out in nuclear medicine, with 123,382 being studies of hospitalized patients.

One of our most interesting activities was transport a radioactive tracer by helicopter in November 1983, to perform a PET scan at the National Institutes of Health Clinical Center with our newly developed tracer, carbon-11 N-methyl spiperone. We flew the tracer to Bethesda from the roof of the Hopkins Childrens' Center in a helicopter because of the 20-minute half-life of the carbon-11 tracer.

The year 1983 was also memorable because I was inducted into the Calvert Hall Hall of Fame on October 4, 1983.

On March 8, 1984, we conducted a dedication program for our Cyclotron and PET scanner. In addition to Drs Heyssel, Donner and Ross, the Dean of the School of Medicine, speakers included David Kuhl from UCLA and Sadik Hilal from Columbia University in New York. David, a Founding Brother of Nuclear Medicine, invented tomography in nuclear medicine.

On February 21, 1984, I had written to Martin Donner: "You have asked me to outline our plans for the positron tomography projects with special emphasis on how and when we intend this to be used in the care of patients . . . Our work continues to support the concept that 'yesterday's research is today's practice; today's research is tomorrow's practice.'"

On July 18, 1984, I wrote to the administrators of Hopkins Hospital: "I want to stress again the urgency of the need for more space in nuclear medicine. We have just received word that we have been given an additional NIH grant for three million dollars for 5 years for PET research. We are moving ahead with the imaging of benzodiazepine, nefedepine and serotonin receptors, as well as dopamine and opiate receptors. We have strong support from all the clinical and basic neuroscientists at Hopkins."

On August 2, 1984, I wrote to Dr. Martin Donner that "PET procedures have not yet been approved for routine use, but are of clinical value in patients with epilepsy and stroke. We hope that within the not too distant future, we will be using them in patients with dementia as well."

On February 8, 1984, I wrote to Martin Donner: "We are about to carry out the first imaging of opiate receptors in human beings. If we are successful, the result may be an important stimulus to Sol Snyder's wining the Nobel prize, which he justly deserves." We succeeded in accomplishing this on May 25, 1984.

On February 15, 1985, I wrote again to Martin Donner to explain why a second PET scanner and increased chemical facilities were needed for our nuclear medicine clinic. In the letter, I wrote that "The competitive edge that the cyclotron gives us will permit not only the survival but the flourishing of Johns Hopkins Hospital. The public will continue to want our brains, not our beds."

"We need space for a second PET scanner, one that will permit an increase in our brain studies, as well as permit the extension into cardiology and oncology, two areas in which

PET is being developed and which are of great interest to Drs. Weisfeldt, Owens and their colleagues."

A major consideration for our choice of a positive- rather than a negative-ion cyclotron was that we wanted to focus on biomedical research, not cyclotron development. As our consultant, Dr. Michael TerPogogossian, said: "The carbon-11 won't care how it was made."

By 1986, funded by the neurosciences grant from the NIH, we had accomplished the following:

1. Selected, purchased, and installed a cyclotron dedicated to biomedicine.
2. Developed a radiochemical and radiopharmaceutical production facility where high specific activity radiotracers for PET could be prepared on a regular basis.
3. Selected, purchased, and installed a PET scanner.
4. Developed tracers for PET studies of dopamine, serotonin, opiate, benzodiazepine, and muscarinic cholinergic receptors.
5. Performed the first successful imaging of a neuroreceptor (the dopamine receptor) in a living human being.
6. Performed the first successful imaging of the opiate receptor.
7. Developed a mathematical model for determining dopamine receptor concentrations in the human brain.

One of the most important discoveries in the late 1970s was that the effect of many drugs affecting behavior was related to their binding to specific receptors in the brain. Before these chemical studies of the living human brain with PET, it was only possible to measure the electrical activity of the brain in order to try to understand the relationship between the mind and the brain. Today we can examine how perception, thinking, feeling and moving affect brain chemistry in health and disease. Histopathology, using the microscope to reveal cellular changes, was able to delineate regional pathological changes revealed by brain biopsy or autopsy. Half a century ago, there was great excitement about the introduction of histochemistry into the field of pathology. Today, we have moved to an even more fundamental level—the molecular level. Process and change can be examined as well as organ and cellular structure.

Whenever a chemical process is involved, there are at least two possible diseases, one in which the process is abnormally slow, the other in which it is abnormally fast. Arthur Koestler said: "In biological systems, what we call structures are slow processes of long duration; what we call processes are fast processes of short duration."

"It is a well established historical generalization that the last thing to be discovered in any science is what the science is really about" (A.N. Whitehead). When I entered the field in 1958, nuclear medicine referred to "radioisotope scanning," with an anatomical orientation. We defined the field as "the visualization of previously invisible organs by means of radioactive tracers." The fusing of anatomical images in x-rays with radioisotope studies has been routine at Hopkins since 1958, an example being the detection of pericardial effusion by showing that the blood pool lies well inside the radiographic contours of the patient's heart. Today, PET/CT is commonplace. We now define nuclear medicine as the medical specialty concerned with global and regional in vivo chemistry and physiology. Molecular nuclear medicine is now closely aligned with pharmacology and genetics. Nuclear medicine defines disease in one or more of four categories: (1) structure; (2) function; (3) biochemistry; and (4) communication.

Until recently, the practice of medicine has been dominated by gross anatomy, histopathology, and the chemistry of the blood and urine. Nearly four centuries ago, Locke wrote: "Anatomy is absolutely essential to a surgeon, but that anatomy is likely to afford any great improvement in the practice of medicine, I have reason to doubt. All that anatomy can do is show us the gross and sensible parts of the body."

After our cyclotron was installed, the physicist in nuclear medicine, who thought he would be put in charge, resigned when I appointed Dr. Bob Dannals, a young chemist who had just received a PhD from our program, to be in charge of the cyclotron and chemistry facility. I didn't want a physicist taking our cyclotron apart to find out why it was working so well.

Figure 107 Bob Dannals (on the right in the front row of the picture). Bob headed the cyclotron unit from the time of its installation in 1981.

We then had to convince others of the value of PET in clinical medicine. In a July 10, 1980, letter to me from the Hospital Director, Bob Heyssell, denying our request to purchase a scintillation camera for the intensive care unit, wrote:

"It was said to me the other day that scanning could be a screen for the decision as to do or not to do arteriography. When the question was put as to whether the cardiologist would be willing to forego arteriography if the scan indicated that one was not necessary, the answer was that some would and some would not . . . The point that I am trying to make to you is that the day has past when we can simply add on more and expensive testing unless we can clearly demonstrate that it makes a real difference in management of the patient and patient outcomes. This is not, incidentally, simply a matter of cost containing, it has to do with good medical practice. That is what we ought to be about in these institutions.

"The question is whether doing the cardiac studies makes a difference in terms of outcome for the patient. The article that you sent me indicates that people with heart attacks are going home earlier. It does not state that every patient who is going home has to have a thallium scan . . . " Eventually in 1981 a portable scintillation camera was purchased for use on the Emergency Room.

We were very fortunate to be able to finance our PET project through NIH grants. Nuclear Medicine often was the Cinderella in the Radiology Department. On October 2, 1978, I wrote to Martin Donner: "For over two years, we have had no major capital funds. Our entire cardiovascular operation was and is based on equipment on loan. We would be a third rate division if we had to depend on the capital funds provided by the hospital. If we are not to lose key people, such as Phil Alderson, we must soon get approval for at least one major piece of equipment to replace out outmoded equipment." (Dr. Alderson subsequently became Chairman of the Radiology Department at Columbia-Presbyterian Hospital.)

In April 1987, I was notified by the Dean that William Brody had accepted the Chairmanship of the Department of Radiology at Hopkins. Brody was subsequently quoted as saying that "if we didn't already have a PET program at Hopkins, we sure wouldn't start one." In an interview shortly after his assuming the Chairmanship of the Department of Radiology in 1987, Dr. Brody asked me what I wanted for nuclear medicine. I replied "independence." Our relationship was always strained thereafter. I erroneously thought that Dr. Brody's opinion of nuclear medicine might make him receptive to the idea of his becoming free of us. How wrong I was. On May 22, 1990, Dr. Brody wrote to all members of the Radiology faculty: "Nuclear Medicine will remain a separate Division although we hope to increase cross-coupling between nuclear medicine with other modalities for each organ system as we gain experience with the new organization . . . Dr. Wagner has informed me of his intention to step down as Director of Nuclear Medicine when a successor has been identified." Dr. Brody left Hopkins in 1994 to become Provost at the University of Minnesota. On September 1, 1996, Dr. Brody returned to Hopkins as President of Johns Hopkins University.

Four years later, Dr. Richard Wahl from the University of Michigan was selected by Dr. Elias Zerhouni, who had become Chairman of the Department of Radiology, as my successor. I often thought that the ten year delay in making the appointment of my successor might have reflected Dr. Brody's opinion about the identity of nuclear medicine within radiology. He did not view nuclear medicine as a medical specialty. During the decade of the 1990s, I was no longer active in clinical nuclear medicine, and directed most of my professional life to the Division of Radiation Health Sciences of the School of Public Health. I continued to direct our NIH Program/Project and Training Grant from 1966 until 2001. In 2004, the Radiation Health Sciences Division ceased to exist, and our few remaining faculty members joined other divisions. Today, I still present the annual Highlights lecture at the annual meeting of the Society of Nuclear Medicine, direct an annual post-doctoral course on nuclear medicine topics at Hopkins, the successor to our old Colby College course, and give lectures throughout the world. Since 1994, the Highlights lecture has been translated every year into the Japanese language and distributed by General Electric Tokogawa Medical Systems, Ltd. Until 2004, it was also translated into Chinese.

I was pleased (and surprised) to read in the weekly radiology journal *RT Image* that I had been chosen on September 6, 2004, as one of the 25 most influential people, institutions, and organizations in radiology. The citation stated: "Without Henry N. Wagner, Jr. you can't help but wonder where the field of nuclear medicine would be. Perhaps still in the dark ages. But because of his accomplishments, a new world of molecular imaging was uncovered . . . as an internationally known and respected expert in nuclear medicine, when Wagner talks, the world listens."

Once we sponsored a dinner at the National Academy of Sciences in honor of Senator Lister Hill during my Presidency of the American Federation For Clinical Research. Vice-President Hubert Humphrey was the principal speaker, and he began his talk: "Lister, I've been told that flattery is OK as long as you don't inhale it. I've seen you taking some mighty big breaths." You can imagine how many deep breaths I took when I read this article, particularly at this stage of my career.

The article in *RT Image* also referred to three honorary degrees which I received, one of which was on May 28, 1980, at Washington College in Chestertown, Maryland, 5 miles down the river from our weekend house.

In my commencement address, I said: "There is no way to escape the challenge of radiation . . . We must learn how to weigh benefits against risks, learn how to deal with problems more scientifically, more rationally, and less intuitively and emotionally. We must learn to balance self-interest and societal interests, between reality and the abstractions of reality . . . Harmony rather than comfort or prosperity should be our goal, the harmony of working with dignity, freedom, and self-respect, realizing our full potential as human beings."

Professor A.M. Harvey often said: "At Hopkins, we select good people and leave them alone." I benefited greatly from this philosophy and tried to follow it in my own professional leadership positions.

Dr. Brody left Hopkins to assume an administrative position at the University of Minnesota. He returned to Hopkins as President of the University in August 1996. In March 2002, he participated in the inauguration of the Henry N. Wagner, Jr. Professorship in Nuclear Medicine at Johns Hopkins School of Medicine. This chair, occupied by my successor, Dr. Richard Wahl of the University of Michigan, was established largely through the efforts of Manuel Dupkin, a trustee of the University and Hospital, a friend for over three decades.

7

The Early Days

Over the years, our faculty and students in nuclear medicine at Hopkins, were the best and the brightest, not only in the United States, but all over the world. Among our young distinguished faculty in the early days was Dr. Arthur Karmen. While a medical student at the New York University College of Medicine, he developed the first spectophotometric method for assaying glutamine oxaloacetic transaminase (SGOT) in the blood, a procedure that became widely used throughout the world. I first met Art when we were at the National Heart Institute of the NIH in 1955–1957. He became a leader in the study of amino acids, sympathetic amines, steroids, and fatty acids. He came to Hopkins in 1963, and subsequently left to become head of Laboratory Medicine at Albert Einstein School of Medicine in New York.

We were very fortunate that Russell Morgan decided to develop a program in Radiological Sciences in the Johns Hopkins School of Public Health. This made possible our having graduate students in our nuclear medicine program, many of whom went on to become faculty members after receiving their PhD degrees. Every year, we had about 20 students enrolled for an advanced degree in the Radiological Sciences Division of the School of Public Health. At the same time, there were three trainees, and ten research fellows in the Division of Nuclear Medicine in the Medical School (seven were foreigners). Many returned to start nuclear medicine departments in their own countries. Medical students from Johns Hopkins and visitors from other medical schools spent a month or more in nuclear medicine. Some of them, such as Paul Worley, returned to basic science departments and eventually became full professors. One example of two generations of nuclear medicine physicians trained at Hopkins is Drs. Phillip Van Heerden and his son, Ben, from South Africa.

An example of academic/industrial teamwork, Dr. Henry Kramer of the Union Carbide Nuclear Company spent a year working with us in 1965–1966 while still a Union Carbide employee. He subsequently made many important contributions to the development of radiopharmaceuticals and was an active participant in the Society of Nuclear Medicine. He developed tracers for the study of the kidneys, a Mo-99/Tc-99m generator, and the fission products Mo-99, Xe-133, and I-131. In 1976, we had a contract with Union Carbide Corporation to develop neutral and positively-charged mercaptoethyl amine complexes of technetium-99m for brain, heart, and pancreas imaging. These were positively charged in contrast to previous agents which were negatively charged. The total budget was $21,238. In 1982 we had a contract with New England Nuclear Corporation to study the relationship between chemical structure to brain uptake of metal complexes. The direct costs were $22,833. The result of these contracts was a publication authored by Susan Lever and colleagues in the *Journal of Nuclear Medicine* in 1985.

Dr. Ephraim Lieberman was a founder of Diagnostic Isotopes, a company responsible for the commercial development of more than 10 radiopharmaceuticals. His company subsequently became known as Cadema Medical Products of Middletown, N.Y. I became Chairman of its Medical Advisory Committee on August 5, 1986. On April 4, 1991, I became a member of the Board of Directors of Capintec, Inc, a company founded by Arthur Weis, a pioneer in the development of nuclear medicine.

Dr. Steve Larson came to Hopkins as an Assistant Professor of Radiology and Radiological Science from his position of Clinical Associate In Nuclear Medicine at the National Institutes of Health where he served from 1970 to 1972. He left Hopkins in 1975 to become Associate Professor of Nuclear Medicine and Director of Nuclear Medicine at the University of Oregon in Portland, Oregon. In 1976, he became Professsor of Medicine in the Laboratory Medicine and Radiology at the University of Washington. He subsequently became Chief of Nuclear Medicine at the VA Hospital in Seattle in 1981. In 1982, he returned to the NIH as Director of Nuclear Medicine. At present he is Chief of the Nuclear Medicine Service at the Memorial Sloan-Kettering Cancer Center in New York. In 2004, he was selected as the Scientist of the Year by the Radiological Society of North America in view of his enormous contributions to nuclear oncology.

Figure 108 Professor Steven Larson.

Another outstanding trainee was Dr. Kenneth McKusick, who completed his residency in nuclear medicine in August 1972 and served as an Instructor at Hopkins through June of 1973 at which time he joined the Peter Bent Brigham Hospital in Boston, and in 1980 was promoted an Associate Professor of Radiology.

Dr. Thomas Mitchell was my teacher of nuclear physics in 1959 at the U.S. Naval Medical School in Bethesda, Maryland. He spent 1956 to 1960 teaching physics to nuclear medicine trainees. He received a PhD in physiology from Georgetown University in 1963, and joined the Hopkins faculty in the School of Public Health in 1969 as an Associate Professor of Radiation Health Sciences. He was recognized every year as the best teacher in our Radiological Science program by graduate students, residents and junior faculty. He collaborated on innumerable research projects.

We were fortunate to have Dr. James Adelstein spend 6 months in nuclear medicine at Hopkins shortly after he had been selected by Dr. Herbert Abrams, Chairman of the Radiology Department at Peter Bent Brigham Hospital in Boston to start and head a nuclear medicine program. He is probably the only person in the world who read from cover to cover the *Principles of Nuclear Medicine* textbook that we published in 1968. Shortly after his arrival in Baltimore, Jim fell and broke his wrist. He turned down my suggestion to perform serial studies with strontium-87m to monitor the healing of the fracture. So much for hormesis. Jim apparently had not been not impressed with our prior study of skeletal growth in chickens as they matured. He carried out pioneering studies of the diagnostic process, emphasizing diseases of the thyroid. He stimulated the interest of Dr. Barbara McNeil, who subsequently became very productive in the field of medical diagnosis. During Jim's subsequent career at Harvard, he concentrated on radiobiology and radionuclide therapy.

Richard Reba, who subsequently became President of the Society of Nuclear Medicine, was a research trainee in 1962, and carried out further studies of the quantitative measurement of Hg-203 chlormerodrin accumulation by the kidneys in the detection of unilateral renal disease. The first imaging of the kidneys with this tracer had been published in 1960.

Over the 50 year life of the Society of Nuclear Medicine, 10 (20%) of the Presidents of the Society had trained in nuclear medicine in the Johns Hopkins Nuclear Medicine program:

1970–1971 Henry N. Wagner, Jr. (The 18th President)
1973–1974 Wil B. Nelp
1989–1990 Richard A. Holmes
1993–1994 Richard C. Reba
1995–1996 Peter T. Kirschner
1997–1998 H. William Strauss
2000–2001 Jonathan M. Links
2001–2002 Alan H. Maurer
2004–2005 Peter S. Conti

The following persons won the Johns Hopkins Nuclear Medicine Distinguished Alumnus awards presented at the Johns Hopkins Nuclear Medicine Alumni party at the annual meeting of the Society of Nuclear Medicine:

1971 Masahiro Iio
1972 Wil B. Nelp
1973 Frank Deland
1974 Donald Tow
1975 Richard Holmes
1976 Yasahito Sasaki
1977 Ephraim Lieberman
1978 Richard Reba
1979 John McAfee
1980 Manny Subramanian
1981 Peter Hurley
1982 Edwaldo Camargo
1983 David Goodwin
1984 Anne Wagner
1985 Steve Larson
1986 H. William Strauss
1987 S. James Adelstein
1988 James Langan
1989 Michael Maisey
1990 Bengt Langstrom and Leon Malmud
1991 Philippus Dr. R. vanHeerden
1992 Mary Lou Mehring
1993 Kenneth A. McKusick
1994 Peter T. Kirchner
1995 Yoshiharu Yonekura
1996 Glenn T. Seaborg (honorary alumnus)
1998 Jonathan Links
2003 Peter Conti

Marianne Chen joined our program as a research assistant in 1971, and entered our doctoral program in 1979, receiving her PhD degree in 1981. She became an Associate Professor of Internal Medicine at the University of South Florida in 1988. Among her many research achievements were the automation of microbiological assays, including studies of tuberculosis.

Dr. Norman LaFrance entered our nuclear medicine training program in 1980 after completing a residency in Internal Medicine in Connecticut. He was a resident from 1980 to 1982, and subsequently became a faculty member in charge of our clinical studies.

Dr. Edwaldo Camargo first came to Hopkins from July 1973 to June 1975 as a research fellow in nuclear medicine. He became an Assistant and then an Associate Professor of Radiology before deciding in 1983 to return to Brazil for personal reasons. When these were resolved, he returned to Hopkins on June 1, 1988, to become Associate Director of our clinical unit. He relinquished his position as Clinical Director of Nuclear Medicine in August 1991 in order to return to Brazil. Other trainees from Latin America included Dr. Jose Moreno and Luis E.M. Tenorio from Panama. Distinguished visitors included

Dr. Julio Kieffer from Brazil, Drs. Alfredo Cuaron, Roberto Maass, Jorge A. Maisterrena, and Carlos Marinez Duncker from Mexico. Dr. Duncker's son was to subsequently spend a year with us as a trainee, one of several trainees from two generations who trained with us. Dr. Juan Touya and his brother Eduardo, of Montevideo, Uruguay also visited us on many occasions.

Edwaldo Camargo was very effective in helping start nuclear cardiology at Hopkins, as well as in the in vitro laboratory, where he was an important contributor to automated microbiology, especially in tubeculosis and leprosy. He helped establish a nuclear medicine satellite operation in the surgical intensive care area. We received a letter on June 4, 1980, from Dr. George Zuidema, Surgeon-in-chief, in which he stated: "I fully concur with your appraisal that we have need for a mobile camera with computer capability for the seventh floor area."

Dr. Norman LaFrance made major contributions to the operation of clinical nuclear medicine from 1982 until 1988, at which time he left to join the DuPont Company as a medical director. He was at that time an Associate Professor, and was ably assisted in his clinical work by Dr. Cahid Civelek, who was in charge of heart studies. Samuel Sostre, James Frost, Dean Wong, and Lorcan O'Tauma joined our nuclear medicine faculty after their period as graduate students.

Dr. LaFrance was also Chairman of the Radioactive Drug Research Committee (RDRC), a position that I had previously held until February 19, 1982, when I resigned as Chairman of the committee after a dispute with the Associate Dean for Research of the medical school.

The Institutional Radioactive Drug Research Committees (RDRC) had been introduced in 1974 when the exemption of radiopharmaceuticals from regulation by the FDA was removed. Until then, the approval of radiopharmaceuticals was by the Nuclear Regulatory Commission (NRC), formed at the time of the breakup of the Atomic Energy Commission into Energy Research and Development Agency (ERDA), and subsequently becoming the NRC and the Department of Energy. Because of the great number of radiopharmaceuticals (over 1,000) that were being developed, a number likely to increase in the future, the FDA had the idea of creating Radioactive Drug Research Committees (RDRCs) in universities or other institutions to approve the proposed use of new radiopharmaceuticals under certain limited conditions. No new tracers could be examined in human beings for the first time. The studies had to be for research rather than for the care of specific patients. It had to be demonstrated that there was no human pharmacological effect. The radiation dose limits were based on the occupational radiation protection criteria established by the Nuclear Regulatory Commission under 10 CFR 20.101. When the radiation exposure limits were exceeded or if the procedures were diagnostic, an Investigational New Drug Application (IND) had to be submitted to the FDA, a limitation described in 21 Code of Federal Regulations CFR 361. The complex toxicological studies that were required cost over $100,000, and pharmaceutical companies were not attracted to diagnostic agents which were given only a few times to a patient. Industry focused on therapeutic agents, especially those that were administered for years in chronic illnesses. Many of the radiopharmaceuticals, such as technetium-99m agents, were administered in nanomolar quantities. It was not possible to obtain enough material to carry out toxicity tests.

Among the most active faculty in the development of nuclear cardiology was Dr. Cahid Civelek who trained in nuclear medicine at Hopkins from January 1982 to August 1984. In March 1991, he was appointed Director of the Section of Nuclear Cardiology in the Division of Nuclear Medicine. He was appointed an Associate Professor in 1994. Before he left in April 2003 to head the Nuclear Medicine Department at St. Louis University Medical School, he published many papers, and trained numerous American and international physicians in nuclear cardiology. He has a worldwide reputation in nuclear cardiology and lectures at medical centers in the U.S. and internationally.

Dr. Helen Mayberg was a post-doctoral fellow in the Division of Nuclear Medicine from 1985 to 1987, and joined our faculty in 1987, first as an Instructor, and then an Assistant Professor before leaving in 1991 for the University of Texas in San Antonio. As a neurologist who recognized early the value of nuclear medicine technology to help solve neurological problems, Helen has made many contributions to PET studies of the brain.

In 1987, we started a long-standing and productive relationship with Dr. Carol Tamminga and Will Carpenter, Professors of Psychiatry at the Maryland Psychiatric Research Center, in the study of patients with schizophrenia.

One of the most important events in the early days of nuclear medicine was the founding of the World Federation of Nuclear Medicine (WFNMB) in 1973. Dr. Iio and his colleagues, Professor Hideo Ueda and Dr. Sadatake Kato, took the risk and responsibility of holding the first World Congress of Nuclear Medicine and Biology in Tokyo and Kyoto in 1974. At the time, it was not known whether the concept of a world federation would be supported by the members of the national nuclear medicine societies throughout the world. The meeting was a huge success and set high standards that were upheld during the next six world congresses in Washington, Paris, Buenos Aires, Montreal, Sydney, Berlin, and Chile.

On the morning of September 17, 1978, the day the Second World Congress was to open in Washington, Anne and I called room service at the Washington Hilton Hotel. If there was an answer, I would know that the impending strike of hotel employees had been averted. Staying at the Hilton (where President Reagan was shot 3 years later) were political leaders, including Senator Ted Kennedy, Menachen Begin, and Egyptian and Israeli political figures who were attending the Peace Conference at Camp David where Egypt and Israeli settled the issue of the Sinai desert, the "Camp David Accords." The

Figure 109 Hideo Ueda, President of the First World Congress of Nuclear Medicine in 1974 in Tokyo/Kyoto, and Henry Wagner, President of the Second World Congress of Nuclear Medicine in 1978 in Washington, D.C.

Figure 110 Masahiro Iio, Secretary General of the First World Congress of Nuclear Medicine, and first Japanese fellow in nuclear medicine at Hopkins.

Figure 111 A social event during the First World Congress of Nuclear Medicine in Tokyo and Kyoto.

Figure 112 Masahiro Iio and Henry Wagner at the gate of Tokyo University in 1964.

sight of these persons in the lobby of the Hotel in the evenings enhanced the international flavor of the Second World Congress of Nuclear Medicine and Biology.

The Second World Congress was held in Washington, D.C. Again there was uncertainty about funding, because there was no source at the start. The Society of Nuclear Medicine (SNM) charged the Congress $25,000 to run the commercial exhibits. It was only after the meeting attended by 1,655 persons was underway that we were able to meet the budget of $300,000. Fifty-eight nations were represented, with 922 registrants from the USA, 135 from Japan, 53 from France, and 55 from West Germany.

At the opening ceremony at 5:00PM on September 17, 1978, greetings from President Jimmy Carter were read, followed by addresses by Donald Frederickson, Head of the NIH, and Nobel Laureate Roslyn Yalow, followed by a wine and cheese reception. On September 18, the Congress sponsored the Preservation Hall Jazz Band from New Orleans at the Kennedy Center followed by a rooftop reception. On September 19, there was a banquet at the Hilton Hotel, and the following day, a private opening of the National Gallery of Art including a concert by the National Gallery Orchestra, playing the Beethoven Overture to Prometeus and Mozart's Symphony #3. There were tours of Washington for $11.00; a Presidential Tour including Mt. Vernon, Arlington Cemetery and the Jefferson and Lincoln Memorials for $18.00: a tour of the Kennedy Center for $13.00; and a tour of the Smithsonian for $6.00.

The organizational meeting at which it was decided to have the 2nd World Congress of Nuclear Medicine and Biology in Washington was held in Philadelphia on June 17, 1975, at the time of the annual meeting of the Society of Nuclear Medicine. An effort was made to have the World Congress held jointly with the 1978 annual meeting of the SNM, but an agreement could not be reached. The commercial exhibits were under the direction of the Executive Director of SNM, Judy Glos. Jim MacIntyre, President of SNM, agree to this arrangement.

Ed Matson of Abbott Laboratories and John Ryan of 3 M chaired the industrial affiliates committee, whose members paid $100–200 each. The affiliates agreed to pre-pay the charges for exhibit space beginning immediately and continuing for 3 years. On July 2, 1975, a group of women called Courtesy Associates agreed to provide administrative services immediately and delay charging us until the year of the meeting. A contract signed on March 24, 1976 agreed to pay them $65,000 to $72,000 depending on how much time their efforts took. The registration fee for the meeting was $100 for attendees and $50 for spouses.

The theme of the meeting was the worldwide promotion of nuclear medicine. The goal was to "glamorize Washington and the USA for overseas visitors," as well as to ensure the survival of the WFNMB, which has now been accomplished, with World Congresses held every four years. The next meetings will be in South Korea in 2006 and South Africa in 2010.

The minutes of the organizing committee for the 1978 meeting stated that the "major thrust is to help nuclear medicine achieve a clearer identity vis a vis the public, government, colleagues in other specialties, and people within the specialty itself."

We in Hopkins nuclear medicine have had the good fortune to have over 50 Japanese physicians and scientists work with us, beginning with Dr. Masahiro Iio in 1961. We helped Mr. Iwabuchi and Mr. Tominaga from Ishikawa Prefecture in Kanasawa, Japan, in their establishment of a nuclear medicine research laboratory including PET, with a budget of $20 million.

The 3rd World Congress was held in Paris in 1982. It was very successful and had a budget of $700,000. The 4th meeting was held in Buenos Aires in 1986, the 5th in Montreal in 1990, the 6th in Berlin in 1994, the 7th in Sydney in 1998, and the 8th in Santiago in 2002.

The Nuclear Medicine Division at Hopkins has always had an international orientation that lasted for many years after our trainees had left Hopkins. For example, in 1983, I joined Dr. Iio in writing a book on Geriatric Nuclear Medicine. We described the ideal radiotracer for studying the aging brain: We postulated that it should:

1. be labeled with carbon-11 or fluorine-18
2. cross the blood–brain barrier to permit intravenous administration
3. have a high affinity and selectivity for neuroreceptors, including receptors for neurotransmitters, glucose, and amino acids.
4. slow dissociation from targets to facilitate imaging
5. little or slow metabolism within the brain
6. rapid clearance from the blood.

Before coming to Hopkins, Dr. Iio worked in the Department of Internal Medicine at Tokyo University, headed by a cardiologist, Dr. Hideo Ueda. When Dr. Julius Krevans visited Professor Ueda, they agreed that Dr. Iio should spend some time at Hopkins in cardiology. When this was not possible, Dr. Iio was accepted to join the new Nuclear Medicine Division at Hopkins. He became enamored about the field of nuclear medicine and dedicated the rest of his life to it, subsequently becoming Chairman of the Department of Radiology at Tokyo University.

Nuclear medicine had developed in the early 1950s in Japanese universities. In Keio University, a program was begun under the direction of Professors H. Yamashita, I. Kuramitsu, and F. Kinoshita. Dr. H. Yamashita was a pioneer in Japanese nuclear medicine. In 1951, he was sent to the United States by the Japanese government to study the use of radioisotopes in medicine. He became friends with many prominent nuclear medicine physicians and scientists in the United States, including Drs. Charles Dunham, R.T. Overman, Marshall Brucer, W. Chamberlain, and Hal Anger.

Figure 113 Some of the attendees at the First World Congress of Nuclear Medicine. To the right of Professor Ueda is Claude Kellersohn, President of the Third World Congress of Nuclear Medicine. To the left of Professor Ueda is Henry N. Wagner, Jr., President of the Second World Congress of Nuclear Medicine.

Figure 114 Some attendees at the Fourth World Congress of Nuclear Medicine.

Figure 115 Mr. H. Seta, President of Nihon-Mediphysics and a strong supporter of nuclear medicine.

Figure 116 Shige Kaihara, third Japanese fellow training in nuclear medicine at Johns Hopkins with his wife at the wedding of his daughter.

Figure 117 Marcus Schwaiger, head of nuclear medicine at the University of Munich, Germany, and a major contributor to nuclear cardiology, starting with his stay at the University of California in Los Angeles.

The early studies of Dr. Yamashita's group involved cobalt-60, phosphorus-32, and iodine-131. Another prominent institution was the program at Nagoya University, under the direction of Dr. T. Nagai. He worked at the National Institute of Radiological Sciences from 1959 to 1974. He spent some time at the Oak Ridge Institute of Nuclear Studies. From 1968 to 1973, he was a Senior Scientific Officer at the IAEA. In 1974, he was appointed as a Chairman of the Radiology Department in Gumma University.

One of the most important persons in the early days of nuclear medicine was Dr. William G. Myers. If Ernest Lawrence is the father of the cyclotron, Bill Myers is its godfather. In 1940, when Bill attended a Sigma Xi lecture by Ernest Lawrence's physician brother, John, at Ohio State University, describing the early studies at UC Berkeley with the first man-made "twinkling" atoms, Bill decided immediately that nuclear medicine would be his medical specialty. His senior medical school thesis was "Application of the Cyclotron and Its Products to Biomedicine." That same year, 1941, Ohio State got its own cyclotron, a 42-inch machine used in physics and medicine. By 1966, Dr. Myers and his colleagues had worked with the positron-emitting radionuclides: carbon-11, nitrogen-13, oxygen-15, fluorine-18, iron-52, and iodine-124. In 1968, he wrote: "The 23% carbon content of man will cause carbon-11 to have a rapidly increasing significance, especially

Figure 118 John Kuranz, founder and President of Nuclear Chicago, first commercial developer of the Anger camera and treasurer of the Second World Congress of Nuclear Medicine in Washington, D.C., in 1978.

Figure 119 Maria and Keith Britton at the Second World Congress of Nuclear Medicine in Washington, D.C. Both were pioneers in the growth of nuclear medicine in Great Britain.

Figure 120 Walter Wolf, a major contributor to radiopharmaceutical development in nuclear medicine.

Figure 121 Henry Wagner with Dr. Svjatoslav Medvedev and a colleague from Leningrad.

Figure 122 Henry Wagner and Michael Welch visiting Kuwait just after the Persian Gulf war in 1991.

Figure 123 James and Beatrice Smith, Vice-President and Secretary, respectively, of the Second World Congress of Nuclear Medicine in Washington, D.C. Jim subsequently became head of nuclear medicine in the U.S. Veterans Administration.

Figure 124 Richard Reba, Chairman of the Local Arrangements Committee of the Second World Congress of Nuclear Medicine, a member of the nuclear medicine faculty of Johns Hopkins, President of the Society of Nuclear Medicine, and head of nuclear medicine at George Washington University in Washington, D.C., and the University of Chicago.

Figure 125 Anne Wagner with Fred George, Chairman of the local arrangements committee during the 1971 meeting of the Society of Nuclear Medicine in Los Angeles during the Presidency of Henry Wagner. He was head of nuclear medicine at the University of Southern California (USC).

in scintigraphy, when small medical cyclotrons become common in most hospitals." How right he was! Historically, carbon-11 was first used in man in 1945 in the Donner Laboratory by Tobias and his colleagues. They used Geiger-Muller tubes to detect the photons coming out of various parts of their bodies after they had inhaled carbon-11 carbon monoxide.

In 1946, Myers traveled to the Pacific to serve as Radiological Safety Officer during the Bikini Atoll testing of the atomic boat. He described his experience:

"We were sitting, Geiger counters in hand, on the deck of a little gunboat when the bomb was detonated. It was awesome, more than a million tons of water were thrown a mile into the air. Battleships anchored near the blast site sank instantly. The *Arkansas*, all 25,000 tons of her, stood on end, and was tossed around like a toy."

When he returned to Ohio State, he became obsessed with the chart of the nuclides. "I would look at the chart and imagine that this nuclide and this and this might be useful in medicine. I would say to my colleagues 'Here's one we ought to take a good look at.'" Bill introduced 10 radionuclides into medicine: cobalt-60, gold-198, chromium-51, iodine-125, iodine-123, iodine-121, strontium-87m, strontium-85m, potassium-38, and carbon-11.

He said: "Some of us among the nuclear physicians still recall the good old days— 'from cyclotron to vein'—during the first decade of nuclear medicine."

From 1949 to 1978, he taught a core course in nuclear medicine at Ohio State University. More than a thousand resident physicians and life-sciences graduate students took this course. I was honored to deliver a talk about Bill when he received the 3rd Hevesy Nuclear Pioneer Award of the Society of Nuclear Medicine. He often spoke of Hevesy's book *Adventures in Radioisotope Research*, published in 1962. Bill quoted Hevesy's Faraday Lecture in 1950: "The indicator chemist is to some extent an historian, highly interested in the past of atoms and molecules . . . He has great concern in the distinction of whether molecules in tissue are old or new."

In June 1948, he reported to the American Radium Society his initial results with cobalt-60 and predicted correctly that thousands of cobalt-60 units would someday be installed in cancer treatment centers throughout the world. Bill developed chromium-51 and iodine-125. Today hundreds of millions of procedures are performed worldwide each year with tracers that he proposed. In 1962, he promoted the use of iodine-123, which has many advantages over iodine-131 for imaging.

Bill was elected Historian of the Society of Nuclear Medicine in 1973, the same year that he received the Paul C. Aebersold Award of the Society of Nuclear Medicine for Outstanding Achievement in Basic Science Applied to Nuclear Medicine.

In 1978, he made closing remarks at the Second World Congress of Nuclear Medicine held in Washington, D.C. He concluded: "My role as historian of the Society of Nuclear Medicine leads me to appreciate that savoring the past enriches the present, and presages the future."

Bill's best friend, Hal Anger, embodied in a single person the convergence of nuclear physics, electronics, optics, and information handling. Hal introduced a completely new and different device for imaging radionuclides in the body. Working in the Donner Laboratory of Medical Physics at the Lawrence Berkeley Laboratory of the University of California, he placed a pinhole collimator in front of a sodium iodide crystal 4 inches in diameter. Behind the crystal he placed an array of seven photomultiplier tubes. The output of the tubes was recorded so that the position of each scintillation detected by the crystal could be localized by grid coordinates and an image of the tracer distribution flashed on a cathode ray tube with the same coordinates. The new "camera" could provide a continuous measurement of the time course of the radioactivity in all regions at the same time. Cassen's rectilinear scanner could not.

Figure 126 Hal Anger, inventor of the scintillation camera.

Figure 127 Hal Anger exhibiting the scintillation camera at the annual meeting of the Society of Nuclear Medicine.

In June 1958, Hal first exhibited the "scintillation camera" at the Fifth annual meeting of the Society of Nuclear Medicine in Los Angeles, and at the American Medical Association meeting in San Francisco the next week. The first images that he showed were of the thyroid gland. For the next three years, he tried to get manufacturers of nuclear instrumentation to begin production of his camera. The first commercial Anger camera was installed in the laboratory of Bill Myers at Ohio State University Hospital in September 1962. The first use of the Anger camera for dynamic studies was presented by Dr. Myers in Montreal at the June 1963 meeting of the SNM. In his citation of the Nuclear Pioneer Award of the SNM to Hal Anger in 1974, Dr. Myers said: "To me a fascinating aspect of all of his innovation and invention is that Hal Anger so clearly saw what needed to be done, and did it, long before the medical profession became cognizant of the need. There lies the essence of true genius."

The introduction of technetium-99m into nuclear medicine by Paul Harper at the University of Chicago in 1963 was a major factor in the eventual success of the Anger camera. On March 24, 1964, Hal wrote to Bill Myers: "Today we took some brain tumor pictures in 10 seconds with 10 mCi of technetium-99m. Our counting rate over the head was 300,000 cts/minute. It looks like a wonderful isotope. Best regards. Hal O. Anger."

In 1969, with $17,000 supplied by Dr. Charles Doan, Bill placed a special order for the first commercial Anger camera with the Nuclear Chicago Corporation. Bill had to threaten to sue the company to get them to deliver the camera, but in September 1962, the first commercial scintillation camera was installed in his laboratory at Ohio State. Two years later at an IAEA symposium on "Medical Radioisotope Scanning," Bill presented the only Anger camera paper presented at the meeting—images of rat kidneys imaged with iodine-131 orthoiodohippurate. He concluded: "The scintillation camera is an elegant instrument for the study of dynamic processes in vivo that are not otherwise demonstrable."

The Chinese philosopher Lao-Tzu said: "Leaders are best when people scarcely know they exist. When their work is done, their aim is fulfilled, the people will say 'We did it all ourselves.'" In 1967, Bill helped Alan Fleisher, a physicist from Berkeley, in the start of a company to build cyclotrons for medicine. Retiring from Ohio State after 46 years, Bill spent the next two years working with the first cyclotron produced by this company for Memorial Sloan-Kettering Hospital in New York, working with carbon-11 and nitrogen-13.

In 1981, I had the honor and pleasure of delivering at the annual meeting of Society of Nuclear Medicine the Nuclear Pioneer lecture, honoring Bill. I concluded: "You can know that there will be continuing benefits long after your efforts have ceased. You have had the pleasure of discovery, the opportunity to spend your life doing what you wanted to do, the freedom to study and investigate, to develop friendships all over the world, and to know that your teaching and research have helped all mankind. All these satisfactions are yours. No one could ask for more."

Shortly after Dr. Iio returned to Japan from Hopkins in 1963, he invited me to visit Japan for the first time. We traveled together throughout Japan for a period of 16 days, visiting universities, temples, shrines, and inns, during which time he introduced me to Japanese science, culture, and art. He imbued me with a love of Japan, the Japanese people and culture that has grown stronger and stronger over the passing years. Many Japanese have become lasting friends over nearly half a century. Among them are Professor Torizuka of Kyoto University, Professor Hisada of Kanazawa University, and Professor Yasuhiro Sasaki, head of the Institute of Radiological Sciences in Chiba, and many, many other leaders of Japanese nuclear medicine.

Figure 128 Henry Wagner with Yasuhiro Sasaki at a conference after work.

8

The Thyroid Paves the Way

The care of patients with thyroid disease illustrates the general principles upon which the field of nuclear medicine is based. Its procedures are used to examine the chemistry of the thyroid by means of radiation detectors outside the body. The information is dynamic, rather than static. Space and time are joined in documenting quantitatively the chemistry of the living human body. The tools of nuclear medicine examine and characterize quantitatively the dynamic state of the processes that are the basis of life. Disease can be detected before the "constancy" of the constituents of the body has become abnormal and is reflected in blood tests. These tests reveal the "steady state" of the abnormal processes that have become abnormal after compensatory biochemical processes become unable to solve the patient's problems.

My extensive contact with patients with thyroid disease began in my work with Dr. Russell Fraser, an endocrinologist at Hammersmith Hospital in London in 1957. I had taken care of patients with thyroid disease under the direction of Dr. Samuel Asper at Hopkins, but my experience had been quite limited. At Hammersmith, caring for a large numbers of patients for whom radioactive iodine played an essential role in diagnosis and treatment made a great impression on me. I could see in vivo biochemistry as the foundation of diagnosis and treatment.

Disease of the thyroid is often detected when a patient or the examining physician first detects an enlargement of the thyroid gland or nodules in the neck, or when the patient complains of symptoms resulting from increased or decreased secretions of thyroid hormones, resulting in hyperthyroidism or hypothyroidism. Using a fine needle to biopsy the thyroid, the physician can detect histopathological findings at the cellular level. Radioactive iodide or its analogue, technetium-99m pertechnetate, make it possible to examine the biochemical processes going on within the thyroid gland itself, and quantify increased or decrease incorporation of iodine into the gland, as well as the release of synthesized hormone secreted by the thyroid gland. Radionuclide imaging and magnetic resonance imaging can provide structural details, such as nodules or enlargement of the gland. Examining the structure, function, and biochemistry of the thyroid and its interaction with other organs not only provides the diagnosis, but also a way to plan and monitor the effect of surgery or drug treatment. Quantitative measurements can help diagnose the patient's problems, predict the future course of the disease, and help plan treatment.

Figure 129 Measurement of thyroidal accumulation of radio-iodine with a Geiger-Mueller tube.

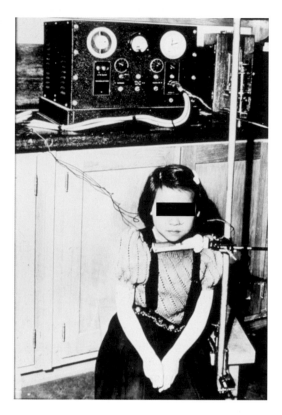

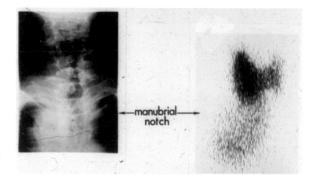

Figure 130 Identification of a substernal goiter with a rectilinear scanner and radioiodine.

The disease hypothyroidism, or myxedema, had been recognized as early as 1827, when the condition was observed to result from the surgical removal of the gland. In 1896, Vassake and Geberaku also described the entity of hypothyroidism after thyroid-ectomy. The relationship of iodine to the thyroid was discovered in 1896 by Baumann.

In ancient times, an enlargement in the neck was not thought to be a manifestation of disease, but rather a sign of beauty. In the first century A.D., Pliny the Elder wrote: "Only men and swine are subject to swellings in the throat, which are mostly caused by the noxious quality of the water they drink." In the 12th century, the ashes of sponges and seaweed were used in the treatment of goiter. In the painting "The Flagellation of Christ," created in 1515, one of the men gazing at Christ has a multinodular goiter. In 1527, Paracelsus lectured to his students that drinking water was a cause of goiter. He postulated that the offending substance was iron oxide. It was only centuries later that iodine was identified as the substance in seaweed and other marine products that could prevent goiter.

The discovery of the element iodine in 1809 offers another example of scientific advances made by wartime research. A saltpeter manufacturer named M. Courtois discovered gaseous iodine, but did not realize that it was an element. In 1813, Sir Humphrey Davy reported that iodine was an element.

In 1824, there was written in the *Edinburgh Medical and Surgical Journal*:

"We learn from the first memoir of Dr. J.C. Coindet, that in the year 1813 . . . he introduced iodine into medical treatment." In 1852, Chatin, a Professor of Pharmacy in Paris, published an extensive series of papers on the geographic distribution of iodine, and recommended iodination of the water supply. His recommendations were ignored by the French Academy of Sciences. Over 50 years were to elapse before there were further studies of the use of iodine in the prevention of endemic goiter. This renewed interest resulted from the work of Baumann, who in 1896 showed that the human thyroid contained large amounts of iodine. In 1915, E.C. Kendell crystallized thyroxine from thyroid tissue, and, in 1926, C.R. Harington determined its chemical structure. In 1954, Gross and Pitt-Rivers identified a related molecule with only three, rather than four, iodine atoms. The compound, called triiodothyroinine (T3), was found to be physiologically more active than thyroxine (T4), and was thought to be a product of thyroxine. The latter hormone is secreted by the thyroid, and then converted to T3 in other tissues, as a result of deiodinathion of T4 to mono- and diiodotyrosine which then combine to form T3.

The clinical manifestations of overactivity of the thyroid called "hyperthyroidism" was first described by C.H. Parry in 1825. In 1835, Graves related the manifestations of enlargement of the thyroid, palpitations, and exophthalmos to disease of the thyroid.

Cretinism was described by Paracelsus in 1603, but it was not until 1878 that W.M. Ord described the clinical manifestations of decreased thyroid function that he called "myxedema."

In 1907, David Marine in Michigan began a series of observations and experiments in dogs that led him to conclude in 1924 that hyperplasia of the thyroid was a result of iodine deficiency and led to goiter. This was recognized as an example of the relationship between nutrition and disease. The process of "homeostasis" was in disarray.

The use of radioiodine to help diagnose disease of the thyroid gland began in 1940. Seaborg and Livingston produced iodine-131 in Ernest Lawrence's laboratory in Berkeley. Measuring the rate of accumulation of this tracer by means of an external radiation detector directed at the patient's neck made it possible to differentiate between decreased, normal and increased function of the thyroid.

In 1950, radioactive iodide began to be widely used to examine patients with thyroid disease, especially those with goiter. In the first imaging with a rectilinear scanner, scintillation detectors with focused collimators localized the distribution within the neck of radioactivity, which was then portrayed by means of a mechanical stylus that would print out the spatial distribution of radioactive iodine in the patient's neck, or delineate one or more nodules within the thyroid. If a single nodule failed to accumulate radioiodine,

Figure 131 Benedict Cassen, inventor of the rectilinear scanner in the early 1950s.

Figure 132 Benedict Cassen and Hal Anger at a meeting of the Society of Nuclear Medicine.

it increased the likelihood that the nodule was cancerous. Hal Anger improved the imaging procedure by replacing the mechanical stylus of Cassen with a flashing light that activated x-ray film, a process called photorecording.

Another early example of the usefulness of rectilinear scanning was the finding of an avid accumulation of iodine-131 in a mass beneath the sternum, which indicated that the patient had a substernal goiter, a noncancerous enlargement of the thyroid that extends down into the thoracic inlet, or arises in aberrant mediastinal thyoid tissue. Sometimes the substernal extension is part of a multinodular gland extending down through the thoracic outlet, compressing or displacing other structures, such as the trachea, esophagus, or laryngeal nerves. Often the patient with a substernal goiter may complain of a chronic cough without a palpable enlargement of the thyroid in the neck. Some but not all patients may need to have it removed by surgery.

At Hammersmith Hospital, one of my responsibilities was to separate iodine-132 from its parent radionuclide, tellurium-132, by the process of distillation. The iodine-132 had a $2\frac{1}{2}$ hour half-life. The physicist at Hammersmith, Dr. John Mallard, told me that after I returned to the United States, I should visit Drs. Stang and Richards at the Brookhaven National Laboratory and learn about the method that they had developed to separate these "daughter" radionuclides from their "parent" radionuclides. The method consisted of attaching the tellurium-132 to a resin column and then separating the iodine-132 that resulted from radioactive decay of the longer-lived tellurium-132 by eluting the daughter radionuclide off the resin column. This "radionuclide generator" was frequently called a "cow." The same principle was applied to other parent/daughter systems, including the molybdenum-99/technetium-99m generator that was to have a major impact on the field of nuclear medicine. Paul Harper at the University of Chicago recognized that the ready availability, short half-life, type of radioactive decay by isomeric transition without particle emission, and 150 kev photons emitted in the process of radioactive decay made technetium-99m ideal for scanning by the newly invented Anger camera with lower radiation dose to the patient than that from radioactive iodine. The molybdenum/technetium generator was advertised for sale by the Brookhaven National Laboratory from 1951 to 1953 before it was recognized by Harper to have such great potential in nuclear medicine. Technetium-99m became the "workhorse" of nuclear medicine.

No one has made a more important contribution to the field of nuclear medicine than Hal Anger. Millions of patients all over the world have had their diagnosis and treatment as a result of the invention of the "Anger camera." In 1951, Hal learned of the invention of the rectilinear scanner by Benedict Cassen at UCLA, and thought: "I can do better than that." He recognized that the mechanical movement of the crystal radiation detector over the patient's body was a limitation of the "scanning" method, and set out to develop a "gamma ray camera" that would record simultaneously the photons coming from the patient's body without any moving radiation detectors.

In 1948, Hal joined the Donner Laboratory in Berkeley, California, part of the Lawrence Radiation Laboratory, named after the Nobel laureate, Ernest Lawrence, inventor of the cyclotron in the early 1930s. John Lawrence, brother of Ernest, was head of the Donner Lab, which opened in 1941. Among the first projects to which Hal was assigned was the modification of the 184-inch cyclotron so that it could be used for irradiation of pituitary tumors with high energy deuterons. During World War II, the 184-inch cyclotron had been converted to a large-scale spectrograph, or "calutron," used to produce uranium-235 for the first atomic bomb.

The public presentation of the first gamma camera by Hal was the pinhole camera. (*Nature* 170, 200, 1952). Gamma photons emitted by I-131 in a patient with metastatic cancer were used to produce images obtained with a pinhole collimator in front of a 2-in × 4-in thallium-activated sodium iodide crystal 5/6 of an inch thick. A subsequent publication was entitled "The Gamma-pinhole Camera and Image Amplifier" (UCRL-2524, 1954).

A description of the results using the first scintillation camera was in an article "A New Instrument for Mapping Gamma-Ray Emitters" in *Biology and Medicine Quarterly Report UCRL-3653*, in January, 1957. The instrument was exhibited at the 1958 annual meeting of the Society of Medicine and later at the meeting of the AMA. The new invention was warmly received, but the next challenge was its commercial development.

John Kuranz, President of the Nuclear Chicago Company introduced a 12-inch crystal scintillation camera at the meeting of the Society of Nuclear Medicine at Berkeley, California, in June 1964. In 1972 Hal received the honorary Doctor of Science award by Ohio State University. Up until then, he had held only an undergraduate degree. In 1974 he received the Society of Nuclear Medicine Pioneer Award.

The property that makes radioactive isotopes, such as iodine-123, iodine-131, or technetium-99m so valuable is that they can be accurately and specifically measured when present in exceedingly small concentrations within the body. The gamma ray photons emitted by the isotopes as they move throughout the body can be accurately counted at the surface of the body by the external radiation detectors and used to produce "images." This fundamental property makes possible their use as "tracers" to examine the "dynamic state of body constituents" and solve problems that could not otherwise be addressed in living patients. Radioactive tracers have biochemical properties and behavior within the body identical to the molecules being "traced." In most cases the mass quantities injected are so small (in the nanomolar range) they have no biological or pharmacological effects.

Why are there so many tests of thyroid function? We can no more characterize the manifestations of thyroid disease with a single measurement than we can characterize the manifestations of heart disease by measuring only the pulse rate, blood pressure, or the oxygen content of blood. An avid accumulation of radioiodine by the thyroid is a manifestation of hyperthyroidism, but may also be observed in patients with iodine-avid non-toxic goiter without hyperthyroidism. In such patients, the thyroid accumulates abnormally high amounts of iodine but produces inadequate amounts of thyroxine. On the other hand, an abnormally low rate of accumulation of radioiodine by the thyroid may be the result of abnormally high serum iodine levels, as well as the result of decreased thyroidal function due to the disease, hypothyroidism. Inflamation of the thyroid (thyroiditis) is often manifest by an abnormally low accumulation of radioiodine.

The most frequent uses for imaging the sites and rates of radioiodine by the thyroid are:

1. Identification of the nature of masses beneath the sternum.
2. Identification of thyroidal tissue beneath the tongue or in other unusual places.
3. Detection of residual thyroidal tissue after an attempt to totally remove the thyroid surgically.
4. Detection of multinodular goiter, when only one nodule can be felt by physical examination.
5. Differentiation of functional ("hot") from nonfunctioning ("cold") nodules to help determine the probability of thyroid cancer.

6. Detection of metastatic thyroid cancer in the lungs, skeleton or elsewhere in the body.

High doses of radioactive iodine can be used to kill hyperactive or cancerous cells. When thyroid cancer occurs in a single lobe of the thyroid, often the treatment is removal of the entire thyroid. Possible complications of surgery include hypoparathyroidism due to the inadvertent removal of the parathyroid glands or damage to the recurrent laryngeal nerves. Many surgeons prefer to remove only the lobe of the thyroid containing the cancer. Radioiodine scans are used to detect the residual thyroid tissue, or to identify distant metastatic thyroid cancer if the metastatic lesions accumulate radioiodine. Most cancers of the thyroid consist of differentiated cells that accumulate radioiodine. The radiotracer, F-18 fluorodeoxyglucose, can detect metastases that do not accumulate radioiodine because the cancerous cells are primitive and undifferentiated, having lost the ability of normal thyroid cells to accumulate radioiodine. After surgery, large doses of radioiodine are administered to remove any normal thyroidal tissue remaining or destroy residual cancer cells. The response to radioione treatment is influenced by the type of cancer (papillary, follicular, medullary, and anaplastic), the patient's age and sex. Most patients with thyroid cancer can be successfully treated and live a long life.

The revolutionary advances in the diagnosis and treatment of thyroid cancer in the 1940s were widely reported in national magazines, including the most popular magazines, *Time* and *Life*. Everyone was happy to see a "peaceful use of atomic energy, a product of wartime research," in which an "atomic cocktail" was given to a patient and the radioactive iodine would search through the body, find the cancer and destroy it. Subsequent reality did not completely live up to the promise, but the hope that more such agents could be developed became fixed in the public's mind. "Atomic medicine" was recognized as a major breakthrough in medicine, and reassured the public that radiation and atomic energy were not all bad. Radioiodine was called a "magic bullet." "Atomic cocktails" provided evidence that wartime research leading to the atomic bomb could provide great benefits to humankind, and help assuage the guilt feelings of many wartime atomic scientists.

9

The Breakthrough to Lung Scanning

In 1960, we had no idea of lung or sketetal imaging as we indulged in blue sky forecasting of what nuclear medicine might accomplish. As is the case with most innovations, both of these developments occurred in response to recognized problems. Lung scanning was invented in 1963 illustrating Arthur Koestler's concept of the "act of creation." Creativity occurs when previously unconnected matrices of experience intersect. The new synthesis occurs at the intersection of two planes of experience. The development of lung scanning resulted from the study of the reticuloendothelial system (RES) with radioactive albumin particles.

The RES, the most important defense against infectious disease, is the ability of white blood cells in the body to engulf and digest particles by the process of *phagocytosis*. The process is assessed by the measurement of changes in the number and type of circulating white blood cells.

In 1955, Benacerraf and his colleagues first used aggregates of human serum proteins labeled with radioiodine to study phagocytosis. When radioiodinated serum albumin is heated, it produces aggregates that are rapidly removed from the circulation by phagocytosis after intervenous injection. George Taplin, a pioneer in nuclear medicine at the University of California in Los Angeles, studied the process of aggregation of human serum albumin.

Dr. Masahiro Iio, the first Japanese trainee, who arrived in 1961 in nuclear medicine at Hopkins showed that the rate of clearance of radiolabeled aggregates of albumin from the blood after intravenous injection can be used to assess the function of the reticuloendothelial system in living human beings. He found the maximum rate of phagocytosis to be $1.07\,mg\,min^{-1}\,kg^{-1}$ body weight. This maximum rate of removal of the aggregates from the circulation after their intravenous injection was found to be increased in persons with bacterial infections and decreased in persons with viral infections. The results were published in 1963. Blockade of the RES phagocytic function in normal human beings could be produced safely and effectively by the administration of large doses of aggregated human serum albumin.

We also found that the phagocytic capacity of the RES decreased significantly with aging, but that the decrease was less than the 15% variability that occurs in normal persons of all ages.

The studies of the RES between 1962 and 1964 were carried out under circumstances that would not be possible today. Prisoners are no longer permitted to volunteer for research studies on the grounds that truly informed consent is not possible.

Figure 133 George Taplin, a pioneer in nuclear medicine at the University of California in Los Angeles.

Patients with typhoid fever, sandfly fever, and tularemia were volunteers in a University of Maryland program to evaluate the efficacy of vaccination as a means of preventing these diseases. The volunteers were recruited from among the inmates of the Maryland House of Correction in Jessup, Maryland. The research was supported by the U.S. Army Research and Development Command.

Once, while Dr. Iio and I were at the prison carrying out experiments on the RES, a Japanese visitor came to the the Nuclear Medicine Division at Hopkins to see Dr. Iio. He was told that Dr. Iio was at the prison. The word that Dr. Iio was "in jail" spread around Japan.

We measured the capacity of the reticuloendothelial system to remove injected albumin aggregates from the circulation before, during, and after the illness. In the course of preparing the human serum albumin aggregates for injection, we occasionally observed that when the particles became larger than optimum for the study of the reticuloenothelial system, they became lodged in the lungs after intravenous injection. What happened next illustrates the role of chance in advancing science, and the intersection of planes in the process of creativity. Attending surgical rounds, I heard Dr. David Sabiston describe a new surgical procedure to remove large emboli that had become detached from the large veins of the pelvis or legs and traveled up the vena cava to produce obstruction of pulmonary blood flow, a condition known as massive pulmonary embolism. The patients were often in shock and critically ill. There had to be an immediate diagnosis of pulmonary embolism if the patients could be put on the newly invented cardiopulmonary bypass instrument and the massive pulmonary emboli removed at surgery. It occurred to me that "lung scanning" of pulmonary arterial blood flow might make possible the immediate diagnosis and make possible the surgical removal of the clots. The patients were often in shock and critically ill. We turned to the experimental animal surgery laboratory, directed by Vivian Thomas, Dr. Alfred Blalock's principal assistant and a major contributor to the development of cardiac surgery. Vivian subsequently received an Honorary Doctor of Medicine degree from Johns Hopkins.

Working with a medical student, Bob Jones, who subsequently became head of cardiovascular surgery at Duke University, we carried out studies in 43 dogs with experimental pulmonary embolism, produced by inserting a balloon filled with radiographic contrast media into a large pulmonary artery. Then, after injection of the albumin aggregates labeled with iodine-131 or chromium-51, we could clearly and immediately see by rectilinear scanning where the emboli were located within the lung. They could be removed at surgery.

After the studies in 43 dogs, I was the first volunteer to have a lung scan. In those days, there was no "Institutional Research Committee" to approve the procedure. After my

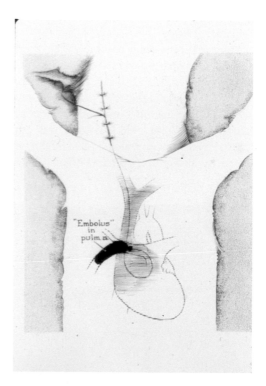

Figure 134 Experimental pulmonary embolism in a dog.

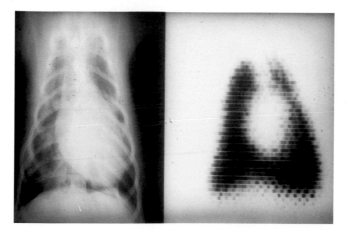

Figure 135 Normal perfusion lung scan in a dog.

initial study, which clearly showed the distribution of blood flow within my lungs, we carried out the procedure in 18 medical students and demonstrated that the method was safe and effective. This recruitment of medical students as subjects for research is prohibited today. After these studies evidencing the success of the method in normal people, we eagerly waited for the first patient to arrive. Soon an elderly woman was admitted to Hopkins in shock. The diagnosis of massive pulmonary embolism was made by lung scanning, and the clots were successfully removed by Dr. Sabiston. The first results of lung scanning to measure regional pulmonary arterial blood flow in 83 patients with pulmonary embolism, lung tumors, cysts, pneumonia, and atelectasis was published in February, 1964.

Lung scanning soon became widely used for the diagnosis of pulmonary embolism, and is still used today, forty years after its invention. An Article in the *New England Journal of Medicine* (271, 376, 1964) describing the development and application of lung scanning in the diagnosis of pulmonary embolism was cited over 245 times before 1980, according to Current Contents, and was selected as "This week's Citation Classic" on March 31, 1980. In 1986, our work was officially honored by the American College of Chest Physicians in a medal presented to me in 1986 at the San Francisco Hotel by its Executive Director, Dr. Alfred Soffer.

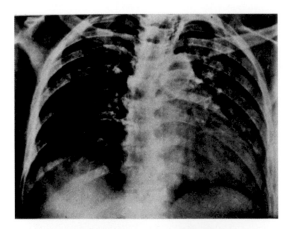

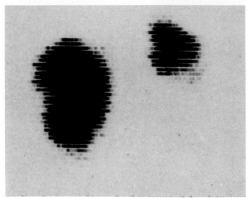

Figure 136 First lung scan ever performed in a patient with massive pulmonary embolism (1963).

The development of lung scanning for the diagnosis of pulmonary embolism moved nuclear medicine from its emphasis on the thyroid into the mainstream of medical practice. Lung scanning with radioactive aggregates and gases, such as xenon-133, is still used to measure regional perfusion and ventilation of the lungs. The techniques are used (1) in the measurement of regional ventilation and perfusion in quantitative terms in locating and characterizing regions of malfunction and (2) in the planning of treatment and monitoring the response.

Before lung scanning was used to diagnose pulmonary embolism, Dr. C. Rollins Hanlon of St. Louis, quoting Dr. Evarts Graham, said that the surgical treatment of pulmonary embolism was inadvisable because "the likelihood of misdiagnosis was so great that the chance of harming the patient exceeded any possible benefit that might accrue." He was referring to the Trendelenburg operation, developed in 1938 for the surgical removal of clots from the pulmonary artery.

In 1961, Sharp was the first to describe pulmonary embolectomy with the use of extracorporeal (heart/lung machine) circulation, which allows pulmonary embolectomy to be performed "deliberately." Dr. David Sabiston of Hopkins emphasized that the advent of cardiopulmonary bypass made the operation much safer. Here again one can see that "progress depends on problems." Lung scanning was developed to make possible rapid diagnosis prior to emergency surgical treatment of massive pulmonary embolism.

Advances developed to solve one problem are soon applied to help solve other problems. Lung scanning is not only used for diagnosis of pulmonary embolism, but also in the care of patients with other lung disease. Lung scanning of the relative perfusion of each lung in patients with different types of lung disease is correlated with the oxygen consumption of each lung. We observed that most lung diseases are characterized by regional decreases in pulmonary arterial blood flow. The finding of characteristic concave perfusion defects along the borders of the lung and the absence of any corresponding parenchymal lesions in the chest radiographs made it easy to diagnose pulmonary embolism. However, the frequent occurrence of pulmonary embolism in patients with chronic obstructive lung disease remains difficult.

In 1968 we published the use of radioactive xenon-133 in the differential diagnosis of pulmonary embolism. Using two stationary scintillation detectors, we observed that it took less than 30 seconds for the radioactive xenon to reach equilibrium in persons without obstructive lung disease. The use of the Anger camera to measure the washout of xenon-133 from the lungs after its equilibrium became routine in the diagnosis of chronic obstructive lung disease.

In patients with chronic lung disease, we produced hypoxia in one lung by having patients breath 100% nitrogen in one lung and 100% oxygen in the other. This resulted in a 42% decrease in pulmonary arterial blood flow to the hypoxic lung as a result of vasoconstriction to the hypoxic lung.

Lung scanning was helpful in detecting the uneven distribution of lung perfusion in patients with emphysema, important in determining the cause of the disease and, in some patients, planning surgical treatment. Chronic bronchitis was uniformly distributed in both lungs.

In 1969, lung scanning was used to select 18 of 179 patients for surgical treatment of bullous disease of the lungs in patients with chronic obstructive lung disease. Assessment of regional lung function helped ascertain whether the lung compressed by bullae was capable of expansion after removal of the bullae. The results of surgery in these properly selected patients was excellent in terms of symptomatic improvement.

Lung scanning of the relative perfusion of each lung in patients with different types of lung disease could be correlated with the oxygen consumption of each lung. We observed that most lung diseases are characterized by regional decreases in pulmonary arterial blood flow to the involved parts of the lung.

The first measurements of regional ventilation of the lung with xenon-133 had been made with stationary scintillation detectors that could measure the time it took for the radioactive xenon to reach equilibrium. It took less than 30 seconds in persons without obstructive lung disease. Subsequently, the Anger camera was used to measure the washout of xenon-133 from the lungs after its equilibrium. Its use became routine in the diagnosis of chronic obstructive lung disease. For decades, lung scanning was used in patients with tuberculosis, other pulmonary infections, chronic obstructive lung disease, emphysema, atelectasis, pulmonary hypertension, and structural abnormalities of the heart or lungs.

In 1964 we showed that the administration of the thrombolytic drug, urokinase, had a significant thrombolytic effect. After studies in 20 dogs, we participated in a large multi-institutional study of urokinase therapy of patients with pulmonary embolism. By 1967 we were able to show that lung scanning could depict the pattern and measure the rate of recovery of perfusion defects in 69 patients with pulmonary embolism.

There were multiple perfusion defects in 65% of the patients when first examined. In 27 patients with small perfusion defects, there was complete recovery of normal perfusion. In 31 patients with intermediate degrees of involvement, 38% had complete return to normal. In 9 patients with severe involvement, 20% became normal, and 70% improved. The rate of return to normal perfusion was significantly greater in patients treated with urokinase. The scans were read subjectively by physicians with no knowledge of whether the lung scans were from control or urokinase-treated patients. This was one of the first multicenter clinical drug trials ever carried out. Lung scanning had made an essential contribution. Thrombolytic therapy has subsequently become widespread in thrombo-embolic and coronary artery disease.

Objective assessment of the perfusion defects in 473 lung scans in 46 patients in the urokinase clinical trial also led to the development of one of the first uses of computers in medicine.

T. K. (Raj) Natarajan created the Hopkins Image Display and Analysis (IDA) computer system. Analysis by IDA of the changes in size of perfusion defects in the lungs of patients with acute pulmonary embolism was compared to the subjective interpretations of the same scans by five expert nuclear medicine physicians. The relative standard deviation in assessing the size of the lesions varied from 59% for small defects and 11% for large lesions. With IDA, the size of the perfusion defects could be measured with much greater precision, and interpretations could be made by inexperienced as well as experienced persons.

The initial results of the randomized, national cooperative urokinase trial were published in the *Journal of the American Medical Association* in December 1970. Pulmonary arteriography showed a decrease in the size of clots in 70% of the urokinase-treated patients compared to anticoagulant treated controls, but the restoration of blood flow to the lesions was slower and less dramatic in the lung scans. In 11 patients, embolectomy was performed with a 73% mortality. This study was the first successful large scale

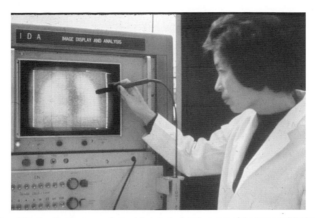

Figure 137 Quantification of the size of pulmonary perfusion defects in the lung by means of the Image and Display Analysis (IDA) computer developed at Johns Hopkins in the late 1960s.

multi-institutional clinical trial, preceding many more to follow over the following decades.

Among the memorable events related to the urokinase study was the periodic gathering of the nuclear medicine specialists involved in the study at our weekend house on the Chester River on the Eastern Shore of the Chesapeake. It was also there that the residency requirements for nuclear medicine were defined by the American Board of Nuclear Medicine.

Lung scanning made a great impact on the field of nuclear medicine and its popularity led us to a develop postgraduate course in nuclear medicine every summer. With a few invited faculty, these postgraduate meetings entitled Topics in Nuclear Medicine continued every summer at Colby College in Waterville, Maine, for 17 years, before we took the course back to the Hopkins Campus. In those early days at Colby, the entire field of nuclear medicine could be covered in one week. All meals were served in Dana Hall on the Colby campus, and the faulty and family lived in the college dormitories.

Lung scanning with macro-aggregates of human serum albumin had become widely used throughout the world by 1969. That year, attempts to use the labeled aggregates to study the systemic as well as the pulmonary circulation and to measure arterio-venous shunting failed because of the irregular size and fragility of the aggregates. Therefore, we substituted human serum albumin microspheres of uniform size that could be labeled with technetium-99m or indium-113m. The advantage of the microspheres was that they can be labeled immediately prior to use with a short-lived radionuclide. The microspheres have a mean diameter of 37.4 microns with a standard deviation of 8.2 microns. Diagnostic doses are from 0.1 to 5 mg, which is one ten thousandth of the toxic dose in animals. Microspheres became routine for studies of the systemic and pulmonary circulations. An example of the early studies was the distribution of cardiac output in experimental shock in dogs. During the irreversible stage of shock, the blood flow to the brain was no longer preserved. Another finding was that in hypertrophic pulmonary osteoarthroapthy (clubbing of the fingers associated with lung disease), most of the increased blood flow to the extremities goes through arteriovenous shunts, stealing blood flow from the capillary beds. Surgical vagotomy decreases this shunting.

Lung scanning was a precursor of nuclear cardiology. In 1969 perfusion lung scanning was performed in 20 patients with cyanotic congenital heart disease. Lung scanning

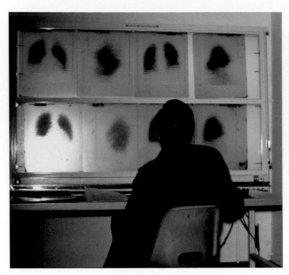

Figure 138 Viewing of a series of perfusion lung scans in a patient in the Urokinase Pulmonary Embolism clinical trial.

Figure 139 Joseph Kriss, head of nuclear medicine at Stanford University and a founder of the American Board of Nuclear Medicine.

Figure 140 James Quinn, a pioneer in nuclear medicine and frequent faculty member of the Hopkins nuclear medicine course at Colby College.

made possible determination of the differential blood flow to the two lungs, particularly in patients with scoliosis in whom this information is difficult to obtain by other methods. Scanning was also used to detect hypoplasia of pulmonary arteries, and the patency of arterial and venous shunts.

Shunting of the blood from the right to left side of the heart could be quantified, as well as the relative amounts of pulmonary and bronchial blood flow, the latter coming from the left ventricle. In patients with total absence of the pulmonary artery, all the intravenously injected microspheres are shunted into the systemic circulation and reach the lungs only via the bronchial circulation.

Xenon-133 and krypton-85 were used to measure muscle as well as pulmonary arterial blood flow. The rate of clearance of the tracers injected into muscle is a function of the blood flow to the site(s) of injection. External monitoring of this clearance was examined in 52 subjects. Clearance was markedly diminished in patients diseases of their circulation. This technique never achieved widespread clinical use.

The clearance of these two inert gases from the brains of dogs after injection of the tracer into the internal carotid artery was used to establish that external cardiac massage during cardiac arrest restored cerebral blood flow to one third of that measured during normal cardiac function in the same dog. Over a blood pressure range from 15 to 150 mm Hg, cerebral blood flow increased in a linear fashion as arterial pressure was increased by external cardiac message.

10

Computers in Nuclear Medicine

A century ago, when machines replaced most activities dependent on human muscles there was enormously increased industrial productivity. We now take for granted that machines will carry us from place to place, produce most of our goods, wash our clothes, and do the dishes.

Research during World War II not only gave us atomic energy, but, even more important, computers. The Information Age followed the Atomic Age, and today computers are essential in nearly all aspects of life, including nuclear medicine. Computers and artificial intelligence are revolutionizing health care.

A human being cannot possibly store in memory all his or her knowledge and experience. No doctor knows enough to single-handedly care for a sick person. Thus, computers are becoming more and more necessary in all aspects of health care, from the emergency room to the physician's office to the radiology and nuclear medicine departments. The greatest effect on the health care system will occur when computer networks can access information equal to that in the National Library of Medicine in a few seconds. Networks have already reached a size and number that far exceed the memory and logical capability of the human brain. Personal computers can be programmed to "think" in a way indistinguishable from the way a human being thinks. A computer network can join physicians all over the world. In the not-to-distant future, networks and databases will be used to answer questions, and extend the subjective interpretations of diagnostic studies.

In the 1960s, it became clear to us in the Division of Nuclear Medicine at Johns Hopkins Hospital that computers would soon play a major role in clinical practice and research, including nuclear medicine. The most revolutionary event at that time was the replacement of "main frame" computers with minicomputers, which took computers out from behind the glass walls of Hospital and Medical School Computer Centers and brought dedicated minicomputers to where we actually worked. Before that time, I often felt like Dorothy in the *Wizard of Oz* being told by those in the Hopkins Computer Center: "Don't pay attention to the man behind the curtain." We envisioned the creation of an international computer health network to help take care of patients and provide an accessible repository of the enormous amount of health-related information that needed to be available at the fingertips of those making life-saving decisions in the care of their patients, as well as in carrying out clinical research.

No one contributed more to the invention of the computer than John von Neumann, who had the vision of a calculating machine that used numbers stored in the machine itself to control its operation. This was the key concept. Until that time, calculations and mathematical operations had been performed by mechanical devices. He also saw that the computer could handle text as well as numbers, that it was possible to "program" a computer, using verbal instructions stored in the computer. The ability to combine sets of instructions telling the computer what to do was truly revolutionary. These instructions, called "software," gave the user the ability to instruct the computer to help solve all sorts of complex problems, including those concerned with health and disease.

In the 1950s, "main frame" computers, such as the IBM 701, paved the way. Hubert Pipberger at the Washington, D.C., VA Hospital and C. Caceres at the U.S. Public Health Services hospitals first began to use these computers to assist in the interpretation of electrocardiograms. By the 1960s, pioneers of nuclear medicine recognized what computers could do in nuclear medicine, especially because the data on which the nuclear imaging was based, was acquired in a readily-digitized form from the start. Newly available "minicomputers," such as the PDP 8, obtained by the Nuclear Medicine Division at Hopkins, could be used to acquire, display, and interpret images, and answer clinical questions.

A dream had been realized. Since World War II, creative physicians and scientists had dreamed of the day when computers would be used to help physicians "make the diagnosis," and then help them make decisions in planning treatment. The computer was viewed as being able to produce a list of possible diagnoses with the probability of each being present and then suggest further diagnostic tests that might help solve the patient's problems. In the 1960s, we visualized the time when every individual's medical record would contain all the information related to the person's health, including biochemical and physiological data. We knew that the amount of information in a database of medical images would require great increases in computational speed and capacity. This has now become widely available.

It is possible to create "personal chips" that each person can carry to serve as the repository of the enormous amount of information that could be used in a medical emergency or illness, or in every visit to a physician. These personal indices of the state of the person's health recognize when the person's "manifestations" of health deviate from normal values stored in huge international databases. Changes can be evaluated by comparison of the patient's database to the normal values of large populations. The normal values from populations of "normal" people can reflect changes with age and be recorded in each person's database throughout his life. Each medical test performed during the person's life can be stored in the person's database. These "normal" values reflect the person's susceptibility to disease. They can also suggest preventive measures.

If computers can track airline and hotel reservations, millions of income tax returns, and bank accounts, why can they not create an international database of manifestations of health and disease? Some may ask whether this will lead to "big brother" government looking over our shoulders, that our privacy will be invaded by employers or insurance companies. These potential problems can be avoided or solved. Many equally sensitive issues about databases are solved every day.

Google and Yahoo can incorporate over 5 billion Web sites into a worldwide system for commerce. Why not do the same thing to promote health and well being? When important events have occured, people do not say: "I was for it, or I was against it, but rather I didn't know it was happening."

Intelligent activities and procedures can be broken down into a well-defined sequence of operations, referred to as "algorithms." The value of an algorithm in medicine, such as the interpretation of brain FDG/PET images, is to help answer questions, such as whether a patient who is beginning to be forgetfull has early Alzheimer disease. Previous limitations of size, storage space and speed of operation that have limited such use of brain image databases no longer exist. Vast numbers of computers all over the world can interact with hundreds of thousands of linked small computers.

The incorporation of an algorithm into a database necessitates that the procedures must be clearly defined and represented as a logical series of operations. In the 1960s, we clearly defined the use of "minicomputers" to help interpret rectilinear scans and gamma camera images. The electronics industry had learned how to make miniature electronic circuits on a chip of silicon. The PDP-5 minicomputer, introduced in 1963 had power, greater than that originally used in space flights. Silicon "chips" made it possible for the minicomputer to do all the work of the central processing unit of the main frame computers. The "chip" did for microprocessing what the DC-3 airplane did for the airline industry.

The Digital Equipment Corporation introduced the PDP-8 minicomputer in 1965. We acquired the fifth model produced by the company, and our engineer, T.K. Natarajan, interfaced it to our rectilinear scanners and the Anger camera. In 1969, we published the design and application of the Image Display and Analysis (IDA) system in the *Journal of Nuclear Medicine* (*Radiology*, vol 93, p 823–827, October 1969).

Our IDA computer system had the following characteristics, which are still the foundation of computers in nuclear medicine throughout the world:

1. The system is interactive with the physician using the computer display to help him or her interpret the studies. There are many advantages in looking quantitatively at the basic data from which the images are created.
2. A light pen is used to flag regions of interest in multiple serial studies.
3. The time-course of the radiotracer in various regions selected with the light pen is displayed together with quantitative parameters calculated by the computer.
4. The rate constant data showing the time-course of the radioactivity in the picture elements of the serial images are displayed as "functional images," first described in 1969 by a trainee in nuclear medicine at Hopkins, Shige Kaihara and others (*Radiology*, vol 93, p 1345–1348, December 1969). Functional images displayed the distribution in the parts of the body of a derived parameter associated with each picture element. Examples included the time-course of wash-in or wash-out of the radioactivity from the regions of interest within the body. Other parameters included time-to-peak accumulation. Since then, thousands of other algorithms for data acquisition and processing have been incorporated into nuclear medicine procedures.

When we presented the results with IDA at the IAEA meeting in Vienna in 1969, we stressed the value of quantification in the objective assessment of the effectiveness of the thrombolytic drug, urokinase, in patients with pulmonary embolism. We also emphasized the value of a color display.

Among the earliest uses of the Image Display and Analysis (IDA) developed at Johns Hopkins Division of Nuclear Medicine was imaging the uptake by the patient's thyroid gland of Tc-99m pertechnetate over a 30 minute period (*J Clin Endocrinol Metab*, vol 34, February 1972). The entire procedure took only 30 minutes from the time of injection of the tracer. Patients with hypo- or hyperthyroidism could be easily distinguished from normal persons. With radioactive iodine, the procedure required measurements 24 hours after injection. Other studies in which the computer played an essential role in the 1970s included the imaging of regional cerebral blood flow in the diagnosis of stroke, renal blood flow and excretory function of each kidney in patients with hypertension, quantification of gall bladder function in patients with right upper quadrant abdominal pain, quantification of hypo- and hyper-plastic regions of the bone marrow in anemic patients, and serial lung scans in patients with pulmonary embolism being treated with the new thrombolytic drug, Urokinase (*Radiology*, vol 97, November 1970). The computer was used to quantify and compare the rate of recovery of treated and untreated patients.

11

From the Lungs to the Heart

From the start, nuclear medicine has had to deal with political issues, particularly because it cuts across so many other specialties. Today, studies of the heart account for the majority of studies in nuclear medicine. Most are performed by cardiologists as well as in well-established nuclear medicine departments. In 1992, the American Board of Nuclear Medicine (ABNM) proposed to the American Board of Internal Medicine that a subspecialty board of cardiovascular nuclear medicine be established. The cardiologists on the board believed that it did not take a year of special training for cardiologists to perform nuclear cardiology studies. Today both the cardiology board and the ABIM require demonstrable expertise in nuclear cardiology.

On December 27, 1978, I wrote to Dr. Lewis Becker, Associate Professor of Cardiovascular Medicine at Hopkins: "At a recent meeting of the Nuclear Regulatory Commision that I attended as a member of the Advisory Committee, we decided that cardiologists should have a minium of two to three months training in nuclear medicine to qualify them for licensing in nuclear cardiology . . . I suggest that we develop a joint program (at Hopkins) in which cardiology and nuclear medicine would be equal partners."

The American Board of Nuclear Medicine was approved by the AMA in 1971. Drs. Merrill Bender of the University of Buffalo and E. Richard King, then Chairman of Radiology at the Medical College of Virginia, were key players in this great achievement, beginning with their concept stated in a letter on July 19, 1968, that "there should be an affiliate certifying board sponsored by the American Board of Radiology and one or more specialty boards to develop training requirements for a residency program in Nuclear Medicine." They outlined the details of the proposed program.

To facilitate our efforts to establish the American Board of Nuclear Medicine, I joined the AMA, and was elected a member of the Board of Directors of the Baltimore City Medical Society on December 3, 1976, and subsequently became its President. I was an Alternate Delegate representing Maryland members of the AMA and the Medical and Chirurgical Faculty of Maryland from 1983 through 1984, and subsequently elected Delegate to the AMA in 1992. I was elected to the Council on Scientific Affairs of the AMA starting on July 2, 1985, and served on the council for 9 years, being re-elected twice. Beginning in March 1990, I participated in the evaluation of the safety and effectiveness of new devices and procedures used in medical devices as a member of the AMA "Diagnostic and Therapeutic Technology Assessment (DATTA)" panel.

I was able to have the AMA accept a resolution that I submitted on behalf of the Maryland delegation that doctors who have a financial interest in facilities where they send patients should disclose this fact their patients. This action was stimulated by the fact that in some states, such as Florida, half of the physicians owned or were investors in clinical laboratories, imaging centers, and other medical facilities. Doctor-owned ventures at times were found to perform more tests per patients with higher charges than similar facilities not owned by doctors. Subsequently there was passage of the Stark bill in Congress which limited the unwarranted use of costly machines and procedures.

Each year hundreds of thousand of patients with coronary artery disease are operated on. There is evidence that at least a third of the operations would not have been performed if the patient selection included diagnostic procedures, including PET and SPECT. These procedures decrease the overall cost of caring for the individual patient, even though the overall cost of providing the service increases total health care costs. What must be avoided is an excessively competitive environment where hospitals feel the need to advertise that they have the newest technology as soon as it arrives on the scene.

Before the development of nuclear angiocardiography, contrast material was injected and x-rays were used to depict the transit and distribution of a radioopaque dye delivered by an indwelling catheter. The computer and the Anger camera made possible:

1. The diagnosis of coronary artery disease as the cause of chest pain.
2. The significance of electrocardiographic abnormalities.
3. The diagnosis of myocardial infarction.
4. The degree of damage to the heart.
5. The function of the right and left ventricles at rest and during exercise.
6. The enlargement or dilatation of the cardiac ventricles.
7. The presence of mitral or aortic valve disease.
8. The existence of intra-cardiac shunts.
9. The failure of the right or left ventricles.
10. The selection of candidates for coronary angiography and possible surgery.

By 1977, it was possible for me to write:

"Cardiologists have long been among the most physiologcally oriented of all physicians, insisting on quantification and validation of the data whenever possible. The increasing use of computers in nuclear medicine has permitted us to meet their strict requirements for quantification of functional information, both regional and global, to help solve their patients' problems."

Nuclear cardiology depends on:

1. the radionuclide ventriculogram to measure the size and function of the ventricles during rest and exercise.
2. Regional myocardial perfusion imaging during rest and exercise.
3. Myocardial infarct visualization and sizing.

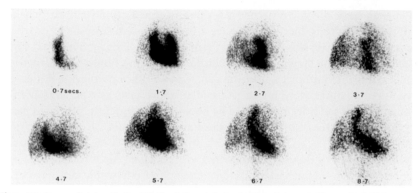

Figure 141 Passage of a bolus of iodine-131 albumin through the heart and vessels, imaged with an early scintillation camera.

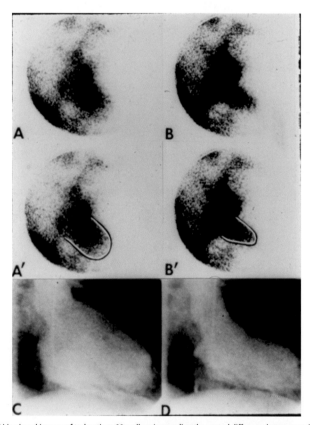

Figure 142 Early gated blood pool images of technetium-99m albumin revealing the normal difference between end-systole and end-diastole.

Extending the advances in pulmonary nuclear medicine, cardiology moved nuclear medicine still further into the mainstream of medicine. Electrocardiography, and subsequently cardiac catheterization, had provided the physiological foundation of cardiology, and served as a model for extending the use of radioactive tracers in the care of patients with heart disease.

Studies of the circulation, the first radioactive tracer studies in nuclear medicine, were carried out in 1925 by Dr. Herman Blumgart, who injected solutions of the radioactive gas, radon, into the arm vein of human subjects and measured the time that it took for the radioactivity to reach the opposite arm. He called this measurement the "velocity of the circulation." The first clinically useful study was the imaging of pericardial effusion in patients after administration of radioiodinated albumin, a procedure first proposed by Rajali and others in 1958. In 1966 we reported the results in 56 patients. By minimizing magnification of chest radiographs, we were better able to compare the intracardiac blood pool with the radiographic image of the heart, using "fused" images since 1958. Technetium-99m albumin was the agent of choice for the detection of pericardial effusion by rectilinear scanning.

The scintillation camera made possible "radionuclide angiocardiography," the continuous monitoring of the passage of a bolus of a technetium-99m labeled tracer, Tc99m-pertechnetate in the earliest studies, and subsequently, Tc99m-human serum albumin, as it passed through the heart and great vessels. The radionuclide technetium-99m made the Anger camera a success. The radionuclide technetium-99 had been first generated by the cyclotron in Berkeley in 1938 by Segre and Seaborg, who found that its decays into Tc-99m, which has a half-life of 6 hours and can be administered safely in millicurie doses to provide larger numbers of photons than had been possible with nuclides such as iodine-131. The great merit of technetium-99m was presented at the meeting of the Society of Nucler Medicine in Berkeley in 1964. This radionuclide remains the most widely used in nuclear medicine throughout the world.

In 1970, we described our results in 76 persons, 60 of whom suffered from congenital heart disease. We were able to diagnose correctly intra-cardiac right-to-left shunting in all 30 patients with cyanotic heart disease. Characteristic findings were seen in

Figure 143 Manny Subramanian, who developed many technetium-99m bone imaging radiopharmaceuticals, and his wife, Kaliyani.

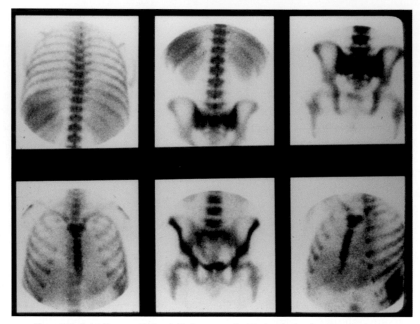

Figure 144 Skeletal images with the scintillation camera after injection of technetium-99m phosphonate.

patients with inter-ventricular shunting, transposition of the great vessels and tetralogy of Fallot. While contrast angiography revealed better anatomical detail, radionuclide angiography "gave sufficient information to establish a diagnosis." Among its uses was distinguishing respiratory distress syndrome from congenital heart disease in newborn infants.

In the 1950s, Saperstein and his co-workers (*Am J Physiol* 193: 161, 1958) measured the distribution within the heart of intravenously injected radioactive potassium-42 to measure myocardial blood flow. In the 1960s, Carr and his colleagues used cesium-131 as a substitute for potassium-42, but found that its extraction from the circulating blood (extraction efficiency) by cardiac muscle was too low. Rubidium-81 was tried next, but the 4.6 hour half-life of this tracer limited its use to a few institutions.

At Hopkins, in 1961, we explored the use of iodine-131 iodide to image regions of myocardial infarction, a procedure introduced by Dreyfuss and his colleagues the year before. We observed that accumulation of the tracer in the stomach greatly limited its usefulness in examining the heart.

In 1971, Hurley and associates (*J Nucl Med* 12: 516, 1971) introduced potassium-43 for imaging regional myocardial perfusion. This radionuclide was not ideal, because its emission of beta particles increased the radiation dose to the patient, and its gamma photon energy (373 and 397 keV) were higher than was optimal for scanning. Nevertheless, clinical studies were able to reveal regions of myocardial infarction. No patient with a proven myocardial infarction had a normal scan. The new procedure was valuable in patients in whom electrocardiographic and biochemical changes were equivocal, and provided regional functional information that could add useful information to the visualization of the coronary arteries by contrast angiography.

In 1973, diagnostic techniques in nuclear cardiology fell into four categories:

1. Screening healthy people in periodic examinations
2. Identifying heart disease in patients with chest pain, shortness of breath, or weakness
3. Follow-up studies in patients with abnormal electrocardiograms
4. Pre-operative decision-making in candidates for coronary artery surgery

Potassium-43 was used to image myocardial infarction. Between 1971 and 1972, 25,000 coronary artery bypass operations were performed in the United States, compared to 9,000 operations for congenital heart disease, and 17,000 open-heart operations for other types of acquired heart disease. Dr. H. William Strauss and I concluded in 1973: "It seems likely that measurement of the spatial and temporal distribution of important cellular elements and chemical substances within the body—the essence of nuclear medicine—will join other non-invasive modalities such as phonocardiography, electrocardiography, vectorcardiography, and echocardiography in filling the gap between the man in the street with undiagnosed coronary heart disease and the very ill patient admitted to the cadiac catheterization laboratory."

With the scintillation camera, the beating ventricules could be seen in a series of 37 patients studied in 1978 using a motion-picture display of the intracardiac distribution of technetium-99m albumin during 16 phases of the cardiac cycle. The images of motion improved the perception of regional dysfunction, and increased the certainty of diagnosis.

Of the many technological advances in nuclear cardiology was the concept of a "functional image." The invention of the scintillation camera by Hal Anger in 1963 made it possible to measure the distribution of a radioactive tracer in a region, such as the heart, at various times after the intravenous injection of the tracer. One can measure the rate of change of the distribution of the tracer as a function of time. One can then use the computer to analyze the time course of the tracer in various regions ("volume elements") to obtain mathematical time-functions, such as exponential accumulation or clearance rates, and portray the numerical results as "functional images." Others subsequently referred to these images as "parametric images." The images portrayed the values of the quantitative rate constants in the clearance of the radioactive gas, xenon-133, from the regions of the heart, called "functional imaging." The first "functional images" were presented by Shige Kaihara and colleagues in 1968. Kaihara was the second of over 35 Japanese research fellows to study nuclear medicine at Hopkins by 1986, starting with Masahiro Iio in 1961.

Monitoring the progression through the heart and lungs after administration of technetium-99m albumin, nuclear angiocardiography, was first used at Hopkins in 1972 to differentiate cardiac and non-cardiac causes of cyanosis in newborn infants. Subsequently, the procedure was also found to be useful the initial demonstration of transposition of the great arteries, truncus arteriosus, and hypoplastic right or left ventricle.

Figure 145 Harumi Itoh, a trainee at Johns Hopkins who subsequently became head of radiology at Fukui University.

A nuclear angiogram begins with the injection of a sharp bolus of human serum albumin or red blood cells labeled with technetium-99m. The bolus of radioactivity peaks in the right ventricle several seconds after intravenous injection, and the transit through the great vessels and chambers of the heart is depicted in rapid sequence cinematically. Computer storage of the data adds the ability to store the data for later manipulation and quantitative analysis. As the radioactive tracer passes through the heart, the activity in each chamber can be displayed as a function of time in graphic form. One looks for delayed transit in the passage of the tracer as a sign of ventricular dysfunction. Abnormal passage from the right to the left ventricle indicates the presence of a right-to-left shunt. When the amount of blood ejected with each heart beat is expressed as a fraction of the end-diastolic volume, the resultant fraction is the "ejection

Figure 146 Normal distribution of thallium-201 in the left ventricle.

Figure 147 Computer image of the right (blue) and left ventricle (red) after injection of technetium-99m albumin.

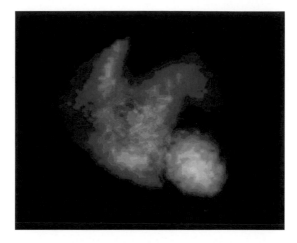

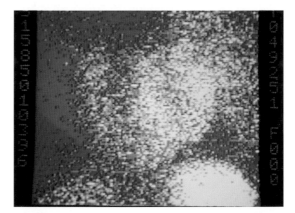

Figure 148 Image of right ventricular hypertrophy with thallium-201.

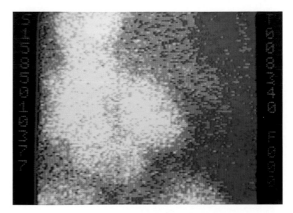

Figure 149 Image of right ventricular dilatation with technetium-99m albumin.

fraction," which is an important measurement of the ventricle's ability to pump blood. When damage to the heart is fresh, the end-diastolic volume is normal, while the amount of blood ejected with each beat (stroke volume) is reduced. Later the ventricle dilates, increasing the end-diastolic volume. Stroke volume may be restored but the ejection fraction remains low.

In addition to its use in patients with coronary arty disease, nuclear cardiology is useful in documenting normal cardiac function in patients with "innocent" murmurs, aneurysms of the great vessels, and arteriovenous fistulas. In 1972, we reported the results in 43 children. Nuclear angiocardiography was found useful in the diagnosis of cyanotic children, including 10 neonates.

Nuclear cardiology has become accepted in the care of patients with suspected coronary artery disease. It now accounts for half of all nuclear medicine procedures in the United States.

A Hopkins nuclear medicine trainee, Malcolm Cooper, with Strauss and Zaret (*Radiology* 108: 85, 1973) first used potassium-43 in rest-exercise testing for detection of coronary artery disease before myocardial infarction has occurred. Soon thereafter, Elliot Lebowitz and colleagues (*J Nucl Med* 16: 151, 1975) at Brookhaven National Laboratory (BNL) showed that thallium-201 had many more favorable physical characteristics. The possibility of using thallium-199 for regional myocardial perfusion imaging had been suggested in 1970 by Kawana and colleagues (*J Nucl Med* 11: 333, 1970). In July 1976, we reported the normal pattern of thallium distribution in the heart of normal subjects at rest and during exercise.

In addition to viewing the images of the distribution of thallium in the ventricles of the heart, the activity of the radiotracer is measured quantitatively. The images obtained when the tracer is injected during exercise is quantitatively compared by computer to the images obtained 4 hours after the cessation of the exercise. Often the images are color-coded to facilitate comparison. Graphing the data facilitates the quantitative assessment of the differences between the exercise and rest images. The images can also be compared visually and quantitatively to those obtained in normal persons. The rates of clearance of the thallium-201 can be shown graphically. SPECT imaging is more satisfactory than planar gamma camera imaging. Significant stenosis of a coronary artery is defined as a reduction of 70% or more from normal values. In patients found to have stenosis of the left main coronary artery or involvement of three coronary arteries, particularly when there was decreased initial perfusion and slow washout of the tracer in the delayed study, there is twice the rate of subsequent myocardial infarction compared to those with normal distribution and washout of the tracer.

The history of how gated blood pool imaging (nuclear venticulography) began is illustrative of how discovery favors the prepared mind. In performing potassium-43 imaging of regional myocardial blood flow, we decided to use the electrocardiogram connected to our minicomputer IDA to divide the cardiac cycle into end systole and end diastole in order to "stop" the motion of the heart in recording the potassium-43 distribution. "Gating" the recording of the data would help eliminate distortion from motion of the left ventricle.

When these images were being shown at our daily nuclear medicine morning conference, one of our trainees suggested that if we rapidly alternated the end systolic and end diastolic images, we could perceive the movement of the blood pool as a result of the beating of the heart. We were fascinated to be able to portray the motion of the left ventricle to reveal the functioning of the heart as it filled and emptied. We decided to do this routinely with all of our potassium-43 images. Eventually we had IDA divide the cardiac cycle into 16 intervals rather than just end-diatole and end-systole. For blood pool and regional ventricular wall motion studies, we used technetium-99m human serum albumin. In the Lewis A. Conner Memorial Lecture in St. Louis at the national meeting of the American Heart Association in 1978, I summarized the measurements made in nuclear cardiology:

1. Right and left ventricular volumes
2. Right and left ventricular hypertrophy
3. Cardiac output at rest and during exercise
4. Mitral value regurgitation
5. Right-to-left and left-to-right shunt quantification
6. Global and regional right and left ventricular wall motion
7. Regional myocardial blood flow

The clinical questions addressed in 1978 were:

1. What is the cause of the patient's chest pain?
2. What is the functional significance of electrocardiographic (ECG) abnormalities?
3. Has the patient had a myocardial infarction?
4. How much of the left or right ventricle is damaged?
5. What is the patient's right and left ventricular function at rest, during exercise, or after drug treatment?
6. Does the patient have a ventricular aneurysm?

Today, thallium-201 and technetium-99m sestamibi are widely used myocardial perfusion agents. The extraction of thallium-201 by heart muscle is greater than that of potassium-43, and its accumulation in the liver is lower. The same instrumentation used for technetium-99m is useful for thallium-201 imaging. Its 73 hour half-life facilitates commercial distribution of the radiopharmaceutical.

On January 19, 1978, an article headlined "Quicker heart test approved" appeared in the *Baltimore Sun*. Reporter Mary Knudson wrote: "A quick method of diagnosing a heart attack and coronary artery disease was approved this week by the federal Food and Drug Administration ... Portions of the heart getting adequate blood will show proper concentrations of an isotope called thallium-201 ... Dr. Allan Green is medical director of New England Nuclear, the Boston-based company that makes thallous chloride, the isotope solution." At the time, we at Hopkins had been using this tracer in patients with suspected or proven coronary artery disease since 1974, and were performing 500 studies per year.

In August, 1979, I wrote in the *Journal of the American Medical Association*:

"If a patient has angina and is not a candidate for surgery, neither a coronary arteriogram nor a thallium-201 study is needed. The patient should be treated medically for angina.

"In an asymptomatic patient with an abnormal ECG, I would recommend an exercise thallium-201 study, particularly in women, who have a large number of false-positive exercise ECG's.

"Our goal is to decrease the number of coronary arteriograms reported as normal in patients with chest pain. In our hospital, 57% of women, 25% of men younger than 50, and 11% of men older than 50 are not found to have coronary stenosis at arteriography.

"We can now determine with 90% certainty those patients without 70% stenosis in one or more coronary vessels."

In studies of the heart, as well as other organs, we can examine the relationship of regional structure, function, and biochemistry in a single study. We can define the regional abnormalities in patients with myocardial infarction and determine whether an area of infarction is not contracting but still alive, and can become the target of aggressive treatment. Previous research indicated that cardiac cells that accumulate FDG are alive, even if they are not contracting. The patients have a far better prognosis when there is evidence that the cardiac cells are still alive. Such patients can go on to interventional procedures that improve their functional recovery and extend their lives.

In addition to detecting coronary heart disease, nuclear cardiology is addressing the problems of heart failure and sudden death associated with arrhythmias. Not only can we examine the regional contraction, but also neuroreceptors on the membranes of the muscle cells. Genetic studies of the heart include the detection of programmed myocardial cell death, called "apoptosis." Annexin-5 is an important tracer for imaging apoptosis both in heart disease and in oncology.

Not all innovations in nuclear cardiology developed at Hopkins have been incorporated widely into widespread clinical practice. The "nuclear stethoscope" is an example. In 1975, we described a simple, single detector system for displaying left ventricular time-activity curves after injection of technetium-99m albumin.

We first presented a description of the "nuclear stethoscope" and its use at the meeting of the American Heart Association on November 17, 1975, in Anaheim, California. The device was invented to extend the use of the Anger scintillation camera and decrease the cost of procedures used to monitor left ventricular function. The device does not provide regional, but provides an overall assessment of the performance of the left ventricle. The left-ventricular time-activity curve is produced by dividing the recorded activity into 48

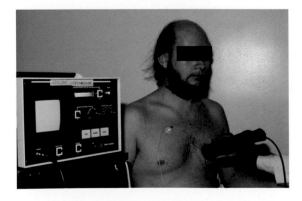

Figure 150 The "nuclear stethoscope" invented at Hopkins to examine the movement of blood through the chambers of the heart.

Figure 151 The "nuclear stethoscope" for measurement of left ventricular ejection fraction with technetium-99m albumin.

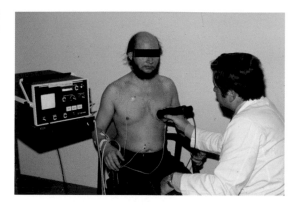

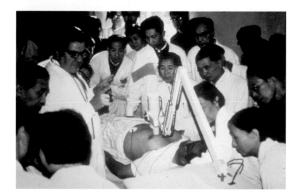

Figure 152 Demonstration of the nuclear stethoscope in Beijing in 1979 using technetium-99m albumin obtained from the first technetium-99m generator ever brought to China.

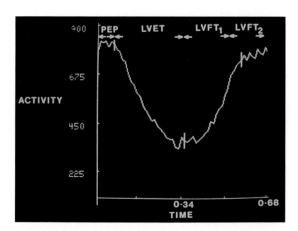

Figure 153 Left ventricular time activity curve obtained with the nuclear stethoscope and technetium-99m albumin.

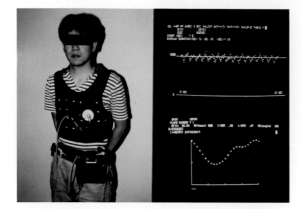

Figure 154 Commercial version of the nuclear stethoscope and left ventricular time activity curve.

Figure 155 Dr. Liu Xiujie with the nuclear stethoscope taken to China in 1979 for use with the first technetium-99m generator, both taken to China by Henry Wagner.

or more intervals during each cardiac cycle. In 1982 we interfaced a miniature cadmium telluride detector (Radiation Monitoring Devices, Watertown, Massachusetts), connected to a microcomputer to monitor left ventricular function. The system was light, comfortable, and easily attached to the patient's chest. It was used chiefly for monitoring cardiac function in patients in the intensive care unit and in assessing the effects of drugs on cardiac function.

In July, 1975, Drs. Jim Adelstein, Carl Jansen and I published a report in the journal *Circulation* from the Inter-society Commission for Heart Disease Resources. The title was: Optimal Resource Guidelines for Radioactive Tracer Studies of the Heart and Circulation. Since April 15, 1972, I had been a member of the Council on Cardiovascular Radiology of the American Heart Association.

In 1979, I was part of a group of 12 physicians who traveled throughout China in a two week program sponsored by the American College of Physicians. The group was led by Dr. Tsung O. Cheng, Professor of Medicine of the George Washington University Medical Center, and Dr. Samuel Asper, Professor of Medicine at Johns Hopkins School of Medicine.

We saw major surgery, including the removal of a pituitary tumor, performed under acupuncture with the patient awake and able to answer questions. We saw a great toe being transplanted to the site of a patient's thumb that had been amputated in an accident. We filmed rooms full of patients receiving acupuncture and moxibustion.

One of our colleagues fell into a pool of night soil to the amusement of peasants working nearby. We stripped him, covered him with a blanket, and gave him shots of gamma globulin. He got to ride in the car with T.O. Cheng, and Sam Asper, instead of in the van with the rest of us. We remarked that we now knew the qualifications needed to ride in the car.

We met several physicians who spoke perfect English that they had not spoken for decades. They had been educated at the Peking Union Medical College, which taught in English until 1930.

I took along a nuclear stethoscope and the first molybdenum/technetium-99m generator ever to enter China, and demonstrated its use in several cities, before leaving the instrument and generator with Dr. Xiu-jie Liu of the Fu Wai Hospital in Beijing. Dr. Cai Rusheng, deputy director of the Cardiovascular Institute at Fu Wai stated in 1979: "Our herbal medicines have thousands of years of use behind them, but still we do not know how they work, even if they work ... Research with this instrument is a major step toward evaluating the old with the new and bringing our medical science up to date without losing what is good and effective from traditional Chinese medicine."

Among the distinguished Chinese physicians whom we met in 1979 was Dr. Wu Ying Kai, a famous cardiovascular surgeon of the Fu Wai Hospital, and Dr. C.L. Tung, a graduate in 1924 from the University of Michigan and subsequently head of cardiology at Peking Union Medical College (the "Hopkins of China."). Dr. Tung was often referred to as the "Paul Dudley White of China." Dr. White was a famous cardiologist who, among other things, was the cardiologist taking care of President Dwight Eisenhower. Dr. Tung's textbook *Practical Cardiology* is the "bible in cardiology in China."

In Dr. Tung's book, he mentions the drug ephedrine which was developed in the 1930s from the Chinese herbal medicine "ma huang." A number of other herbal medicines have been refined for modern use to treat heart disease, malaria, hypertension, and cancer. Dr. Cai Rusheng said: "There are more than 100 herbal drugs that are supposed to have a cardiac effect, but our knowledge of them is not objective, not really scientific, but just based on clinical observations and impressions."

Before the Communists came to power in 1949, Hopkins had a close relationship with the Peking Union Medical College, the country's leading medical school, subsequently called Capital Hospital, and we hoped that a renewed Sino-American collaboration could be developed. In 1983, Dr. Liu Shouchi Tao and I published a paper in the *European Journal of Nuclear Medicine* entitled: "Measurement of Effects of the Chinese Herbal Medicine higenamine on Left Ventricular Function Using a Cardiac Probe." The effects of higenamine were similar to the response with isoproerenol. This provided objective evidence of the pharmacological effect of an herbal medicine.

In 1987, Dr. Liu, his colleagues, and I published in the *European Journal of Nuclear Medicine* the use of the nuclear stethoscope to monitor the effects of the human natriuetic polypeptide on left ventricular function in patients with hypertension. The study exemplified the value of the nuclear probe in interventional nuclear cardiology.

Figure 156 Henry Wagner with a "barefoot doctor" in Beijing in 1979.

On October 7, 2004, exactly 25 years after my first trip to China, my wife, Anne, and I took a non-stop flight from Chicago to Beijing, quite a change since 1979 when our first group visiting China had to go through Los Angeles and Tokyo, changing airlines there. Changes in Beijing, Shanghai, and Tai Yuan where I lectured were beyond belief, strikingly contrasted with the ancient city of Ping-Yao, founded in 1270 B.C. that we also visited. In Shanghai, for example, there are 300 buildings over 100 meters high, all of different architectural form, blending together beautifully, and illuminated like a fairyland at night. It brought to mind the words of Napoleon: "Let China sleep, for when she awakens, she will shake the world."

The blue Mao jackets and caps that were everywhere in 1979, three years after the end of the cultural revolution, were no longer seen. Automobiles and traffic jams were commonplace. In 1979, we wondered what it would be like when the Chinese replaced their bicycles with automobiles. Now we knew. The new multi-lane highways were beautifully landscaped with rows of trees and greenery lining the roads and median strips. There were no private cars but only governmental cars in 1979. The picture of Mao was still prominent at the front of the Imperial Palace (Forbidden City) where 24 emperors had reigned on 175 acres of beautiful buildings and palaces across from the Great Hall of the People on Tiananmen Square.

Johns Hopkins was well represented at a 2004 meeting, with Dr. Liu as President and Dr. P.F. Kao, who trained at Hopkins, receiving a prize for the best scientific poster, entitled: "Tc-99m/I-131 simultaneous dual-isotope brain striatum phantom SPECT study for simultaneous Tc-99m Trodat/I-131 IBZM pre- and post-synaptic dopamine imaging."

Dr. Mao-Song Jiang from the Shanghai Jua Dong Hospital was our first Chinese trainee in nuclear medicine at Hopkins, coming in 1980. Dr. Liu came later in 1980, and then returned again as an IAEA fellow for 6 months in 1987. He worked primarily in studies with the "nuclear stethoscope" and published over 20 papers on its use, primarily to study the effect of herbal medicines on the heart. Dr. Ma Jixiao, Chief of Nuclear Medicine at the Sixth People's Hospital in Shanghai, was a visiting Professor in nuclear medicine at Hopkins in 1987. All of these Chinese visitors became friends over the past quarter of a century.

The visit between October 9 and 13, 2004, was to attend the 8th meeting of the Asia and Oceania Congress of Nuclear Medicine and Biology held every four years since 1976. The principal meeting was in Beijing, with satellite meetings in Shanghai and Hong Kong. Two hundred and fifty attendees from 35 countries came from overseas. One evening there was a Johns Hopkins party at a restaurant in Beihai Park. Dr. Cheng Mo Zhu of the Shanghai 2nd Medical University gave a brilliant Highlights lecture in Beijing, in a style very similar to those that I have presented for 28 years as highlights of the American Society of Nuclear Medicine meeting.

I had the opportunity to visit the department of Dr. Gang Huang at the 1,000 bed Renji Hospital where there is a modern PET/CT scanner. I asked Dr. Chang, who was Vice President of the Hospital how he had obtained the funds for the scanner and he replied: "If you have power, everything is easy." When I asked the head nurse if any patients having PET/CT examinations were worried about the radiation, she replied: "No, they only worry about the cost of the study."

The airports in all three cities—Beijing, Shanghai, and Tai-Yuan—were beautiful, efficient, very large and modern. Security was as tight as in the United States.

In 2004, there were PET Centers in 10 Chinese cities, most of the instruments being PET/CT devices. There were 3 medical cyclotrons.

An article in the Johns Hopkins magazine in April 1980 stated: "Last year Henry Wagner, '48, MD '52 took his nuclear stethoscope to China to measure the effectiveness of traditional herbs on the heart . . . Screening these herbal medicines is a little like drilling for oil. We're not sure what will prove effective. The stethoscope was left on permanent loan, and the work continued." The year 1980 also marked the dissolution of the Johns Hopkins Medical Society which had been a major part of Hopkins since its founding in the 19th century. I had served as President of the Society in the past.

Both in the United States and China, nuclear cardiology developed as the result of close collaboration between cardiologists and nuclear medicine physicians. In July 1978, in response to a questionnaire sent to the heads of nuclear medicine residency programs and directors of adult cardiology training programs in the United States, I learned in a survey that over 90% of both groups said that nuclear cardiology studies should be performed by both groups, preferably working together as a team. The expertise of the nuclear medicine physicians was of greatest value in the technical aspects of the study. Collaboration was the key to success.

On December 7, 1981, we obtained $8,000 from the Women's Board of Johns Hopkins Hospital to purchase a high quality video taping and playback machine that we used in our morning conferences to train nuclear medicine physicians and cardiologists to "read" the nuclear cardiology procedures, although it was used tape all types of dynamic studies, including cerebral blood flow, renal blood flow, and other studies.

On February 2, 1981, we entered into an agreement with a company CARDIOKINET-ICS INC to obtain software for computer assistance in the diagnosis of coronary artery disease. This was a very successful early approach to "computed assisted diagnosis" (CAD), which is now very well accepted throughout the world.

Molecular imaging with PET, computed tomography and magnetic resonance imaging developed in parallel. Today most nuclear medicine departments require that nuclear medicine residents spend at least 2 to 3 months each in the computed tomography (CT) and magnetic resonance imaging (MRI). Nuclear Magnetic Resonance (NMR) imaging at Hopkins began when Dean Richard Ross established a Radiology Research Conference on May 4, 1977. Experts in NMR from other institutions were invited to review the medical applications of NMR. On March 20, 1978, at the request of Dean Richard Ross, I began the development of an interdisciplinary program for NMR imaging at Hopkins. Eventually, Dr. Jerry Glickson was recruited by our committee to head the NMR imaging program at Hopkins.

On April 27, 1981, I was invited to become a member of the editorial board of the journal *Magnetic Resonance Imaging*, published by Pergamon Press, having been invited by Sharad R. Amtey of the University of Texas Health Science Center in Houston, who was Editor-in-Chief. Other members of the editorial board included Paul Lauterbur at that time from the State University of New York at Stony Brook, who subsequently won the Nobel prize for his work. I began service with the Editorial Board on March 15, 1982. This was the first journal to cover the clinical applications of "an exciting new field."

On April 21, 1982, Dr. Myron Weisfeldt, at the time Director of the Cardiology Division and now Chairman of the Department of Medicine at Hopkins, wrote me: "I am most excited that you plan to establish a Nuclear Magnetic Resonance Imaging Center facility within this institution. I have long awaited this event with every anticipation that it would lead to a marked enhancement of our imaging capabilities for the heart and circulation."

On December 7, 1983, Dean Richard Ross wrote me: "I am pleased that you have accepted the position of Chairman of the Interdisciplinary Committee on Nuclear Magnetic Resonance. Establishment of the ICNMR and acquisition of three NMR systems by the JHMI should provide an excellent opportunity to apply this exciting new technology to a wide range of research projects."

On December 13, 1983, under my Chairmanship, we held the first meeting of the Interdisciplinary Committee on Nuclear Magnetic Resonance (ICNMR), whose members included Dean Richard Ross, Martin Donner, Chairman of Radioloy, and Myron Weisfeldt, Head of Cardiology. "The goal of this committee is to encourage the development of shared resources which are now the leading edge of this exciting area of research."

In 1986, the new MRI facility opened in the basement of the Houck Building at Hopkins with three new scanners under the leadership of Dr. Jerry Glickson, director of magnetic resonance research, who had been recruited by the ICNMR committee from the University of Pennsylvania. In the announcement, we stated: "investigators in the departments of radiology, ophthalmology, psychiatry, neurology and neurosurgery and from the John F. Kennedy Institute plan research with the new equipment."

12

Growth Out of Control

Since 1948, the early detection of cancer has been a major goal of nuclear medicine. The direction of research has been to develop tracers that accumulate in cancer cells because of their biochemical or immunological properties. Another approach has been to detect cancerous lesions because they produce defects in normal processes, such as phagocytosis in the liver or regional blood flow in the lung. In 1975, among the first agents was selenium-75 selenomethionine, an amino acid that accumulates during the process of protein synthesis. This tracer did not achieve widespread clinical use because of the high body background and slow accumulation in the lesions. The same was true for iodine-125-iodoquine, a chloroquine analogue, and indium-111 or cobalt-57 bleomycin, all developed in 1975.

With the discovery of antigens on tumor cells, research in nuclear medicine was directed to the possibility that radiolabeled antibodies might be imaging agents. In 1965, Gold and Freeman first isolated carcinoembryonic antigen (CEA) in colon cancer. Their initial results in rodents looked promising, but the optimistic results could not be reproduced in human cancer, chiefly because of a lack of specificity in the binding of the tracer to tumors.

In 1969, Edwards and Hayes began to use gallium-67 citrate to detect bone lesions in patients with Hodgkin's disease. They observed that there was accumulation of the tracer in non-osseous cancer as well in the cancerous lesions. The high incidence of "false negatives" prevented the use of gallium-67 citrate in mass screening of patients for cancer, but the finding of a positive accumulation was helpful in staging the degree of involvement of Hodgkin's disease and in assessing the response to treatment.

In 1978, we at Hopkins reported our results in gallium-67 studies in 46 children with cancer. Forty-seven percent of the studies revealed avid gallium-67 accumulation in the involved lesions. Hodgkin's disease or non-Hodgkin's lymphoma accounted for 44 of the 83 (53%) positive scans in our series. We concluded that the technique was helpful to assess whether cancerous lesions in patients with lymphomas are gallium-67 avid. If they were, we carried out serial follow-up studies to monitor, and, when necessary, to alter treatment.

Bone scanning was very successful in early studies of patients with cancer and soon became widely used throughout the world. The detection of metastatic lesions in bone was an important finding in decision-making. In 1979 we reported the results of bone scanning with technetium-99m pyrophosphate or diphosphonate in 367 patients with

Figures 157 and 158 Gallium-67 (and indium-111) images of the liver indicating that a lesion seen in the technetium-99m sulfur colloid image is malignant.

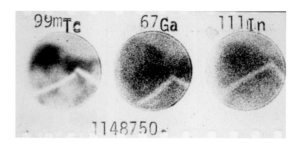

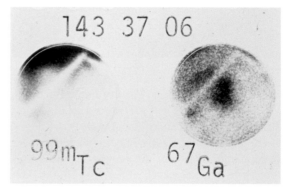

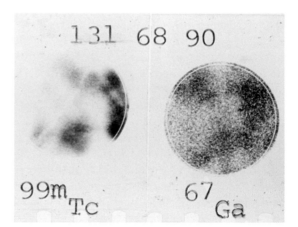

Figure 159 Gallium-67 image of the liver indicating that a lesion seen in a technetium-99m image is not malignant.

cervical cancer. Only 4 patients were shown to have bone metastases. Since there were so few bone metastases detected, in patients with cervical cancer, screening for metastatic disease did not seem warranted.

While bone scanning with phosphonates achieved an important role in detecting bone metastatic disease, the search continued for tracers that would detect the transformation of normal to cancerous cells, and be useful in early diagnosis, at a time when treatment

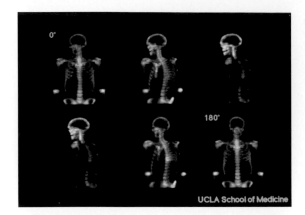

Figure 160 Whole-body bone imaging in a normal person.

would be most effective. The candidate tracers were almost limitless, because cancer is not a single process, but is characterized by hundreds of biochemical manifestations in different types of cancer. Each made possible a different approach to the development of clinically useful tracers in patients with cancer.

Cancer results from the progressive accumulation of genetic mutations that affect specific locations in a cell's DNA, resulting in uncontrolled growth of cancer cells. These cancer cells escape the regulatory mechanisms for controlling growth and continue to multiply. Cancer can arise from a single cell that has undergone these genetic changes that eventually results in these abnormalities in billions of cells. The genes that induce this uncontrolled cell division are called *oncogenes*. The growth is counteracted by the actions of genes that restrain the cell's division. When these become non-functional, the oncogenes are locked in an "on" state. When the cancer becomes large enough to be detected, its cells have undergone numerous changes subsequent to the initial event. In current thinking, cancer is considered a disease of the molecules that control the synthesis of the genetic DNA. Hence, the term "molecular biology" led to the term "molecular medicine" to characterize the use of radiolabeled tracers to detect and treat cancer.

For over a quarter of a century, the concept of "molecular medicine" has been the basis for the design and development of new therapeutic drugs. In 1983, soon after our first imaging of neuroreceptors in human beings, we began collaborative studies with scientists in pharmaceutical companies. Today, several worldwide pharmaceutical companies have extensive molecular imaging divisions, including Merck, Glaxo Smith Kline, and Pfizer. Molecular imaging first in animals and then in human beings has resulted in a decrease in the time required to develop and obtain approval of New Drug Applications (NDAs) from the FDA for pharmaceuticals through the use of positron emission tomography in experimental animals and human beings. These have achieved wide use in all four phases of drug design and development. The use of other drugs, such as Gleevec, designed to treat leukemia and stomach cancer and Herceptin to treat breast cancer has benefited from molecular imaging.

Lengauer and Vogelstein proposed that "chromosome instability" is one of the earliest events in cancer cells, preceding the genetic mutations. "Master" genes that control growth are affected early as a result of chromosomal abnormalities. The cell "stumbles each time it attempts the carefully choreographed dance of cell division." (W. Wayt Gibbs, *Scientific American*, July 2003).

Peter H. Duesberg and Ruhlong Li of the University of California have proposed another theory. The chromosomes themselves are damaged, and become "aneuploid," characterized by an abnormal number of chromosomes, or the chromosomes become truncated, extended, or contain abnormal segments. An example of a chromosomal abnormality is the tripling of chromosome 21 rather than a doubling. Each cell contains three chromosome 21s rather than two. As long as a century ago, Theodor Boveri suggested that these aneuploid chromosomes might be the cause of cancer, but his ideas were dropped in favor of the search for oncogenes. Today, the list of mutations linked to cancer have increased to more than 100 oncogenes and 15 tumor suppressor genes. The question remains: "which comes first, aneuploidy or mutations?"

Genomics is a major component of "molecular medicine." Genomics can be thought of as the approach to the patient from the "bottom up," while the search for abnormal genes by molecular imaging of phenotypes, is "top down."

Louis Sokoloff developed "quantitative autoradiography" at the NIH using C-14 labeled glucose. After administration of the tracer to animals, they were sacrificed and their brains frozen and sliced for autoradiography. Deoxyglucose is an analogue of glucose through the first steps of cell metabolism—accumulation and phosphorylation—but then goes no further. This makes it possible to measure the regional ATP-generating activity of the cell, that is, its energy-generating activity. Measuring the optical density of the autoradiographic film shows the distribution and quantity of glucose utilization in different parts of the brain. Among the early studies was the mapping of visual pathways in the monkey brain, showing the passage of neuronal activity through the newly identified multiple pathways of the brain, including the amygdala, a part of the brain concerned with emotions. Not long afterward, fluorine-18 fluorodeoxyglucose made possible analogous studies in experimental animals, and eventually, in living human beings in health and disease.

Giovanni DeChiro at the NIH began using fluorine-18 deoxyglucose (F-18 FDG) to identify those regions of the brain that became activated during mental activity. In the course of these studies, he found that brain tumors avidly accumulated F-18 deoxyglucose (FDG) to a degree greater than the activation of the normal brain. He was aware of the previous studies by Warburg, who found that anerobic metabolism was increased in cancer. DeChiro's first studies with F-18 FDG were in 23 patients with gliomas, reported in 1982. He found that the more malignant tumors had the highest FDG accumulation.

What is a possible mechanism for the increased accumulation of FDG? The earth in its primitive state contained no oxygen, but eventually in the process of chemical evolution, oxygen, hydrogen, nitrogen, and carbon became the most abundant elements found in all forms of life. Cells are composed of organic molecules, including amino acids, sugars, and the purine and pyrimidine bases of the nucleic acids that make up genes. All living things depend on chemical synthesis by green plants that use sunlight as an energy source in the process called photosynthesis. Ultraviolet and visible light were responsible for chemical evolution and are involved in the sustaining of life as we know it today. Photosynthesis depends on the pigment molecule chlorophyll to produce oxygen which profoundly modified the earth's atmosphere and provided the conditions necessary for the evolution of life. Photosynthetic phosphorylation is the process by which light forms the high energy phosphate bonds of adenosine triphosphate (ATP) to serve as the principle carrier of chemical energy in the cell. DeChiro found that the accumulation of FDG correlated with the histopathological degree of malignancy. (In 1976 there were 2 scientific presentations using FDG in oncology at the annual SNM meeting; by 1983, there were 25, with a steady increase every year since. Today there are over 500 FDG presentations in oncology alone.)

Figure 161 Abass Alavi was the volunteer for the first study of the brain with F-18 fluorodeoxyglucose (FDG) at the University of Pennsylvania, where he subsequently became head of nuclear medicine.

The greater accumulation of F-18 fluorodeoxyglucose (FDG) reflected increased anaerobic metabolism. The less malignant cells manifest a greater accumulation of C-11 acetate, which reflects aerobic metabolism. FDG accumulation by cancer cells is the result of increased expression of the enzyme, hexokinase II, an ancient enzyme present to archeobacteria that existed even before there was oxygen on earth. FDG accumulation reflects the de-differentiation of cells to a more primitive, pre-aerobic state, characterized by uncontrolled growth. In late 1930s in Germany, Otto Warburg was the first to postulate that tumors are similar to archeobacteria in the metabolism of glucose.

Hexokinase is located inside the cytoplasm surrounding the nucleus of the cell, and plays a key role in the energy supply of the cells. After active transport of glucose through the cell membrane, hexokinase catalyzes the phosphorylation of glucose to glucose-6 phosphate, the rate limiting step in glycolysis. F-18 fluorodeoxyglucose (FDG) is transported into cells by the action of two protein transporters, Glut-1 and Glut-4. FDG-6 phosphate is not further metabolized because, being negatively charged, it remains trapped within cells. This makes possible PET imaging of the process of glucose utilization by the cancer cells, a classic example of "molecular medicine."

Chromatin contains the genetic code in "chromosomes," which are easily damaged by hypoxia, resulting in impaired regulation of gene expression. Hypoxia within a tumor is an early event leading to subsequent genetic disarray in cancer. The degree of hypoxia varies in different types of cancer.

In 1942 Bauer and Wendeberg detected high concentrations of calcium-47 and strontium-85 at the active sites of Paget's disease, as well as in fractures and cancerous bone metastases. In 1959 William Fleming began using strontium-85. We used strontium-85 at Hopkins until 1969, when it was replaced with strontium-87m, another isotope of strontium which can be obtained from its parent, yttrium-87, in a radionuclide generator, making it readily available. Today fluorine-18 fluoride, introduced in 1969, and F-18 FDG are used for PET imaging of the skeleton, as well as technetium-99m phosphonates.

Diseases in which bone imaging is useful include: (1) metastatic lesions, (2) sarcomas, (3) multiple myeloma, (4) Paget's disease, (5) surgical osteotomy, (6) osteomyelitis, (7) arthritis, and (8) fractures. Detecting bone metastases is particularly useful in planning subsequent treatment.

More and more is being learned about the mechanism of the accumulation of different tracers in different cancerous lesions. Thyroid cancer cells express the protein "sodium/iodide symporter" that concentrates radioiodine. The cancerous cell loses this ability with increasing degrees of malignancy. Thyroid cancer metastases can be detected by FDG, even though they no longer accumulate radioiodine. Prostatic cancer, bronchioalveolar lung cancer, early breast cancer, and renal cancer often do not accuulate FDG to a large degree, reflecting their low grade degree of malignancy. These types of cancer may express somatostatin receptors which can be used to detect the lesions. Other cancers, such as prostate cancer, can be detected by C-11 acetate imaging, which reflects aerobic oxidation via the Krebs cycle. C-11 choline accumulation reflects synthesis of cell membranes, and is also used to detect cancer.

Not only is glycolysis increased in many types of human cancer, but is also increased in the myocardium during stress. At rest the myocardium relies chiefly on fatty acid metabolism. Chest pain developing during electrocardiographic stress testing is associated with lactate production as a result of anerobic metabolism related to myocardial ischemia. Increased FDG accumulation is also seen in hypoxic tuberculous lesions.

Nuclear medicine has made enormous advances in the care of patients with cancer as well as other diseases throughout the world over the last half century. Its proof of principle is well established, and it is apparent to all physicians and the public. Fluorodeoxyglucose (FDG) imaging to measure whole body or regional glucose utilization has become routine in patients with cancer. In developing countries without access to cyclotron-produced FDG, gallium-68, with a 68-hour half-life, can be helpful in oncology. The tracer can be obtained on a routine basis from a germanium generator which has a 271-day half-life. One needs only to buy one of these generators every 6 months. This makes possible extension of PET into areas of the world where cyclotron-produced tracers are not readily available.

Another tracer, 18-FLT (a fluorinated analog of thymidine) is used to image cell division while F-18 and C-11 choline are used for diagnosing prostate and bladder cancers, which do not usually accumulate FDG. Sodium fluoride is used for imaging skeletal lesions. At times the same or modified tracer molecule used in diagnosis can be the basis of targeted radiation therapy. Examples include Zevalin and Bexxar, which target the CD-20 antigen on non-Hodgkin lymphoma cells.

Many types of cancer express molecular receptors that can be imaged in patients with cancer. For example, somatostatin receptors are expressed in tumors affecting the gastrointestinal tract. Other tracers that bind to other receptors can characterize cancerous lesions.

In the care of patients with cancer, it is helpful to combine the anatomical information from computed tomography (CT) with that obtained by PET or its ally, single photon emission tomography (SPECT). The use of PET/CT and SPECT/CT in so-called hybrid instruments is increasing all over the world, and is likely to replace dedicated PET instruments. Special imaging devices can be used for specific purposes, such as the detection of breast lesions or during surgery by means of simple "probe" detectors or small field of view imaging devices. The complexity of the imaging device or probe should be matched to the complexity of the problem being addressed, rather than "one size fits all" imaging.

These are but a few of the great advances being made in trying to solve the mysteries of cancer, and helping patients to escape its curse. More and more is being learned about genetic and molecular factors that result in uncontrolled malignant transformation of the cells of the body, and how they are spread through the body. Andrew C. von Eschenbach, director of the National Cancer Institute (NCI) of the NIH, has made the goal of the NCI "to eliminate suffering and death due to cancer by 2015." This may seem a dream to many, but it does provide a tribute to current research and a worthy target.

13

Molecular Communication

In 1949, it was discovered that the Indian herb, *Rauwolfia serpentina* helped calm patients with schizophrenia. Reserpine, the active component and a related drug, chlorpromazine, reduces the concentration of the neurotransmitter, dopamine, in the brain and various other organs. These drugs act by blocking the binding of dopamine to the dopamine receptor, leading to the hypothesis that schizophrenia and possibly depression might be definable at the molecular level. Schizophrenia, classified as a psychiatric disorder, might then be considered a "medical" illness. Autopsy studies of patients with schizophrenia revealed increased numbers of dopamine receptors in the caudate nucleus and putamen of the brain. Until the development of PET, knowledge of the chemistry of the human brain could be obtained only from neurochemical and neuropathological examination of experimental animals or from studies of the human brain at autopsy. PET was as important a landmark in the study of the human brain as the discovery that neuronal activity was associated with measurable electrical activity within the brain of living persons.

The imaging of receptors in the human brain was extended to the field of oncology when it was discovered that many types of cancer express receptors that influence tumor growth. Attention was directed not only to the development of radiopharmaceuticals targeted to image these receptors in the brain, but also for diagnosis and treatment of cancer. Neuroendocrine tumors, such as islet cell tumors of the pancreas and carcinoid, were successfully imaged with analogues of somatostatin receptors. The peptide, somatostatin, was discovered by Roger Guillemin of the Salk Institute in San Diego and Andrew V. Schally of the Veterans Administration Hospital of New Orleans, both of whom received the Nobel prize in 1978. Interestingly, the same year Rosalyn Yalow of the Veterans Administration Hospital in the Bronx, N.Y, shared the prize for the development of radioimmunoassay of peptide hormones.

Other receptor ligands in oncology include those for imaging endothelial growth factor (EGR) receptors for non-small cell lung cancer, estradiol receptors for breast cancer, HER-22/neu receptors for breast and ovarian cancer, gastrin-releasing peptide receptors for small cell lung cancer, interleukin 2 receptors for lymphoma T-cells, p-glycoprotein receptors for leukemia, colon and renal cancer, glucose transporter receptors for gliomas, lung and breast cancer, and transferring receptors for lymphoma and lung cancer.

Figure 162 Rosalyn Yalow receiving the Nobel prize in Stockholm in 1977.

We set out to accomplish the first imaging of receptors in the human brain when we learned of the work of Snyder, Kuhar and colleagues in identifying neuroreceptors, particularly dopamine and opiate receptors by autoradiography in rodents. We decided that the time had come for us to try to get a cyclotron and PET scanner. This was accomplished in 1978.

In November 1979, I described our acquisition of our first PET scanner to a reporter from a Philadelphia newspaper: "What we are doing is like a mountain-climbing expedition, where we are all assembled at the base of a mountain. Our goal is to try to find out how abnormalities in brain chemistry are involved in certain major neurological and mental diseases."

We have managed to climb a long way up the mountain, making possible a new era in the study of the relationship between brain chemistry and behavior. This would not have been possible without the invention of single photon emission computed tomography (SPECT), positron emission tomography (PET), and magnetic resonance imaging (MRI). Today it is possible to quantify brain chemistry in the different regions of the brain in meters/kilograms/second (MKS) units. The most significant practical results of this research has been in drug design and development. We measure the effects of drugs on the chemistry of the brain, such as the state and activities of neuroreceptors and transporters. Measurements of receptor availability to naturally-occurring chemical messengers within the brain can now be related to animal and human behavior, and the effects of therapeutic and abused drugs examined in health and disease.Ninety percent of the scanners being purchased in 2004 to carry out these studies with positron-emitting tracers were hybrid PET-CT machines.

The spatial resolution of proton MRI images to reveal structure is far greater than that of PET images. Most MRI studies are based on the imaging of hydrogen atoms. In MRI studies of other atoms, such as phosphorus, the spatial resolution is less than that of PET. Studies of regional phosphorus metabolism can be performed with MRI, while glucose, oxygen, and other molecular processes can be measured by PET with a spatial resolution of 2–4 mm.

The development of FDG to study the brain began in 1954 when Sols and Crane discovered that deoxyglucose-6-phosphate, the product of the reaction deoxyglucose with the enxyme, hexokinase, is trapped after accumulation in neurons, because it cannot participate further in the subsequent metabolic steps of glucose metabolism. Sokoloff and his colleagues at the National Institutes of Health developed a quantitative auto-radiographic method to examine regional glucose utilization within different regions of the brain, and, for the first time, were able to relate levels of regional glucose metabolism to the degree of neuronal activity in specific regions of the brain. The synapses utilize glucose as the source of energy in neuronal transmission. Al Wolf, working with Reivich, Kuhl, Ido and others prepared deoxyglucose labeled with fluorine-18 making possible the extension of the autoradiographic method of Sololoff to imaging glucose utilization in the brain of living human beings.

In the 1960s, a frequent subject of lectures and debates by us and others was the relative merits of rectilinear scanners and the Anger gamma camera in nuclear medicine. The camera was the winner, but the debate continued about the relative merits of single photon versus positron-emitting radiotracers. The key argument for PET was that carbon-11, oxygen-15, nitrogren-13, and fluorine-18 tracers could be detected, while single photon agents were primarily dependent on iodine-123 and technetium-99m tracers.

An example of the state of nuclear medicine in 1970 was reflected in a Hickey Lecture that I delivered at the Wayne State University College of Medicine in Detroit: "Brain scanning (with radioactive tracers) is one of the most widely accepted procedures in nuclear medicine today. Yet we still have not taken the time and trouble to find out the effect that brain scanning has had on the practice of medicine." At that time, brain scanning could not detect lesions less than 2 cm in size, and the procedures took a long time. Yet it was predictable even in 1970 that: "The bringing together of structure and function is a major accomplishment in nuclear medicine. Functional images are two aspects of the same thing. As Koestler has stated 'Structure can be considered as function frozen in the present time.'" What we call structures are slow processes of long duration, while functions are quick processes of short duration.

One of the early findings was that nuclear brain scanning significantly decreased the time between the onset of symptoms and the time patients came to surgery for treatment of brain tumors. This finding provided the first objective evidence of the clinical value of radionuclide brain scanning. Compared with other diagnostic modalities at that time, brain scans were abnormal in 80–95% of patients with brain tumors. The study of the cerebrospinal fluid in patients with normal pressure hydrocephalus was another important advance in those early days.

In 1971, we reported the use of the Anger camera in 55 patients with brain tumors. In 1968 at an IAEA Symplosium in Salzburg, Austria, we concluded: "Our results suggest that tomography of brain imaging with the Anger camera may be of significance in increasing the diagnostic certainty and characterizing the size, shape and position of a lesion." In 44% of the patients with brain tumors, tomography was helpful when the rectilinear scan results were equivocal. At an IAEA symposium in 1971, we reported that the scintillation camera was useful in the diagnosis of middle cerebral artery thrombo-embolic disease and internal carotid artery occlusion.

In the discussion period, I concluded that "the use of high resolution stationary imaging devices for studying the cerebral circulation has resulted in more progress than the use of the earlier multi-detector systems with xenon-133." The beloved physicist in nuclear medicine, Norman Veall, author of the classic textbook of nuclear medicine that I read avidly in 1957 at Hammersmith, developed another radioactive gas method of measuring blood flow, the rubidium-81/krypton-81m generator system, a method still in use today.

The invention of x-ray computed tomography (CT) with its exquisite spatial resolution soon led to a progressive decrease in radionuclide brain scanning. This was fortunate because it resulted in a change of direction for nuclear medicine. Radionuclide brain scanning would change focus from simply detecting brain cancer to examining brain chemistry.

The first use of positron-emitting radionuclides, including carbon-11 and fluorine-18, in the study of the brain was by Wrenn and his associates in 1951. Their work was extended by Brownell and Sweet in 1956. Studies with positron-emitting tracers at that time were superceded by single photon emiting radionuclides, such as mercury-203 and eventually by technetium-99m, which became the workhorse of nuclear medicine because of its availability, inexpensiveness, half life, and mode of decay, the latter being optimal for external imaging of the human body. Positrons, used since the late 1930s, had been put on a back burner until 1973 when Brownell and his associates introduced a multi-crystal positron camera. In 1975, Ter Pogossian, Phelps, and Hoffman developed a posi-tron camera based on a hexagonal array of crystal detectors.

Anger's scintillation camera remained the main focus of attention for brain scaning until Sokoloff and his colleagues reported the use of quantitative autoradiography to measure regional glucose metabolism in the brain. DeChiro and colleagues at the NIH showed that human brain tumors avidly accumulate a glucose analogue, F-18 fluorode-oxyglucose *(vide supra)*.

In the 1950s, endocrinology was the "hot topic," competing with infectious diseases for the interest of faculty and students. Circulating molecules, including insulin, ACTH, hydrocortisone, thyroxine, epinephrine and other hormones were shown to have great effects on the brain as well as other tissues.

When I was a Clinical Associate in the Laboratory of Kidney and Electrolyte Metabo-lism of the National Heart Institute in Bethesda in 1955, I learned of the exciting discov-eries of Bernard Brodie and his colleagues who showed that the drug reserpine depleted the neurotransmitter serotonin from body tissues, including the brain. This was my first "inkling of the linking" of brain chemistry and behavior. These results strongly supported what the philosopher, William James, had written in 1890, that: "Chemical action must of course accompany brain activity but little is known of its exact nature."

James' classic book (1890) "Principles of Psychology" was to be my "bible" for decades thereafter.

Today, one third of all the prescription drugs taken in the United States are given because of their effect on mental function. Many affect the mental state of normal persons, while other more potent drugs alleviate the symptoms of diseases such as schizophrenia. The hypothalamus lies at the top of the brain stem, and is the meeting place of the brain and the endocrine system. The pituitary gland is regulated by molecules produced in the hypothalamus, and called "releasing factors" because they cause the secretion of pituitary hormones. One of these, somatostatin, was discovered by Andrew Schally and Roger Guillemin in Houston, Texas. They received the Nobel prize for this work in 1977, sharing it with Dr. Roslyn Yalow, for her invention of radioimmunoassay.

Otto Loewi received the Nobel prize in 1936 for his discovery that acetylcholine was released when he stimulated the vagus nerve of a frog, slowing of the heart rate. Acetyl choline is but one of more than 100 neurotransmitters, chemicals that are released from neurons when they are activated. Most drugs that affect mental function act by affecting neurotransmitter systems. Examples are opiates that affect pain perception and feelings of well-being, and drugs that help alleviate the symptoms of patients with schizophrenia, anxiety states or depression. Alzheimer's disease is related to degeneration of acetylcholine-containing neurons in certain regions of the brain, and is characterized by plaques of the chemical, beta amyloid, and tangles of the neuronal connections. There is considerable evidence that Alzheimer's disease (SDAT) is characterized by the accumulation in the brain of "plaques" made of the protein, beta amyloid. These amyloid deposits are related to one of the characteristic pathological findings of SDAT, so-called neurofibrillary tangles. Others believe that amyloid is the brain's response to the damaged neurons. They postulate that the production of amyloid sops up neurotoxic proteins and poisonous metals, such as zinc and copper. Neurologist Peter Davies of Mount Sinai School of Medicine in New York has said: "Amost all aged brains have amyloid depostion, even in people who die with no symptoms of Alzheimer's." The debate about whether amyloid is the cause or an effect of SDAT continues. Nevertheless, targeting of amyloid deposits in the brain for early detection of SDAT by PET and SPECT is a major target of several pharmaceutical companies: Eli Lilly & Co. in Indianapolis, Neurochem in Laval, Quebec, and Avid Pharmaceuticals in Texas. Price et al. at the University of Pittsburgh reported in vivo quantification of amyloid binding in SDAT patients using the compound referred to as PIB. Kepe et al. from the University of California at Los Angeles (UCLA) and the University of Ljubljana (Slovenia) are focussing on the accumulation of ^{18}F-FDDNP, another tracer that binds to amyloid plaques. Their results showed that global or overall accumulation of this tracer is greater in patients with AD than in patients with minimum cognitive impairment or in control groups. As yet, no existing drug reverses the inevitable cognitive and memory decline in SDAT but many are in the pipeline of pharmaceutical research.

These companies and several academic institutions are using PET and SPECT imaging as a major focus of their research on the role of amyloid plaques in SDAT. Imaging of β-amyloid plaques is proposed to be used in selecting patients and validating results of clinical trials of drugs designed to target plaques. These new tracers enhance research in drug development. Toyama et al. from Fujita Health University (Aichi, Japan) and the

National Institutes of Health (NIH; Bethesda, MD) have used small animal PET to image β-amyloid with the tracer ^{11}C-6-OH-BTA-1 in a mouse model of AD. They found that the binding was minimal in mice that were 20 months old but was abnormal in mice that were 32 months old.

The existence of specific molecular receptors in the brain had been postulated as early as 1905 by Langley (*J Physiol London* 33: 374–413) and by Paul Ehrlich (*Readings in Pharmacology*, Little Brown, Boston, 1962). Much of our knowledge about neurotransmitters originated from research at the National Heart Institute of the NIH based on the development of spectrophotometry by Sydney Udenfriend. This technique was used by visiting scientist Arvid Carlson in 1957 to demonstrate that another amine, dopamine, was a neurotransmitter. He was the first to suspect that the behavioral effects observed in animals given anti-schizophrenic drugs was due to depletion of dopamine.

Figure 163 Arvid Carlsson, winner of the Nobel prize in 2000 for his discovery that dopamine is a neurotransmitter.

In 2000, the Nobel prize was awarded to Carlsson, then at the University of Göteborg, together with Paul Greengard of the Rockefeller University in New York, who showed that dopamine, after binding to neuronal membrane receptors, resulted in phosphorylation of proteins. Eric Kandel showed that protein activation was involved in memory.

The neurotransmitter dopamine is synthesized in the midbrain in an area called the substantia nigra. In patients with Parkinson disease, these dopamine-transmitting neurons degenerate, resulting in an absence of dopamine in the basal ganglia, especially the putamen. The administration of the dopamine precursor, L-Dopa, greatly relieves the symptoms, an illustration of chemical characterization leading to molecular therapy.

In 1978, Kuhar and his associates at Hopkins showed that dopamine receptors in the basal ganglia could be visualized by autoradiography after administration of drugs that were bound by dopamine receptors. Creese, Burt, and Snyder (*Science* 192: 481–483, 1976) showed that the anti-psychotic effect of drugs in schizophrenic patients was directly related to their degree of blockade of dopamine receptors. It was theoretically possible to label these drugs with carbon-11, and image their distribution in the brain, first in

baboons and then in normal human beings and patients with schizophrenia. We resolved
to try to image dopamine and opiate receptors in the brain and establish a Mind/Brain
Imaging Program.

The path was not always smooth. We faced many doubters in our efforts to examine
"the molecules of the mind." For example, on February 8, 1979, after I had had an interview with Jon Franklin, a Pulitzer Prize winning reporter from the *Baltimore Sun*, I
received a letter from Dr. Bob Heysell, Director of Johns Hopkins Hospital. He wrote:

"I happened to see the article in the Sun in which you are quoted extensively concerning expansion of your "brain study program" for thought imaging . . . We have enough
problems as it is without our faculty and others helping to make us look like idiots or
incompetents or worse."

In Franklin's article, he had written: "Johns Hopkins scientists say they will install a
scanner capable of watching the chemical process of thought as it flickers through the
brain . . . It is designed to help map the circuitry of thought and allow scientists to understand how information is processed by the brain . . . Once brain specialists understand
the normal flow of energy through the brain, they should be able to pinpoint the problem
in patients with certain diseases, including schizophrenia, senility, Huntington's disease,
Parkinson's disease, manic-depressive psychosis, drug addiction and manganese
poisoning . . ."

Franklin added: "Dr. Henry Wagner, the scientist in charge of the project, expects the
device (the PET scanner) to pinpoint chemical malfunctions in the brain and, for the
first time, allow scientists to tinker directly and rationally with the chemical engines of
thought . . . If we find something that runs too fast, we can treat it by slowing it down. If
it's too slow, we can speed it up."

Four years later, we were able to carry out the first successful imaging of a neuroreceptor by PET in a living human being with carbon-11 N-methyl spiperone (C-11 NMSP)
on May 25, 1983. I was the subject in the first study. The labeling procedure was developed
by Bengt Langstrom and colleagues at Uppsala, Sweden, who developed the process of
methylation with carbon-11 methyl iodine, which eventually became widely used for
labeling a host of drugs with carbon-11. We saw the high concentration of the tracer in
the basal ganglia, specifically in caudate nucleus and putamen.

By 1986, studies with C-11 N-methyl spiperone had been performed in more than 50
normal persons and 150 patients with various neuropsychiatric disorders at Hopkins.

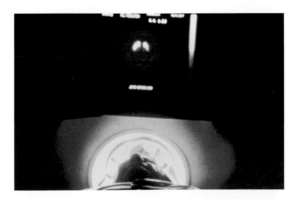

Figure 164 First imaging of a neuroreceptor in the brain
of a living human being. HNW was the subject, and the
tracer was C-11 N-methyl spiperone. The date was May 25,
1983.

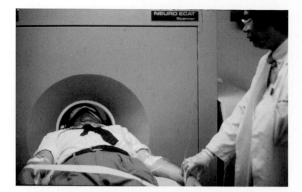

Figure 165 Dean Wong, a trainee in nuclear medicine who subsequently became a Professor is feeling the pulse of the subject, HNW.

Figure 166 The Johns Hopkins nuclear medicine faculty in the 1980s.

Figure 167 Bengt Langstrom, a Swedish chemist, was the developer of C-11 N-methyl spiperone, the first agent used to image a neuroreceptor (the dopamine receptor) in a human being.

On September 23, 1983, Peter Kumpa, a reporter for the *Baltimore Sun*, wrote: "We should be able to learn how those tiny messengers work or don't work. In a variety of perplexing illnesses, the function of those receptors is believed critical."

The study of neuronal receptors in the brain began with the use of psychoactive drugs, such as reserpine, to treat mental illness. Most such drugs interfere with specific chemical processes within the brain that are controlled by monoamine peptides, serotonin, dopamine, norepinephrine, and epinephrines, involved in the transmission of electrical impulses from one neuron to another or to effector cells.

Of all mental diseases, schizophrenia has been thought to be associated with the neurons that use dopamine as a stimulant of specific dopamine receptors, including the D1, D2, and D4 dopamine receptor systems. For decades the dopaminergic system has been guiding drug treatment of patients with schizophrenia. This emphasis was the result of the discovery in the 1950s that a class of drugs called phenothyazines could control the so-called "positive" symptoms of schizophrenia, namely, agitation and hallucinations. This class of drugs helps only about 20% of the patients. They act by blocking the D2 receptors located chiefly in the basal ganglia of the brain, that are involved in movement and emotions. Dopamine normally acts as an inhibitor, blocking the neurons that express D2 receptors. Too much dopamine results in symptoms involving mental functions related to the basal ganglia, while too little dopamine leads to symptoms related to functioning of the frontal lobes.

We believed that PET could help explain how our brains work, and help guide drug treatment. We postulated that many diseases, including Parkinson's disease, dementia, and schizophrenia might be more successfully treated if the diagnosis was based on regional brain chemistry. If the treatment was to be chemical, the diagnostic characterization of the patient should be chemical.

In April 1980, I wrote that "despite the many current obstacles and problems, it seems highly probable that emission computed tomography (ECT) will one day be routinely employed in many institutions." Later that year, I wrote: "The University Hospital cyclotron is an idea whose time has come ... There are now four major producers of hospital cyclotrons and at least eight different positron tomography devices now in use throughout the world." In 1981, we received an NIH grant that made it possible to acquire a cyclotron for brain research.

Figure 168 The Swedish cyclotron being installed in nuclear medicine space at Johns Hopkins in 1981.

We requested $450,000 from the Hospital, which would provide half the funds needed to acquire a PET scanner. The justification for clinical funds was the proposed use of F-18 FDG studies to select the sites of origin of epileptic seizures that could be removed at surgery, a use of PET first carried out by Dr. David Kuhl of UCLA. Dave delivered a lecture at Hopkins to present his results. He began his career at the University of Pennsylvania, but moved to UCLA because they had a cyclotron. He subsequently became head of nuclear medicine at the University of Michigan.

While conventional drug therapy is effective in controlling seizures for most of the 2 million people with epilepsy in the United States, 200,000 persons suffer from epileptic seizures that are not controllable with drugs. Half of these patients, if identified, are candidates for surgical removal of the brain regions where the seizures begin. At surgery, it is important to avoid removing any brain tissue which control vital language or other mental functions. The offending focus creating the seizures can be identified by inserting 4 to 6 electrodes deep within the brain. PET, which is less invasive, can often be of great help in selecting regions of the brain that can safely be removed. On July 31, 1981, we were able to purchase a NeuroECAT scanner from EG & G Ortec in Oak Ridge, Tennessee, having worked out the details of payment with Terry D. Douglass, President. Our journey had begun.

In the mid-1930s, Henry H. Dale had demonstrated that the process of transmission from nerve to muscle was not electrical as everyone had thought, but was chemical. For this work, he received the Nobel prize in 1936. Neurotransmission involves the movement of chemical "neurotransmitters" across a space between nerve endings that are about 150 to 200 angstroms apart. One angstrom is one ten-millionth of a millimeter. A neurotransmitter, such as dopamine, binds to specific receptors on the next neuron, which stimulates the firing of the neuron containing the receptor on its surface. Drugs can block the receptor on the post-synaptic neuron, thus preventing the transfer of information. Curare, for example, acts by blocking the receptor sites for the neurotransmitter acetylcholine at the junction between the nerve endings and muscle cells.

Our goal was to develop chemical "markers" to detect vulnerability to disease before deterioration of the brain has begun. After we had successfully imaged dopamine receptors in the brain of a living human being, we set out to try to answer these questions:

How do the manifestations of mental disease relate to brain chemistry?

How is dopamine receptor activity related to perception of the outside world?

How do perceptions become transformed to personality?

Is there a relationship between brain chemistry and violence?

How does the world of images relate to the world of mental functioning?

Can we tell the difference between the brain chemistry of depression and dementia?

Will assessment of dopaminergic function by PET become as common in mental illness as radioiodine uptake studies in thyroid disease?

Will patients with mental illness be characterized by radiotracer studies?

Can some mental illness be correlated with changes in dopamine receptors?

Will stimulation or blockade of dopamine receptors with drugs be related to observable behavior?

On June 7, 1988, I answered some of these questions in a lecture entitled "From Black Bile to Dopamine: The Search for Sanity" on the occasion of being awarded an honorary doctoral degree from the Georg August University of Gottingen. The laudation was given by one of our former visiting Professors, Dr. Dieter Emrich, one of the most distinguished nuclear medicine physicians in Germany. One year after the first imaging of the dopamine receptor in the human brain, we had been able to image opiate receptors, a decade after opiate receptors had been first identified in mammalian tissue by Pert and Snyder, Simon et al., and Tereniuo.

The tracer carbon-11 carfentanyl (4-carbo-methoxy-fentanil) was the first and carbon-11 diprenophine the second tracer to successfully image opiate receptors by PET. Carfentanil binds chiefly to the mu type of opiate receptor, which is involved in perception of pain. The potency of binding of carfentanil is 7,000 times greater than morphine, which explains why carfentanil, and not radiolabeled morphine, was successful in imaging the mu receptors. Naloxone and related drugs were able to block the accumulation of C-11 carfentanil.

Imaging of opiate receptors by PET has made it possible to address problems that cannot be addressed in experimental animals. The body's own opioid peptides or their

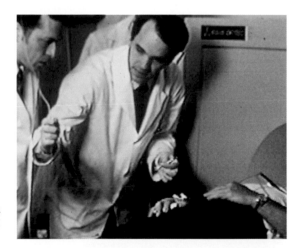

Figure 169 First imaging of opiate receptors in May, 1984 in a living human being. HNW as the subject and C-11 carfentanil the tracer. James Frost made the injection.

Figure 170 Distribution of opiate receptors in the living human brain, imaged with C-11 carfentanil in May, 1984.

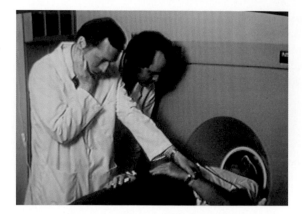

Figure 171 Solomon Snyder and James Frost during the first imaging of opiate receptors in the brain of HNW.

Figure 172 Faculty of nuclear medicine at Johns Hopkins in the 1980s.

receptors seem to be involved in depression, mania, bipolar illness, Parkinson disease, schizophrenia, senile dementia, pain, and drug addiction.

On March 15, 1985, Mary Knudson, a reporter for the *Baltimore Sun*, wrote: "Out of these studies, the scientists said they hoped to be able to develop drug therapies tailored to an individual's needs."

In 1988, James Frost and colleagues reported that mu opiate receptors are increased at the site of origin of temporal lobe epilepsy. This could be an anticonvulsant response that limits the spread of electrical activity to other parts of the brain.

Claude Bernard said: "Every time a new and reliable means of analysis makes its appearance, we invariably see science make progress in the questions to which this means of analysis can be applied."

Figure 173 Claude Bernard.

In 1985, it was possible to say: "Just as radioimmunoassay revolutionized the study of peptide hormones because of its exquisite sensitivity and specificity, so also does the use of positron emission tomography permit quantitative assay of the picomolar quantitites of neuroreceptors within the living human brain. Now that we can locate and quantify these low concentrations, we can begin to probe the chemistry of the human mind."

In 1986, we described a simple two detector system for neuroreceptor studies in the brain. The dose of radiotracer used was 1/10 that needed in PET imaging studies, so that multiple studies could be carried out in the same persons. The system was used in a definitive study to show the duration of action of a new opiate antagonist, Nalmefin, in blocking opiate receptors. The study played a major role in approval of this drug by the FDA. Although the dual-detector system was patented, it never achieved use beyond Hopkins.

Another neurotransmitter, glutamate, helps explain some of the manifestations of schizophrenia. The abused drug, phencyclidine (PCP), known as "Angel Dust," can mimic some of the manifestations of schizophrenia. Normal people given PCP have slowing of movement, decreased speech, and difficulty in cognitive function. PCP and related drugs block the NMDA receptor, which plays a critical role in mental function. The NMDA receptor is related to both the positive and negative manifestations of schizophrenia, while the dopamine system is related chiefly to the positive manifestations.

In 1986, we were able to image benzodiazapine (GABA) receptors in human beings using C-11 suriclone. These receptors are involved in epilepsy, anxiety, and neurodegenerative disorders. In 1987, we imaged serotonin receptors in human beings with a C-11 LSD tracer. The studies were extended to patients with Alzheimer's disease, depression, schizophrenia, and anxiety disorders.

While we still do not know the effect of dopamine on various neuronal pathways, it seems to have a "dampening" effect on neuronal responses to other neurotransmitters, and may be involved in servo-responses, analogous to the effects of lithium.

In early studies, we found that the availability of D2-dopamine receptors in the basal ganglia decreased with advancing age, perhaps related to decreasing motor performance and emotionality. Cerebral blood flow in general, but not cerebral oxygen metabolism, declines with age.

Subsequent to our work with C-11 N-methyl spiperone, other tracers that bind to dopamine receptors include C-11 raclopride, iodine-123 iodalysuride, and several others have been developed. Of greatest interest are serotonin, benzodiazapine, opiate, nicotinic, muscarinic, and sigma receptors. Today, assessment of the state of these receptors can help in the early diagnosis of diseases such as Alzheimer and Parkinson disease.

In June 1986, I concluded: "Positron emission tomography has until now been primarily a research tool in the quest for knowledge about the human mind. It is on its way to become part of the decision-making process in the care of patients with epilepsy (to find a focus in the brain of the seizures), dementia (to distinguish early Alzheimer's disease from benign forgetfulness), brain tumors (to locate and determine their extent), and stoke (to assess the severity of brain involvement)."

The greatest benefit of PET studies of neurotransmission is to help develop more selective therapeutic agents. Knowledge about the process by which information is transmitted from one nerve to another provides insight into the effects of widely used drugs, as well as drugs under development, and helps our understanding of mental diseases. This knowledge in a specific patient can help select, plan and monitor treatment. Drugs can act by blocking a receptor, amplifying the transmission process, or blocking the reuptake of the neurotransmitter into the presynaptic neuron. For example, the widely used drug levodopa improved the lives of patients with Parkinson disease, and analogues have improved effectiveness while decreasing unwanted side effects. The D2 subtype of dopamine receptors on the post-synaptic neurons of the basal ganglia of the brain, particularly the putamen, are defective in Parkinson disease. Stimulating the D2 receptor improves patients with Parkinson disease; antagonizing the receptor makes the patient worse.

Over $25 billion a year, equal to the amount spent by the National Institutes of Health, is spent by the pharmaceutical industry in research and development. Some complain that much of this effort results in "me too" drugs and too few genuinely new ones. Producing new drugs to the point of approval by the Food and Drug Administration and commercial availability is a complicated, expensive and risky business. For example, GlaxoSmithKline has invested $4.5 billion and the efforts of 16,000 people to develop 82 new drugs, 44 of which are moving into the late stages of clinical trials. Both the number and originality of new products has fallen in recent years. In after spending billions of dollars on research, the cost of development to achieve a successful new drug has been estimated to be $100 million for drugs to fight infectious diseases or rare conditions, and over $800 million producing drugs used for treating chronic diseases. Many pharmaceutical companies are looking to molecular imaging, especially PET, to cut costs in terms of time and money. PET not only can reduce the cost of drug design and development, and help select those patients who are likely to benefit by specific drug treatment, but also help demonstrate how more rational drug treatment can help reduce the costs of care. We and others in nuclear medicine continue to work with industry and the FDA to obtain approval for surrogate markers, that is, molecular markers that correlate well with symptoms, such as memory loss or movement disorders. The ability to validate such surrogate markers will greatly facilitate and decrease the cost of drug development. regulatory approval for their use.

On June 22, 1987, Dr. Frank E. Young, Commissioner of the Food and Drug Administration, introduced the concept of a "treatment IND" to address the problem of caring for patients with life-threatening illnesses for which no effective therapy exists. It implemented a fast-track approval process. On March 1, 1988, I proposed to the FDA to extend the concept to "radiotracer INDs" in order to provide a mechanism for drug developers to provide promising experimental drugs used in the diagnosis of patients with life threatening diseases or serious conditions before complete data on the drug's efficacy are available to enable full approval for full commercial distribution. Efforts are being made by the FDA to simplify regulations for radiotracers used in medical diagnosis and research.

Figure 174 One of many meetings of the Johns Hopkins/Japanese Nuclear Medicine Society in Tokyo.

Peace Through Mind/Brain Science

In 1986, I was visited by Mr. Teruo Hiruma, who with three colleagues had founded a company called Hamamatsu Photonics, K.K. (HPK) in Hamamatsu City, Japan, and had introduced television into Japan. Mr. Hiruma recognized the increasing focus on photonics technology in medicine. He believed that in the last half of the 20th and in the 21st century, light would be what electronics had been in the early 20th century. With 1,500 employees in his company today, and with sales of hundreds of millions of dollars, HPK has been an enormous success, creating pioneering designs and manufacturing most photomultiplier tubes used in nuclear medicine imaging. In 1986, Mr. Hiruma had set his sights on designing and building the most advanced PET scanner in the world, and he asked my help in establishing a Japanese/American Mind/Brain Imaging Program (MBIP). The program would be "based on the application of photonics technology to the study of the relationship between the brain and the mind in living human beings." At the time, the company was in the process of developing a high resolution positron emission computed tomography scanner, incorporating many improvements over state-of-the-art PET scanners. HPK was not at that time a manufacturer of PET scanners, but was a major supplier of components of PET scanners being built all over the world.

At about the same time, there was considerable economic tension between the United States and Japan. Harold Brown, chairman of the Foreign Policy Institute of Johns Hopkins University, promulgated the idea that better relations in the commercialization of new technology "would be very helpful in alleviating economic tensions." A "Human Frontier Science Program" (HFSP) was founded to help the Japanese communicate with and acquire the expertise of foreign countries in the basic sciences of biology throughout the world. At that time, the United States and Europe were far more advanced than Japan in biochemistry, pharmacology, and brain research.

In January 1986, Mr. Hiruma met with his Hamamatsu colleagues, leaders of the National Institute of Radiological Sciences in Japan, and nuclear medicine physicians and scientists at Johns Hopkins to discuss the establishment of a joint Mind/Brain

Figure 175 Tereo Hiruma, founder and President of Hamamatsu Photonics K.K., and Henry N. Wagner, Jr., in Hamamatsu City, Japan.

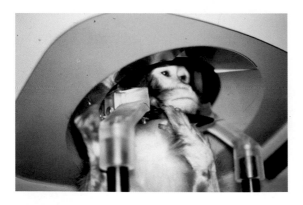

Figure 176 PET scanner designed by Hamamatsu Photonics for the study of unanesthetized monkeys.

Figure 177 Richard Wahl, first recipient of the Henry N. Wagner, Jr., Professorship in Nuclear Medicine at Johns Hopkins.

Figure 178 Sydney Brenner, winner of the Nobel prize for his work on deciphering the genetic code. Sydney continues to play a major role in the symposia on Peace Through Mind/Brain Science sponsored by Hamamatsu Photonics K.K. beginning in 1987.

Imaging Program. Mr. Hiruma was attracted to the idea that the suspicions and fears of the Cold War might be alleviated by collaborative efforts of scientists in the study of the brain chemistry of fear, violence, and war, particularly with the use of positron emission tomography (PET), which was able to examine neuroreceptors in the living human brain in health and disease. Questions that would be addressed harken back to my days working with Dr. Richter. We found major differences in the the endocrine system between wild and domesticated Norway rats. Today, it would be interesting to search for potential psychotropic drugs that might be able to transform the violent behavior of the wild rats, and make them act more like their domesticated cousins.

I told Mr. Hiruma in 1987 that we at Hopkins would do all that we could to help Hamamatsu scientists design a new generation PET scanner, and work with HPK to develop a plan to submit to the Japanese Ministry of Science, proposing the establishment of a Mind/Brain science program in Japan. This collaboration resulted in the founding of the Research Foundation for Opto-Science and Technology in Japan on December 23, 1988. Its goal was "to provide assistance and recommendations to support basic investigations and research in optical science and technology in an effort to promote greater understanding of light and the creation of new science."

We decided to sponsor a yearly Symposium in Japan entitled: "Peace Through Mind/Brain Science" in order to meet the goal of developing and improving photonics science and technology. The foundation would faster the exchange of ideas among scientists and scholars from all over the world. These symposia have continued to be held every year or eighteen months. The first was in May, 1988. Discussion at these conferences emphasized new imaging technologies, including positron emission tomography (PET), that make it possible for the first time to relate human thought, emotions, and behavior to measurable chemical reactions within the living human brain. At these meetings the goals of the program were established:

1. To develop the concept that peace will depend on a better scientific understanding of the human brain and its role in behavior.
2. To facilitate scientific and technological advances in the peaceful applications of photonics—the science and technology of light—to the solution of human problems.
3. To facilitate international communication among basic and clinical neuroscientists; manufacturers and users of photonics technology; scientists, the public, and political leaders.
4. To test the hypothesis that mind/brain science can help explain violent behavior. Can violence be explained not only as socioeconomic and political phenomena, but also as a form of mental illness, subject to investigation by means of biochemical, electrical and psychological methods?

The great psychologist/philosopher, William James, stated in his book *Principles of Psychology* that "chemical activity must of course accompany mental activity, but little is known of its exact nature." PET made it possible to explore the relationship between brain chemistry and behavior. Claude Bermard stated, "Every time a new and reliable means of analysis makes its appearance, we invariably see science make progress in the questions to which this means of analysis can be applied."

In a letter to his friend, Nobel Laureate in Physics Alexander Mikhailovich M. Prokhorov, in Moscow on July 19, 1988, Mr. Hiruma wrote: "I see still big suspicion between country to country. Especially between USSR and Japan... If the world continues to develop the science to kill people, in long future, the risk to destroy whole world may increase year by year ...

"When we say 'peace activity' in the past, this means propaganda for or against something like nuclear bomb. This is indirect ... We want to make peace scientifically ... Our peace activity is completely different from the past one ... Please help me and guide me if you agree with my dream."

On October 19, 1989, Mr. Hiruma, President of Hamamatsu Photonics, sponsored a dinner at the National Academy of Sciences in Washington, D.C., to honor former Senator William Fulbright. Dr. D. Miwa, Chairman of the Board of Directors of the Research

Foundation for Opto-Science and Technology said: "A special word of thanks is due to Senator David Pryor for making it possible for us to express our great appreciation for what Senator Fulbright has done for human beings all over the world . . . We owe our great thanks to Senator Fulbright for putting his great idea (Fulbright scholarships) into practice in 1946, a time when we Japanese needed the help of persons, such as Senator Fulbright."

In 1939, when he was just 34 years old, William Fulbright had been named President of the University of Arkansas. He was elected to Congress in 1942, and to the U.S. Senate in 1944. In 1945 he introduced legislation, revised in 1961 as the Fulbright/Hays Act, that sponsored academic exchanges to promote better understanding between Americans and people of other countries. At the time of the dinner in 1989 over 170,000 scholarships involving 120 countries had been awarded.

The dinner was attended by the Ambassadors of Japan and the Soviet Union and their science advisors, as well as invited scientists from the U.S., Japan and the Soviet Union, including Dr. Nataliya Petrovna Bekhtereva, Director of the Leningrad Institute for Experimental Medicine (a position formerly occupied by Pavlov).

The shared vision of the Research Foundation for Opto-science and Technology was that: "Radioactive tracers may be what it takes to increase our understanding of the emotions of fear, violence, and destructiveness so that we can diminish the dangers of nuclear war. PET (positron emission tomography) studies of the human brain can help us understand better the chemistry of fear, aggression and violence, so that we can direct our energies in safe and constructive directions toward further human progress rather than a nuclear holocaust." The Cold War ended in 1991, but the same vision of peace through science can help solve the problem of terrorism, so rampant today throughout the world.

On March 16, 1987, Mr. Tereo Hiruma, Mr. Manuel Dupkin, a Hopkins trustee, and I visited Steven Muller, President of the Johns Hopkins University, to obtain his approval to establish a foundation in Japan to include a collaborative effort of Johns Hopkins and Hamamatsu Photonics in the establishment of a Hamamatsu/Hopkins MIND/BRAIN SCIENCE IMAGING PROGRAM. It would include the development of a greatly improved positron emission tomography (PET) scanner, better than any in existence at that time. The program was enthusiastically approved, but a problem arose because of its focus on the Johns Hopkins School of Public Health. In reviewing the proposed agreement, the Dean of the Johns Hopkins School of Medicine wrote that "this machine should not be used for patient care and/or clinical research . . . All patient care and/or clinical research should be done under the aegis of the School of Medicine (unless, of course, we are explicitly asked to do so otherwise by our colleagues in Medicine)."

At that time, it was jokingly said that "the widest street in Baltimore was Wolfe Street," which was between the Johns Hopkins Hospital and the School of Public Health. Today, there is close collaboration between the Medical School and School of Public Health, and biomedical imaging has become a major focus of the Department of Environmental Health Sciences. A Mind/Brain Institute was established on the Homewood campus, with the emphasis on magnetic resonance imaging (MRI) rather than PET.

The year after our visit, our progress in establishing a joint Mind/Brain Institute between Hamamatsu and Hopkins was hampered by the politics on the Homewood Campus of Hopkins during 1988. After his being on the top of the world as President of

Figure 179 Manuel and Carol Dupkin, responsible for the establishment of the Henry N. Wagner, Jr., Professorship in Nuclear Medicine at Johns Hopkins.

the University for 16 years, President Muller faced the problem that the College of Arts and Sciences was running in the red. This necessitated cutbacks in faculty, suspension of building commitments, and austere budgets. The rich, money-making divisions—the schools of medicine and public health—were taxed to make up deficits.

At the time of our visit to President Muller in his Homewood office, we had to walk through a group of about 15 students lying on the floor at the entrance to his office. They were protesting the administration of the budget of the Hopkins School of Arts & Sciences. The establishment of a joint Mind/Brain Institute was put on hold. Two deans on the Homewood campus tacitly criticized President Muller's administration to the Trustees of the University. President Muller's predecessor Dr. Milton S. Eisenhower had warned Dr. Muller before he took on the Presidency that if he had not made strong enemies—on the faculty, within the administration, among the students—after five years, he was not doing his job.

In 1990, a building dedicated to PET imaging research in animals was built at Hamamatsu Photonics, followed by another building with PET scanners to examine patients. Because Mr. Hiruma wanted to strengthen his relationships with the United States, his company, Hamamatsu Photonics, and Queen's Hospital in Honolulu agreed to introduce the first PET scanning program in Hawaii, using the new Hamamatsu PET scanner. This instrument is today used primarily for clinical studies in oncology, but also has provided research opportunities for Hamamatsu engineers. The refinement of the new PET scanner was a great help. This research was in the context of an agreement described in a letter to Mr. Hiruma on May 8, 1996. In the agreement was the plan to deliver a newly designed

Hamamatsu PET scanner to Queen's Medical Center in July 1997. The agreement was signed with Governor Ben Cayetano of Hawaii in attendance.

Figure 180   (l to r) HNW, Governor B. Cayetano, Tereo Hiruma, and Mark Coel, head of nuclear medicine at Queen's Hospital.

In November 1986, an International Conference on High Technology and the International Environment was held in Kyoto. Shortly thereafter, we in nuclear medicine at Hopkins proposed on January 27, 1987, a Mind/Brain Imaging Program to advance the application of the technology called photonics (the science of light) to biomedical research. Included among the goals was to develop independent Mind/Brain Institutes in Japan and at Hopkins. Positron emission tomography (PET) was to be a major focus, drawing on the innovative work being carried out at Hamamatsu Photonics K.K. The response of President Steve Muller to the initial discussions of this proposal had been quite encouraging on October 28, 1985.

On October 6, 1987, Mr. Hiruma invited 20 scientists and engineers from the United States, the Soviet Union, and Japan to the first Mind/Brain symposium in Hamamatsu City. In his opening remarks, Mr. Hiruma said: "Recently, there has been a growing argument that Japan should be responsible for its own security, and increase its military budget . . . I cannot agree with this . . . What we are trying to do is establish a program called 'Mind/Brain Science' as a means to a solution . . . By undertaking a Positron Emission Tomography project, as well as creating various photonic technologies, we hope there will be a generation of key scientific knowledge to elucidate why mankind must be in conflict at all times." Among the attendees at this meeting were Dr. Donald A. Henderson, Dean of the Johns Hopkins School of Public Health, Academician Natalya

Bekhtereva, who succeeded Pavlov as Director of the Leningrad Institute for Experimental Medicine, and two other scientists from the USSR. Dr Bechtereva had previously visited Hopkins to see our "PET lab."

In 1987, an $80 milllion PET Center and Mind/Brain Imaging Program (MBIP) were established in the Hamakita Research Park, a 40-acre site owned by Hamamatsu Photonics, near existing company facilities in Hamamatsu City. In addition to the technical advances made since then in PET and optical imaging systems, a series of conferences entitled "Peace Through Mind/Brain Science" have been held over the past two decades. The sponsor of the symposia is the Research Foundation for Opto-science and Technology.

Among the numerous breakthroughs originating in the international symposia was the design (and subsequent construction in several countries, including the United States and Japan) of an optimal animal scanner, which took place between February 8 and 13, 1989. Among the participants was Terry Douglass of CTI, who with Simon Cherry, subsequently played a major role in designing and building animal PET scanners, which have been of enormous value to the development of pharmaceuticals, and have been embraced by the pharmaceutical industry throughout the world.

Figure 181 Professor Natalya Bekhtereva, Director of the Leningrad Institute for Experimental Medicine in Leningrad, who was influential in the Peace Through Mind/Brain Symposia at Hamamatsu Photonics, K.K.

In a hand-written letter to me Dr. Bechtereva wrote was: "What are the physiological changes in the brain when an emotional reaction is developing in an emotionally balanced as well as disbalanced (sic) brain?"

"... war for me has the ugly face of the blockage (siege) winter of 1941–1942 in Leningrad where thousands of people died every day of bombing, hunger and cold. I developed then a very intensive hatred towards any kind of war and this feeling never extinguished."

When Dr. Bekhtereva and her colleagues introduced PET imaging into their institute, I helped them get a COCOM waiver from the U.S. State Department for a PET scanner, which required a waiver in order to export a computer to the USSR. On April 30, 1990, Dr. Bekhtereva wrote to me: "I hope that when (and if) we will be ready to celebrate the opening of our Brain Center of the Academy of Sciences of the USSR you will be able to visit us ... The building for PET is nearly ready, many of the PET components are already installed ... Dealing with the living human brain, one has to keep in mind the first law—to help patients."

In accomplishing this, we worked in collaboration with Simon Serfaty, Executive Director of the Johns Hopkins School of Advanced International Studies. Interestingly, when I described the activities of our Peace Through Mind/Brain Science program, begun in March 1987, to him, he responded in a letter to me dated March 15, 1989: "A basic premise of your endeavor—that brain research is a key to international peace—will be met with a great deal of skepticism by my professional colleagues ... To convince a good number of my colleagues that the key to world peace lies in greater understanding of the human brain would require some rather substantial demonstration."

This brought to mind the aphorism: "The sale begins when the customer says no." We in nuclear medicine accept the fact that the origin and function of the human mind will remain mysterious, but believe it is worthwhile to try to use PET, SPECT and molecular imaging to relate mental functions to the molecular changes that we can now examine in the living human brain. Both the state of molecules and molecular processes can be examined in relation to violence and beneficence just as they can now be related in movement and cognitive disorders. Is it not worthwhile to at least view the possibility that paranoia, insecurity, fear, rage, violence, and aggression may have molecular manifestations in the brain? Perhaps molecular imaging of the chemistry of the brain and its relationship to behavior can, in the words of Albert Einstein, "change our modes of thinking in order to rid ourselves of the war mentality".

To quote Woody Allen: "More than at any other time in history, mankind now faces a crossroads. One path leads to despair and utter hopelessness. The other, to total extinction. Let us pray we have the wisdom to choose correctly." In 1990, the U.S. Congress declared the 1990s the "Decade of the Brain."

Positron emission tomography (PET) continues to increase its applicability to biomedical research and health care. Intercellular communication is now a major focus of medicine, with PET playing a major role in its characterization. PET images integrate molecular and cellular interactions in the living human body. Two genetic principles focus the attention of those of us who work to relate phenotypes to genotypes: pleotropism—a single abnormal gene can affect many parts of the body, and genetic heterogeneity that is, different abnormal genes can result in the same clinical manifestations (clinical phenotype). Measuring regional molecules and molecular processes serves as a bridge between genotype and phenotype. Genes can cause disease, but molecules determine the manifestations of disease. A person's genome is his map; PET examines his territory.

Radioactive tracers, such as glucose, fatty acids, amino acids, peptides and nucleic acids characterize or regulate the growth and development, regeneration and repair, and the response to treatment. Particularly when the treatment is chemical, the characterization of the patient's problem should be chemical. A reasonable question is: what the

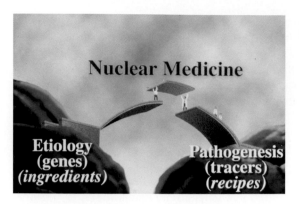

Figure 182 Nuclear medicine bridges genotype and phenotype.

relationships of PET to magnetic resonance imaging? When I first learned about magnetic resonance imaging, I was fascinated. Here was a technology that was: (1) molecular; (2) not associated with ionizing radiation; (3) capable of very high spatial resolution. I communicated with Paul Lauterbur, who was subsequently to receive the Nobel prize for his work, visited him, and was invited by him to join the new Society of Magnetic Resonance Imaging in Medicine, of which Paul was the first President. The first meeting was held August 16–18, 1982, in Boston.

Upon my return, I helped Dean Richard Ross get NMR imaging started at Hopkins, chairing an MRI committee that recruited Jerry Glickson as the first director of NMR. I eventually became somewhat disillusioned with MRI when I learned of the insensitivity of the method for assaying molecules in vivo, and realized that the high spatial resolution was only achievable for hydrogen imaging. I directed my interests back to full time nuclear medicine.

I still ask: what is the brain chemistry of fear, disgust, despair, and aggression? Fifty years ago, as a medical student working in the laboratory of psychobiologist, Curt P. Richter, I would put wild Norway rats captured on the streets of Baltimore in the same cage with Norway rats bred for laboratory research. They did not disturb each other until we applied electric shocks through the floor of their cages. The wild rats then killed the laboratory rats. We found differences in their endocrine systems, strikingly in the adrenals. Today we can begin to look at the chemistry of the brain related to violence and war. Perhaps the greatest impact of molecular imaging of the brain in the 21st century will be to help bring about world peace.

The time has come to consider violence a public health disease, and search for its causes and treatment. Over one million persons die each year in the United States as a result of suicide or homicide. The leading cause of death in teenagers is gunshot wounds. We can now characterize chemical manifestations of neuropsychiatric diseases. Radioactive tracers are playing an increasing role in drug design and development. The era of histopathology as the dominant tool in medicine is coming to an end. A single biopsy specimen reveals only the state of molecules within cells a small sample of tissue. Radiotracer studies make it possible to go far beyond structure into the domains of function and regional biochemistry. We do not just provide new tests for old diseases, but new ways of defining and detecting disease.

14

The Fight Against Infectious Disease

When I was an assistant resident on the Hopkins house staff in internal medicine decades ago, serious infectious diseases were very prevalent. Fortunately, penicillin and broad spectrum antibiotics, such as streptomycin, tetracylines, chloramphenicol, erythromycin, and vancomycin, had become available. The single most important examination was the Gram stain. One could determine immediately whether the offending organism was Gram-positive or Gram-negative. In patients with pneumonia, antibiotic therapy was begun immediately on the basis of clinical examination and sputum smear, and then altered in the light of the culture reports and clinical response.

One of my first clinical research projects was the study of staphylococcus endocarditis, i.e. infection of the heart valves. During the 20-year period 1933 through 1953, there were 104 instances of staphylococcal bacteremia. None of the patients survived. Thirteen patients with staphylococcus endocarditis were treated during the five year period from 1949 through 1953. The incidence of penicillin resistance of the organisms was 42%. Fifty-four percent of the patients survived.

My colleagues and I carried out studies in patients with pneumococcal pneumonia admitted to the ward services of Hopkins and Baltimore City Hospitals between October 1, 1954, and May 31, 1955. We examined the effect of hydrocortisone on the course of pneumoccal pneumonia treated with penicillin. Fifty-two patients were treated with hydrocortisone and penicillin, and 61 control patients with pneumonia were treated with penicillin only. There was faster disappearance of fever, greater comfort and faster improvement in symptoms in those patients who received hydrocortisone as well as penicillin. There were no complications resulting from the hydrocortisone administration.

One of the early studies in nuclear medicine was the use of gallium-67 citrate in patients with infection. Gallium is bound by *transferrin*, a protein circulating in blood, moves into sites of infection, and is bound by lactoferrin in white blood cells. The tracer accumulates in the liver and bone marrow, and is excreted into the intestines. Gallium-67 had undesirably high photon energies, and high radiation doses, and subsequently was replaced by other tracers. Human immunoglobulins (HIG), labeled with technetium-99m and indium-111, accumulate at the site of infections.

Radiolabeling of autologous leukocytes were first used to locate infections by McAfee and Thakur in 1976, because there is considerable accumulation of the *ex vivo* labeled leukocytes at the sites of infection and inflammation. Other tracers include labeled antigranulocyte antibodies and antibody fragments, Tc-99m labeled liposomes, such as HMPAO and HYNIC, receptor-binding peptides that bind to leukocytes, chemotactic peptides, cytokines, interleukins that bind to granulocytes, monocytes and lymphcytes, platelet factors, and anti-E-Selectins.

In 1990, in the United States, at least a million people were infected with the human immunodeficiency virus (HIV). This virus destroys a type of white blood cell, called "T-cells," and macrophages critical to the functioning of the immune system. This results in the patient's increased susceptibility to infections, in a disease called acquired immune deficiency syndrome (AIDS). Many of these patients can be helped by localizing the site of infections by positron emission tomography (PET) with F-18 fluorodeoxyglucose (FDG) and other tracers.

Although its primary use has been in oncology, FDG-PET is beginning to be used more and more in the care of patients with infectious disease. FDG is an agent for imaging regions of increased glucose utilization, which includes infection and inflammation. Much of the accumulation of FDG is due to the increased numbers of granulocytes and macrophages at the sites of infection and inflammation The high degree of accumulation and the increased spatial resolution and quantifiability of PET provide important advantages over imaging with labeled leukocytes, and is gradually replacing the use of labeled leukocytes. More than 5,000 different types of viruses, 100 species of fungus, more than 300,000 species of bacteria, and innumerable parasites threaten human health. More than 14 million people worldwide die each year from infections. Tuberculosis and malaria account for hundreds of thousands of death each year. Often a patient will have fever and other signs and symptoms of an infection, but its site will not be known. F-18-FDG PET can detect hidden infections, such as tuberculosis or abscesses. In fevers of unknown origin, the site of infection can often be detected by F-18-FDG PET.

In 1969, we published a new radiometric method for the detection of bacterial growth in blood cultures, based on incubation of the blood samples in media where the only source of carbohydrate was carbon-14 glucose. The evolved carbon-14 dioxide was a sign of positive bacterial growth. By 1971, we had examined 2,967 blood cultures in 1,280 patients suspected of bacteremia. The radiometric method detected 102 positive cultures from 40 patients. Of these the routine microbiological method detected 98 positive cultures from 40 patients. Seventy percent were detected first by the radiometric method, 65% on the day of inoculation. Under the label "Bactec," this method became widely used throughout the world. One of its most important contributions is in the detection and sensitivity testing for *Mycobacterium tuberculosis*, now a major international problem. The conventional techniques require two weeks for detection of tuberculosis, while the Bactec method takes only 48 hours.

We also developed the Limulus test for pyrogens based on the coagulation of lysates extracted from circulating cells of the Limulus horseshoe crab, prevalent all over the beaches on the Atlantic coast. These lysates form gels in the presence of bacterial endotoxin in exceedingly low concentrations. We developed a test for pyrogenic endotoxins that was 10 times as sensitive as the rabbit pyrogen test in rabbits officially approved by the U.S. Pharmacopeia (USP).

The development of this test is an example of how problems are often the stimulus for scientific advances. The problem we were addressing was that the rabbit test for

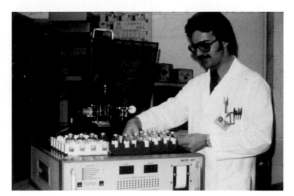

Figure 183 Tomas Guilarte with an early instrument for the automation of microbiology by monitoring C-14 carbon dioxide release.

pyrogens approved by the FDA takes 24 hours. We were routinely producing radiopharmaceuticals with half-lives as short as 20 minutes. Limmulus not only solved our problem, but it was eventually accepted for testing all pharmaceuticals, and is now used routinely for all intravenously injected pharmaceuticals. The development of this test shows how "exotic" basic science studies often have important practical benefits.

15

A New Approach to Disease

In his book, *The City of God*, St. Augustine wrote: "Some medical men who are called anatomists, have dissected the bodies of the dead to learn the nature of disease. Yet those relations which form the concord, or as the Greeks call it, the harmony of the whole body, no one has been able to discover because no one has been audacious enough to seek for them."

We who have spent our professional careers in nuclear medicine have looked at "slices of life," examining the regional function and chemistry of the living human body in health and disease. Our work is leading to new definition of disease.

Despite its birth over a half century ago, nuclear medicine is now based on the concept that diseases can be characterized by *in situ* molecular abnormalities. Our nanoDx probes search out the molecular manifestations of disease. We then use nanoRx radioactive drugs to correct them. In the late 1960s and early 1970s, Drs. Merrill Bender, Joseph Ross, David Kuhl, W. Newton Tauxe, and others struggled to have nuclear medicine recognized as an independent specialty. We succeeded in founding the American Board of Nuclear Medicine in 1971.

Questions that had to be addressed included: (1) Should nuclear medicine be a subspecialty of radiology? (2) Will the many technological advances of nuclear medicine become a part of organ-oriented specialties, such as cardiology, oncology and neurosciences? (3) Should nuclear medicine be an independent specialty, based on the tracer principle, molecular biology, genetics, and biochemistry?

In the United States today, many young radiologists spend an additional year beyond their radiology residency requirements in fields such as neuroradiology, pediatric radiology, cardiovascular radiology, or interventional radiology. We live in an age of increasing specialization because it is impossible for a person to keep up with advances in all imaging specialties. An orientation to a single organ or organ system is based on an out-of-date concept of medicine and disease. Most often, disease does not involve a single organ, but involves the integrated biology of the entire body. For example, cancer is not viewed solely as a disease of a specific organ, but as an imbalance or failure of growth-promoting or growth-suppressing factors, including somatostatin and other proteins. Modern scientific medicine is moving away from an organ orientation, advancing by leaps and bounds by revolutionary advances in genetics, pharmacology, oncology, endocrinology, neurosciences and molecular biology, all of which provide holistic views of disease. The new, integrated, global approach of nuclear medicine goes beyond gross anatomy and pathology, even beyond histopathology, to physiology and molecular nuclear medicine. Localizing disease to specific organs is likely to remain predominantly the domain of surgeons, but many patients' problems will be viewed as regional molecular dysfunction. Molecular imaging will increasingly join histopathology to characterize disease.

Many radiology chairmen still orient their departments into organ systems, "dismantling"sections of ultrasound, magnetic resonance imaging, and computed tomography. In 1989, Dr. James S. Todd, Executive Vice President of the American Medical Association, described the typical medical practice at that time.

"The experienced doctor spends a long time listening to the patient describing in his or her own words the nature of the illness. The taking of the medical history points the doctor towards the diagnosis—especially if he or she is an experienced clinician who can compare the current problem with similar consultations.

"In many cases, your doctor will have a strong suspicion of what is wrong at the end of the consultation, and the experienced clinician will compare your problem with similar patients, but will want to obtain some clear-cut evidence—if possible—to clinch the diagnosis."

The relationship between symptoms and disease has been examined for over two centuries. Symptoms are subjective; signs are objective evidence of disease. Symptoms are usually subordinate to signs, with a continuing effort over centuries to elicit signs of abnormal organs, parts of organs, tissues, cells, and today, molecules. Psychiatric diseases are often classified according to the response of the patient to treatment. For example, people with manic-depressive or bipolar disease respond to lithium therapy, which helps legitimize it as a specific disease entity. Problems arise when several "diseases" including depression, insomnia, and psychosocial disorders respond to the same drug. The breadth of response undermines the specificity of the diagnosis.

The diagnostic process in the practice of medicine today is based on the assumption that you are not suffering from a new, unique disease, but that others have had "your" disease in the past. You feel reassured if your doctor "makes the diagnosis," that is, tells you the name of the disease that you have. This tells him "what is wrong with you." He or she is able to recognize your disease because it is similar to the same problem suffered by others, characterized by "manifestations" similar to yours. Your doctor makes the diagnosis of your illness on the basis of the manifestations of the disease as defined in medical taxonomy. In contrast to what was thought in the past, disease is not some THING inside of you. "Diseases" are abstractions defined by manifestations and created because of their usefulness in predicting what is going to happen, what can be done about it, and what caused the problem.

In a sense, making a diagnosis is based on pattern recognition, in much the same way as recognizing the face of a friend or acquaintance. Historically, an experienced doctor was often able to make the diagnosis solely on the basis of your appearance, and your answers to a few pertinent questions. Today, we realize that the process is much more complicated and difficult, especially when the patient does not recover spontaneously as the result of the body's defense mechanisms that have evolved over hundreds of thousands of years.

The great physicist Werner Heisenberg, creator of the uncertainty principle in physics, said: "What we learn about nature is not nature itself, but nature exposed to our methods of questioning." When we think about the diagnostic process, we need to remember that:

"What physicians learn is not about us as patients, but as patients whose diagnosis is revealed by our methods of questioning and testing."

In the words of Gottfried von Leibgnitz in 1703: "Nature has established patterns but only for the most part." Thus, a working diagnosis is not certain, but must be stated in terms of probability. No two patients are exactly the same. What happens to you will not be the same as what has happened to others with the same diagnosis. Even experienced physicians often have differing opinions about the "correct diagnosis."

In 1968, I proposed the application of Bayes' theorem to the diagnostic process in nuclear medicine. Its essence is that "we live forward but we can only think backward" (Kierkegaard). We must consider the *a priori* probability of a disease as well as the *a posteriori* probability based on its present manifestations.

This approach was first suggested by Ledley and Lusted in 1959. We selected the uptake of radioiodine by the thyroid in 90 patients in Hopkins Hospital as an example. The uptake of radioactive iodine indicating hyperthyroidism was incorrectly predicted by the clinical findings in 21% of the patients when Bayes' theorem was applied by computer, and in 40% when the predictions were made subjectively by physicians. This early study was a step in the development of a completely objective model for assessing a new diagnostic procedure.

Electrocardiographers took the lead in applying Bayes' theorem to the diagnosis of coronary artery disease. In 1979, Diamond and Forester combined the pretest probability of coronary artery disease, based on age, sex, and symptoms with the data from stress electrocardiography, thallium scintigraphy, and cardiac fluoroscopy, using Bayes' theorem. The use of Bayes' subsequently became very important in nuclear medicine studies.

The type of treatment depends on our approach to disease. The rise of pathological anatomy at the turn of the eighteenth and nineteenth centuries reflected the growing influence of surgeons. The genetic approach to disease today is closely aligned to the revolutionary advances in pharmacology. While surgeons searched for localized pathology, pharmacologists emphasized diseases that could be treated by specific drugs. The growth of bacteriology, immunology, pharmacology, and genetics, today, has greatly influenced the orientation of medicine. Also increasing in modern medicine is the role of public health and preventive medicine. Determination of the patient's genome and its phenotypic expression during life often represent the ultimate in specificity of diagnosis, prevention and treatment. The principle of pleiotropy and environmental influences support the essential role of quantitative assessment of phenotypes.

Monitoring of regional biochemical processes by means of the tracer principle will play an ever-increasing role in the molecular medicine of the 21st century. The science and technology of nuclear medicine will continue to lead the way by the application of tracer methods, including fluorescent as well as other stable and radioactive tracers.

Fluorescent compounds are labeled with fluorescent side chains attached to the biologically active molecule. They are used for in vitro and imaging studies in living animals or patients. The sensitivity of using tracers with fluorescent labels is a thousand times greater than with radioactive tracers labelled with iodine-125. This increase can be used to greatly increases spatial resolution in imaging studies. Fluorescent tracers can be used for spatial resolution within individual cells to study compounds that address specific biochemical pathways in pharmaceutical development. The applications of fluorescent molecules in drug discovery give powerful insights into cellular imaging, target characterization, receptor biology, protein-protein interactions, screening assays, and in vivo diagnostics. Fluorescent proteins have proven instrumental in monitoring the effects of RNA and genes in mice whose genes have been inactivated ("knock-out mice"). Their use will become increasingly important in nuclear medicine.

The use of radiolabeled and fluorescent tracers provide quantitative, specific molecular information about the patient's illness. They define the characteristics of the individual patient, a process referred to as "personalized" medical care. Today, in making a diagnosis, the physician ignores the characteristics that make you and your illness unique, and focus on those manifestations that are thought to be most helpful in solving your problems. Most characteristics are unique to you as a patient. The physician today searches for manifestations of your illness that are similar to those of other patients. Your manifestations of disease are the most important. The best physicians are those who look at you both as an individual and as a person manifesting common characteristics of a specific disease. Nuclear medicine, by its continual eliciting of more and more molecular processes, is able to characterize a person's health by more and more characteristics. Some day your diagnosis will be simply your name, defined by multiple molecular manifestations measured objectively and quantitatively, throughout all parts of your body.

Claude Bernard, a French physician from the late 19th century, is the father of scientific medicine. The driving interest of his life was "the search for an understanding, in terms of physics and chemistry, of those processes by which we live, by which we become ill, by which we are healed, and by which we die." He believed that as more is learned about the fundamental sciences of physics, chemistry, and biology, the greater will be our need to modify prior definitions of disease in the light of the new knowledge.

In his classic book, *An Introduction to the Study of Experimental Medicine* in 1865, Bernard described how he went about his research. He looked at: (1) problems; (2) ideas; (3) hypotheses; (4) experiments; (5) principles; (6) predictions; (7) control; and (8) new problems. He distinguished between "observing" physicians and "experimental" physicians.

When I was in medical school, we were taught to begin the diagnostic process by first asking the patient: "What brought you to the doctor?" or "What seems to be the trouble?" The response was called the "chief complaint." Next, the patient was asked to describe details of the present illness, beginning with the time that the patient first began to feel ill. The more symptoms that are commonly found in a specific disease, the greater the likelihood that the doctor would be able to make a "working diagnosis," and then proceed to confirm or refute the diagnosis and search for another disease.

The physician would then "take" a complete medical history, to be sure that no other significant symptoms or events were overlooked. Then, a physical examination was performed, focusing on the parts of the body likely to be involved, in the light of the "working diagnoses." Laboratory procedures were then performed to further confirm or refute the "working diagnosis," and to systematically exclude other possible diseases. The diagnosis determined the treatment. Diagnoses, like theories, were judged by how well they work. The best diagnoses are the ones that result in the best outcome, that is, to bring about a return to previous health.

Figure 184 Graduating class at Johns Hopkins Medical School in 1952. HNW third from left in third row from top.

At times, a single diagnosis cannot be made immediately, and the physician creates a "differential diagnosis," that is, a list of all possible diseases. It may not be possible to make a single diagnosis at this stage of the diagnostic process.

The diagnostic process depends on the existing concepts of disease, which change over time. The physician elicits and then relates your manifestations of disease in the context of the concepts of disease at that time. How diseases are defined is fundamental to the diagnostic process.

The so-called "ontological" concept views diseases as self-sufficient entities, running a regular course and a recognizable natural history. The biographical approach rests on the history of the patient, and the manifestations of disease in the patient as an individual person, based on anatomical, physiological, and biochemical manifestations of disease. The historian Owsei Temkin has quoted Hippocrates: "The art (of medicine) consists of three things—the disease, the patient, and the physician."

Scientific advances today make it necessary to take a new look at the diagnostic process, and therefore in the definition of disease. These include: (1) advances in medical imaging, including computed tomography (CT); magnetic resonance imaging MRI), single photon emission tomography (SPECT), positron emission tomography (PET) and ultrasound (US); and (2) advances in human genetics. These advances affect not only present illnesses, but also the detection of disease in asymptomatic persons at risk of developing disease in the future.

Most physicians and medical students believe that diseases exist, and would exist if there were no physicians to observe them. The reality is that diseases are constructs of our perceptions of disease, of our observations, of the manifestations of disease. Medical diagnosis depends on categories and definitions. They are expressed by words, and therefore are imprecise.

For thousands of years, diseases were defined by the patient's symptoms or abnormal findings during the physical examination of the patient. For example, a description of symptoms is still used to diagnose "pleurisy, and angina pectoris." Some believe that when the physician gives a name to the patient's illness, he has solved the problem.

In 1543, the anatomic studies of Vesalius (1514–1564), whose monumental work *De humani corporis fabrica*, published in 1543, resulted in anatomy becoming the basis for classifying disease. Subsequently, physiology, born with the work of Claude Bernard, and organ pathology became the focus of the diagnostic process. Estimation of the clinician's skill in medical diagnosis was judged by the ability to correctly predict what would be found at autopsy. The Clinical Pathological Conference (CPC) was the most important educational exercise for students in the second year of medical school. We learned to be diagnosticians, specialists in "making the diagnosis" of the patient's illness.

Today, diseases are still defined by location and cause. Their causes include genetic abnormalities, nutritional deficiencies, infectious microorganisms, environmental toxins, and trauma. Research into the cause of disease remains a dominant factor in medical research today, even though most diseases are now recognized as having multiple complex "causes." For example, in the case of infectious organisms, host factors are of great importance. Another example is liver cancer, where an environmental toxin, afla-toxin, and a herpes viral infection, and certain trace elements all interact before the disease is manifest.

There is not a specific disease for each microorganism, for each specific element result-ing in a nutritional deficiency, nor for every type of gene expressing a hemoglobin abnormality. No manifestation can stand alone, apart from its associations with other manifestations. The manifestations of disease, both inherited or acquired, are grouped in categories, called "syndromes", or "patterns." The connections are more important than the individual manifestations themselves. For example, there are fewer diseases than there are abnormal genes or abnormal molecular products of these abnormal genes. The ability to recognize the differences among patients is one of the most essential qualities of an expert physician, as well as of the diagnostic process. Although each patient must be considered as an individual, as stated by Lukacs, "science must depend on regularity."

Sydenham wrote that "A disease is nothing more than an effect of nature, who strives with might and main to restore the health of the patient by eliminating the morbific matter." Diseases should not be considered as "things" themselves, but as a collection of manifestations, constituting a pattern, called the "diagnosis," and based on the usefulness of such a classifications in terms of prognosis or treatment.

In the "International Classification of Diseases," six digits are used to define diseases: The first three digits describe the location of the abnormality. The next three define the disease as being in one of the following categories, each of which is designated by a three-digit number:

Infectious and Parasitic Diseases (001–139)
Neoplasms (140–239)
Endocrine, Nutritional, and Metabolic Diseases and Immunity Disorders (240–279)
Diseases of the Blood and Blood-forming Organs (280–289)
Mental Disorders (290–319)
Diseases of the Nervous System and Sense Organs (320–389)
Diseases of the Circulatory System (390–459)

Diseases of the Respiratory System (460–519)
Diseases of the Digestive System (520–579)
Diseases of the Genitourinary System (580–629)
Complications of Pregnancy, Child Birth, and the Puerperium (630–676)
Diseases of the Skin and Subcutaneous Tissue (680–709)
Diseases of the Musculoskeletal System and Connective Tissue (710–739)
Congenital Abnormalities (740–759)
Certain Conditions Originating in the Perinatal Period (760–779)
Symptoms, Signs and Ill-defined Conditions (780–799)
Injury and Poisoning (800–999)

The next three digits are based on the cause of the disease, such as a bacterium. This approach is conceptually flawed because we now recognize that most diseases involve multiple factors. It is difficult if not impossible to determine the single "cause" of a disease.

A person's genetic makeup or genome, interacting with environmental and cultural factors, result in the manifestations of disease, called the "phenotype." A person's genome is both a blueprint and template for the production of proteins and other molecules from which the person's body will be constructed and repaired continually from conception to death. Knowledge of the person's genetic makeup provides an enormous amount of information about what is likely to cause future health problems or be the cause of present problems. Mutations can be present in the genes that survived the evolution of the human species or as a result of subsequent mutations occurring later in life, called "somatic" mutations. The relationship of specific genes to diseases and the risk of certain diseases is of intense research interest. Certain diseases, such as sickle cell disease, are clearly correlated with an abnormal gene and its product protein. In the cases of diseases in which genetic factors play a major role, the National Institutes of Health provides a database in which 10,000 distinct genetic conditions are listed; some are "recessive," the result of two abnormal genes, one from each parent; others are "dominant," that is, the result of a single abnormal gene; or sex-linked abnormalities on the X chromosome.

An important question is whether a person without symptoms, but with a documented genetic abnormality, should be considered as having a disease. Clearly, if a person has an elevated blood pressure or blood sugar, even without symptoms, that person is considered to have a disease. Identification of a genetic "disease" may lead the affected person to change life style, avoid specific environmental or occupational hazards, avoid cultural stress, and make dietary changes. The effect of such a finding on insurability remains a political problem.

Today, most diseases are defined on the basis of abnormalities in gross pathology and histopathology, as well as microbiology and biochemistry. Advances in medical imaging in the past few decades are now being incorporated into the defining of specific diseases. The characterization of regional molecular abnormalities provides a way to define disease at the molecular level, as well as at the level of the whole body, organ, tissue or cellular level. The task now is to define the quantitative parameters of a specific regional molecular process, that is, to set the limits of normality, and define a molecular disease quantitatively. The totality of these molecular phenotypes varies from person to person, and can define the person's state in health and disease. Manifestations of disease are increasingly elicited by our outstanding new tools.

One of the important characteristics of molecular medicine is the ability to express these manifestations of disease quantitatively. This makes it possible to establish quantitative criteria for the definition of normality and disease. Most measurements among different persons, healthy or ill, fall along a so-called "bell-shape" curve, first described by Lambert Quetelet and Carl Frederick Gauss.

Another major characteristic of nuclear medicine is that it measures processes, i.e., physiology, rather than states, i.e., anatomy or histopathology. An essential part of most languages is that the verbs have a tense. They distinguish between past, present and future. Anatomy can be thought of as Q, while physiology and nuclear medicine are concerned with Q/t, where t = time. Thus, nuclear medicine changes the way in which we define disease. Since the work of Vesalius in 1543, the concept that the body is made up of organs provided a directing paradigm for the diagnostic process. The challenge to the physician then was to locate the patient's problem in a specific organ. This resulted in the development of organ-oriented medical specialties, such as gynecology, cardiology, urology, and so forth. Later, function, or "animated anatomy" was assigned to organs, modifying the diagnostic paradigm, while retaining the anatomical orientation. Diseases of the skeleton were connected to diseases of locomotion. In the case of the heart and blood vessels, structure seemed quite appropriate to function. Claude Bernard, pointed out how difficult it is to deduce the function of an organ from its structure. Functional paradigms, such as endocrinology, evolved, as well as feed-back models in neurological diseases, such as Parkinson's disease. Other models are concerned with energy or information transfer, involving receptors, effectors, and reflex circuits. The electrical model achieved great popularity in neurology. This occurred even before the evolution of the cellular theory of disease, i.e. histopathology, which even today retains a dominant position in the process of medical diagnosis. Today, the new paradigm is "molecular medicine."

The number of molecular processes that can be measured by PET and SPECT is enormous. What is needed now is a systematic grouping of these molecular processes in a manner analogous to the classification of elements in the periodic table. We need a system of molecular "taxonomy," defining what is normal and quantifying what is abnormal. We need not find for "ultimate" causes of disease, but define abnormalities that are progressively antecedent, in order to make possible earlier detection of structural or functional abnormalities. Antecedence should not be confused with "cause." For example, an increased rate of accumulation of iodine is a manifestation of hyperthyroidism, but is not its "cause."

Molecules can be classified as proteins, carbohydrates, and fats. They can be classified according to their involvement in (1) energetics; (2) communication; (3) structural integrity; or (4) molecular transfers. A patient's disease(s) can be defined as involving one or more of these categories. For example, in diseases of the thyroid, the abnormalities involve genetics, i.e. increased expression of the sodium iodide transporter protein, increased accumulation of iodine, and increased synthesis of thyroxine or triiodothyronine. Another abnormality in hyperthyroidism is increased consumption of energy. Thus, the diagnostic process involves abnormal structures, physiological and biochemical abnormalities. Arthur Koestler has said: "Where there is life, it must be hierarchically organized."

This is the revolution called "molecular medicine." Only by understanding the basic biological processes involved in disease can we most intelligently and successfully intervene in the disease process. The first step is to document the molecular processes that are abnormal; then put these manifestations into a pattern of disease. It is remarkable how many basic scientific advances resulted from the introduction of radioactive tracers into biology and medicine, the essence of nuclear medicine.

At the high level of complexity of the human body, syntheses analogous to those in physics are beyond the powers of our imagination. Complexity theory, defined as the search for algorithms, is in its infancy. Nevertheless in medicine it can often be helpful to be able measure a single link in the chain of human illness in order to affect favorably what happens to the patient. Specific findings, reflecting only a very small part of the whole organism, can continue to provide important benefits in patient care.

Unfortunately at times, tests and treatment are carried beyond rational stopping points. More than one third of all medical expenditures in the United States are incurred during the last six months of life. Technological advances can give doctors and patients options that are expensive, often with sub-marginal benefits. In many cases, physicians are not knowledgeable enough to make every decision correctly. Increasing technological complexity makes it even more difficult.

The nuclear medicine specialist is a physician, not a technologist. In 1974, I wrote: "I believe that the nuclear medicine specialist should have the opportunity to speak with and examine every patient prior to the nuclear medicine study. He should know the patient's problems and the key findings. He should list the *a priori* diagnoses with the probability of each. He should plan the studies and instruct the technologists. The patient should not leave the department until the results have been reviewed and a judgment is made about how nuclear medicine techniques help solve the patient's problems. The interpretation (of the results) of the studies should be reported in terms of diagnoses and probabilities. Vague statements such as 'compatible' or 'consistent with' should be avoided. Follow-up should be carried out on a routine basis to ensure consistently adequate performance."

16

The Genetic Revolution

My first contact with genetics was inauspicious. In our undergraduate genetics course at Hopkins, there was an experiment in which we were to count and characterize fruit flies that were reproducing in flasks with cloth stoppers. Returning to the lab one day, I found fruit flies all over the outside of the flasks containing my flies. Thinking they had escaped from the stoppered flasks, I counted their characteristics and recorded the results, not realizing at the time that these flies on the outside of the flasks had gathered there after they escaped from the flasks of my classmates. Needless to say, I was unable to confirm Mendel's laws.

In 1952, while an intern at Hopkins, I attended a lecture by Francis Crick in Hurd Hall, an amphitheater in the hospital that was the scene of Medical and Surgical Grand Rounds, guest lectures, and many other activities for the faculty, house staff, and medical students. He had not yet won the Nobel prize. When I heard his conclusions about the structure of DNA based on x-ray crystallographic images, I thought that he was making the whole thing up. Perhaps this skepticism at that time and the subsequent revolution that was to follow in molecular biology account for my being so humble today.

How far we have come in the last half century can be illustrated by the results recently obtained in treating patients after surgical treatment of breast cancer. Twenty to thirty percent of the breast cancer lesions removed at surgery express a specific protein called HER2 on the surface of the cancer cells, as a result of a genetic mutation in the cells. These tumors are aggressive and often recur after surgery. The drug Herceptin blocks the action of the HER-2 protein in its action to stimulate the growth of the tumor. The HER2 can be identified in the specimens of tumor removed at surgery and the patients can be given Herceptin therapy.

Another new drug based on genetic characterization of the disease is Avastin, which is used to prevent the recurrence of colon cancer after surgery.

Genetics and pharmacology are fields with increasing interactions with nuclear (molecular) medicine. All three are shifting medicine from being concerned only with treatment toward the prevention and early detection of disease. It is predictable that identification of phenotypic characteristics that are consequences of genetic actions will be guided by the findings in an individual's genotype. Genomics, proteomics and phenome assessment will be the hallmarks of clinical medicine of the 21st century.

The genome of an individual contains the history of the human race, the specific population from which the person descended on both parental sides, and the mutations that have occurred over the course of the person's life. It is the blueprint and template for all the protein synthesis that will chacterize the person's life. It will become routine to determine whether a person's genome contains deviations from the "normal" genome, or contains mutated genes that increase the risk of disease. These findings will provide the indications for examination of the associated phenotypes at the molecular and functional level. It remains to be determined which molecular targets or processes are likely to be monitored on a routine basis in healthy persons.

A number of genetic abnormalities have been found to increase the risk of neuro-degenerative diseases, such as Alzheimer and Parkinson diseases. Mutations in three different genes (amyloid precursor protein, presenilin 1, and presenilin 2) are involved in early onset, dominantly inherited Alzheimer disease (SDAT). An important finding in SDAT is increased production and aggregation of beta-amyloid protein.

Mutations of the alpha-synuclein gene are related to Parkinson disease. Many mutations that have developed in germ cells or, after birth, in somatic cells can provide information to help predict the individual's risk of developing diseases in the future, and then serve as a guide to treatment.

Genetics involves not only diagnosis but also therapy. A major event took place on September 20, 1990, when scientists at the National Institutes of Health in Bethesda, Maryland, treated a 4-year-old girl with a disease called "adenosine de-aminase (ADA) deficiency." They injected her with some of her white blood cells that had been infected with a virus carrying a normal version of the gene which had mutated and become inactive. The mutation of this gene normally involved in the metabolism of the amino acid, adenosine, resulted in the abnormality in her immune system that resulted in her being susceptible to infectious diseases. This was the first example of "gene therapy, a new approach that holds promise in the treatment of diabetes, rheumatoid arthritis, many different kinds of cancer, and other diseases."

At 10:00 AM EDT on June 20, 2000, there was great enthusiasm when President Clinton announced to the public that the first draft of the human genome had been deciphered. This major landmark in the history of science and medicine was the result of a journey begun by the U.S. Department of Energy in the mid-1980s directed toward the elucidation of the effects of radiation on living organisms. A long term goal of the Human Genome Project (HGP), was to identify the sequence of the entire human genome by the years 2003–2005. Success came much sooner. The identification of gene targets in disease represent a major advance in the discovery, design and development of new pharmaceuticals. It remains a challenge to determine the role that specific genes and the proteins that the genes express provide as manifestations of health and disease. It did not take long for biotechnology and pharmaceutical companies to recognize the great opportunities to use genetics and nuclear imaging to help develop effective drugs. These would be used in:

1. Providing targets for developing small molecules as drugs to simulate or suppress biological processes.
2. Providing antigens as targets for antibodies or fragments of antibodies.
3. Using genomic technology, such as reporter or promoter genes, expediting the process of drug development.

The Human Genome Project depended heavily on related studies of the genomes of bacteria, yeasts, fruit flies, mice, rats, and other plants and animals. A most striking biological principle is that the genomes of living organisms are encoded by four basic nucleotides: adenine, cytosine, guanine, and thymine. In the human being, there are 3 billion pairs of these "letters" that speak the language of the genes. Only about 3% of all the DNA is known to contain information; 7% is related to regulatory functions, 90% of the DNA is not related to genes at all. The human genome consists of 34,000 genes, not many more that in the fruit fly (14,000 genes); the worm *C. elegans* (19,000); yeast (6,000); or *E. coli bacteria* (4300).

Many genes are the same in organisms as different at fruit flies, yeasts, bacteria, and even plants. Abnormal mutated genes related to human disease are similar to those found in other organisms. Human beings and mice have 200 stretches of genes in their chromosomes that are in the same sequence.

Most diseases, including heart disease, diabetes, cancer, mental illnesses, and age-related disorders, are complex, resulting from many factors in the environment and many different genes. At least 1,000–3,000 diseases result from abnormalities of a single gene, some of which have been related to single molecular abnormalities, such as sickle cell disease.

Molecular nuclear medicine involves the discovery and developing of phenotypic markers that make possible measurement of regional biochemistry in the living body, relating genotypes to molecular phenotypes in vivo. In 2001, 1,000 abnormal genes had been related to diseases (for example, familial PD, breast cancer, etc.). Patients expressing the BRAC1 and BRAC2 genetic abnormalities have an increased risk of breast cancer. Proteins, such as P-glycoprotein, increase resistance to treatment with drugs as a result of the extrusion of the drugs from cancer cells. The use of a radioactive tracer, Tc-99m sestamibi can detect p-glocoprotein expression in cancer and other cells. Molecular phenotypes can serve as therapeutic targets, for example, anti-CD20 monoclonal antibodies are now widely used to treat non-Hodgkin lymphoma.

How far we have come! In 1806, President Thomas Jefferson congratulated Edward Jenner on his invention of vaccination. Jefferson wrote that William Harvey's discovery of the circulation of the blood was a great contribution to knowledge, but he did not see how knowing about the circulation of the blood could be translated into improvement in human health.

Epilogue

Only he who keeps his eye on the far horizon can find the right road. Today, it is widely accepted that the earlier a diagnosis is made, the greater will be the benefit of treatment. Characterization of the extent of disease within the body, called "staging," helps predict what is going to happen, and helps plan effective treatment. Early demonstration of the effectiveness of treatment soon after it has begun improves care and decreases its cost. Early detection of recurrence at a later time makes possible further treatment and often prolongs the life of the patient.

The first use of a radioactive isotope for the treatment of disease was by Dr. John Lawrence on Christmas Eve, 1936. The first patient suffered from leukemia and was treated with phosphorus-32. The most successful use of P-32 therapy was in the treatment of polycythemia vera, a disease characterized by an increased production of red blood cells. If P-32 was "the first child of radionuclide therapy, treatment with radioiodine in thyroid cancer was the midwife that delivered atomic medicine" to the world. Drs. Joseph Hamilton and Mayo Soley of the University of California San Francisco Medical School were the first to treat hyperthyroid patients with iodine-131, discovered and produced by Dr. Glenn Seaborg at Berkeley. Radionuclide therapy is today having a renaissance in "molecular nuclear medicine," the new name of nuclear medicine. As long ago as 1978, I wrote: "Because of its chemical and physiological orientation since its beginning, nuclear medicine has had an orientation to drug therapy ... When it was discovered that ionic gallium-67 accumulated in neoplasms, people began to investigate whether stable gallium might have an anti-neoplastic effect."

Therapy has been linked to diagnosis since the first human studies were performed with radioactive tracers. In 1936, Karl Compton and his colleagues at the Massachusetts Institute of Technology (MIT) and the Massachusetts General Hospital (MGH) produced small amounts of the radionuclide iodine-128 and examined its accumulation by the thyroid in rabbits. Their results were published in 1938. In 1939, the group working at the University of California at Berkeley used iodine-131 to show that the normal thyroid accumulated radioiodine. It was found that in some patients the accumulation of the tracer, radioactive iodine, was occurring at an abnormally fast rate. This could be reduced by administering larger doses of the same radioisotope which would kill some of the thyroid cells and cure the disease, the birth of radionuclide therapy. Whenever specific molecular processes are defective, as in the basal ganglia of patients with Parkinson disease, drugs such as L-Dopa, can stimulate the deficient process.

Figure 185 Ernest Lawrence, inventor of the cyclotron.

After World War II, there was great excitement about the use of radioiodine for the treatment of thyroid cancer and hyperthyroidism. The press and public greeted the beginning of the post-war "Atomic Age" as a welcome relief from the atomic bomb. Only a few pioneers believed that some day nuclear medicine would achieve the widespread acceptance that it has today, when, every year, new tracer procedures make it possible to examine molecular processes and physiology in every part of the living human body, characterizing the patient's problem at the molecular level of organization, more basic than the cellular level of histopathogy.

In the United States, researchers from medical institutions, universities, and industry present the results of their research at the annual meeting of the Society of Nuclear Medicine. With strong government support that began immediately after World War II, American companies have made major contributions, and continue to provide leadership. Today, all over the world, hard-working, dedicated and creative people make major contributions.

To be sure, problems persist. One of these is the dependence of the United States on Canada for the production of reactor-produced radionuclides used in medicine and biomedical research. When privately-owned domestic commercial suppliers of radionuclides began operation in the 1960s, the Atomic Energy Commission ceased its production. Private domestic sources in the United States are sorely needed.

Most of the larger companies in nuclear medicine today are multi-national, reflecting the diversity of people in the field since its inception. Mainstream and entrepreneurial instrumentation companies continue to develop "novel imaging" devices. There is ever-increasing capital investment, benefiting patients and the entire health care system. Industry is increasing its collaboration with academic and private research programs, and cooperating with large pharmaceutical companies.

Nuclear medicine is playing an increasingly important role in drug design and the development of pharmaceuticals. Companies rely on radioactive traces and molecular imaging to decrease the time and cost of developing new drugs from the large number of possible compounds developed in basic sciences, including genetics. From the tens of thousands of candidate drugs, nuclear medicine technology helps select the most promising early in the new drug development pathway, and accelerates the extensive developmental and regulatory process until they become accepted as approved drugs. The same imaging techniques used to validate the diagnostic agent are often used subsequently to select those patients who are likely to benefit from the specific treatment. Then, during the course of treatment, radiotracer studies help monitor the patient's response.

Nuclear medicine is today a multi-billion dollar industry in the United States. Yet more patients could benefit from nuclear medicine procedures than are now receiving them. It has been said that nuclear medicine is "the best kept secret in medicine," but this is no longer the case. The information provided by nuclear medicine is greatly improving the efficiency in the delivery of health care, for example, in the better selection of patients for surgery.

In a report, "Crossing the Quality Chasm," the Institute of Medicine stated: "Our attempts to deliver today's medical capabilities are the medical equivalent of manufacturing microprocessors in a vacuum tube factory. The costs of waste, poor quality and inefficiency are enormous." Nuclear medicine can define diseases at the molecular level for every person. These measurements can help plan specific treatment, monitor its effectiveness, and direct adjustments to treatment when ineffective. This results in more effective, less costly surgical and pharmaceutical therapy.

What we need to do now is decrease the time required to carry out nuclear imaging studies. Fortunately, current research is resulting in procedures and instruments that take less time without reducing quality. Still, on average, it takes about 90 minutes to perform a PET scan, which amounts to about 1,000 studies per year for each scanner. With the new scanners under development, it is possible to perform greater numbers of studies in the same time. For example, several companies, including Hamamatsu Photonics, K.K. have recently built PET scanners that can image the entire human body in less than 10 minutes. Such devices are particularly useful in the screening of large numbers of persons with increased risk of specific diseases, for example, because of a genetic predisposition.

Today, we need to continue to try to relate molecular processes to specific genes (e.g., radioiodine uptake by the thyroid is the result of overactivity of the sodium-iodide "symporter" gene). In sickle cell disease, a genetic abnormality results in an abnormal hemoglobin molecule, which causes increased hemolysis, and the risk of thrombosis. Resistance to chemotherapeutic drugs in the treatment of patients with cancer results from expression of the protein, P-glycoprotein, which transports the chemotherapeutic agent out of the cancer cells. In depressed patients, we can block the sites on the presynaptic neurons that remove the neurotransmitter, serotonin, from the synapse. We can block the dopamine transporter in patients with Parkinson disease, and block the CD20 antigens on malignant cells in patients with non-Hodgkin lymphoma. Radiotracers can define biochemical processes in the brain in order to homogenize patient groups with various mental disorders who are chosen to participate in clinical trials of new drug therapy.

We can assess the effectiveness of gene therapy by examining the effectiveness of viruses used to carry therapeutic genes into injured heart muscle of patients with myocardial infarction. We can quantitatively assess the survival and subsequent division of transplanted cells.

One of the major characteristics of nuclear medicine is that teams of people bring their different technologies to solve patients' problems. Disciplinary boundaries disappear, and are replaced by hybrid domains in which cooperation is the basis of success.

Each specialty brings its own focus of research, and advances are multi-focal. For example, only recently has "molecular imaging" with radioactive tracers been expanded to include optical imaging with fluorescent tracers. Visible light is only a small part of the entire electromagnetic spectrum, comprising wavelengths of 400 to 700 nanometers, part of a spectrum that ranges from gamma rays trillions of times shorter than visible light to radiowaves trillions of times longer.

PET, SPECT, and optical imaging provide new eyes in a world where there is always more to be seen. The greater sensitivity of optical imaging results in a 1,000-fold increase in spatial resolution, which makes possible the application of the tracer principle to study isolated cells in vitro. Positron emission tomography (PET) and single photon emission computed tomography (SPECT) make it possible to examine regional function and biochemistry in large experimental animals and human beings in health and disease, while optical imaging provides intracellular imaging. The higher energy photons of PET and SPECT penetrate the human body to permit measurement of radioactive tracers deep within the body.

The president and CEO of General Electric Medical Systems, Joseph M. Hogan stated recently:

"In the years to come, we envision a health care system that uses molecular medicine to diagnose and treat patients before symptoms appear and treatments that are tailored to an individual based on his or her genetic makeup."

What challenges lie ahead? Most of all we need to simplify the regulatory process for the approval of new radiopharmaceuticals. In November 1987, Dr. John Palmer, director of the division of oncology and radiopharmaceutical drug production of the FDA, called attention to the "regulatory gray area" for the products of in-house cyclotrons. An important difference between diagnostic radiopharmaceuticals and other drugs is that radiotracers are designed to have no biological effect. Therefore we do not need the same regulatory criteria used to assess drugs designed to have biological effects. The term "outcome" has a different meaning when applied to a diagnostic agent, where the product of the procedure is information, not the achievement of a therapeutic effect, as is the case for conventional pharmaceuticals. Often the goal of a diagnostic procedure is to help decide which treatment is likely to counter the disease process or relieve symptoms. Until the late 1970s, the use of radiotracers in medicine was exempt from FDA regulations, without there being any untoward effects during this pre-FDA era. Their use was under the control of institutional Radioactive Drug Research Committees, following the guidelines of the Atomic Energy Commission, and subsequently the Department of Energy. More tailor-made regulations need to be created and implemented by the FDA. Efforts in this direction are already underway. On January 24, 1986, Dr. Ephraim Lieberman, President of Cadema Medical Products, Inc., proposed that for many radiopharmaceuticals, especially in-house PET tracers, the review process could be carried out by selected, qualified academic institutions with all of the credentials and supervision necessary for an effective, timely assessment.

In June 1979, faced with the resignation of Judy Glos, Executive Director of the Society of Nuclear Medicine, who made enormous contributions to its growth and success, I made the following proposal in the *Journal of Nuclear Medicine*:

"I believe the time has come to form a Federation of Societies of Radiological Sciences ... an analogous structure to the Federated Societies of Experimental Biology

(FASEB) . . . the identity of the individual societies would be retained, as they are in FASEB.

"We must convince the public and political leaders that our technologies, although often expensive, serve to reduce the 'guesswork' in medical diagnosis and therefore reduces the costs of delayed or wrong treatment.

"There comes a time when one's future depend on being aware of one's problems, being imaginative enough to see solutions, and being courageous enough to take bold steps. This, in my view, is such a time."

Few people have been given the opportunity to witness the birth and growth of a new medical specialty, increasingly being called "molecular medicine." No specialty is better suited to meet the challenge of addressing the basic questions that make up the practice of medicine:

1. Where within the body is the patient's problem?
2. What are the molecular changes within the body related to the patient's problem?
3. What is going to happen to the patient if the problem is not addressed?
4. What can be done about it?
5. Has the problem persisted after treatment?
6. Has the problem recurred after apparently successful treatment?

Today, molecular medicine is having an enormous effect on health care and is becoming widely known by all health care professionals and an increasingly large part of the public. The specialty has extended the foundation of medicine for over a century, from a time when anatomical and histopathological abnormalities defined disease.

The combination of structural and molecular manifestations by "hydrid" imaging of CT and MRI with PET and SPECT is increasing every day. Today, radiologists and nuclear medicine physicians are becoming expert in structural and molecular imaging, and are becoming qualified as experts in their interpretation.

We are living in what has been called the "Information Age." It is now possible to have a patient play an increasing role in his or her care as a unique individual. As one molecular process after another can be imaged in every organ of the body, advances in information technology are needed more than ever before. Physicians and patients need full access to the ever-increasing "molecular" information, specific for each patient, in order to be able to make more intelligent use of the health care system. Better knowledge about each patient will lead to better, more effective and more efficient health care. Perhaps within several decades, there will be a vast national network of electronic health records.

While the past half century has witnessed revolutionary advances in chemistry, physics, biology, and medicine, the delivery of health care to the individual patient has changed little. There is increasing use of electronic health records (HER's), the foundation for a system that can provide all the information about an individual patient's health and diseases. In the future, each person will have access to a complete, up-to-date review of his or her past and present medical history, manifestations of disease, and medications.

Figure 186 Alan Maurer, President of the Society of Nuclear Medicine and a former trainee in nuclear medicine at Johns Hopkins, awarding Henry Wagner a citation on the occasion of his presenting the 25th annual Highlights talk at the annual meeting of the Society of Nuclear Medicine.

Today, in 2005, only about 13% of hospitals have adopted electronic health records, (EHR), and this number will undoubtedly increase.

Within the next few decades, everyone will have a periodically updated computer chip (EHR), containing lifetime manifestations of his or her state of health. The individual's health information will be periodically entered into an International Database of Health Manifestations (manifestations include symptoms, physical signs, lab tests, imaging, etc.—all the aspects of life related to health or disease). He or she will be alerted if abnormal manifestations are identified. The health chip will periodically search an International Health Manifestations Database (IHMD) that will help answer the following questions in language that can be easily understood: (1) Is anything wrong? (2) What is going to happen? (3) What can be done about it? And (4) Is the treatment helping? Rather than simply giving a name to a person's disease, the health chip will reveal all aspects of that person's health and illness. The health manifestations on the computer chip will search the IHMD to characterize a disease that may develop, predict what is likely to happen, and suggest possible treatments.

Molecular imaging makes such a system especially necessary because of the coming revolution in health care. Using statistical averaging, a series of PET images can be integrated to yield computer images of "normal" individuals, or patients with different diseases. Okada et al., in a joint effort by Hamamatsu Phototonics (Japan), the University of Washington (Seattle), and the University of Michigan (Ann Arbor), studied an existing cohort made up of 551 normal individuals and 31 patients with Alzheimer disease (AD). They derived a composite, statistically analyzed distribution of F-18-FDG in the brains of patients with AD.

In vitro imaging will be used together with *in vivo* imaging. Genetic profiles will provide information about the risk of the patient's developing one or more diseases in the future, and will be able to suggest possible steps for prevention. Relating genetics to physiologic processes, called "functional genomics," will become increasingly possible, characterizing the relationship between one or more genes and specific homeostatic processes, which can vary greatly from person to person. We will be able to display composite images of different regional molecular states and processes. An example is the imaging of regional distribution of dopamine receptors in different parts of the brain and relating the findings to mental function and diseases.

We need to establish whether the variability among normal persons and patients with different diseases follow a single or multimodal distribution. We need to identify those molecular phenotypes that follow Mendelian genetics, characterize complex homeostatic processes, and identify diversity among normals as well as patients with disease. Study of abnormal molecular processes will identify abnormal genes and proteins.

To create an International Health Manifestation Database (IHMD), we need to continue to create well-defined subsets of images of disease manifestations. We need to enter them into an Internet database, analogous to the "open source" philosophy in computer programming, so that normal and disease databases will be available to all. Eventually, we will be able to compare the patient's anatomical, functional, and biochemical imaging studies with images in the IHMD through co-registration or precise overlay of the database image. Already, in the case of an F-18-FDG PET scan of a patient, the physician can go online and relate the images to a dataset of normal people. This capability will be extended to all diseases.

Molecular imaging, the process of identifying, localizing, and quantifying normal and abnormal regional molecular processes and relating these processes to genotypes, histopathology, and the patient's clinical problems (phenotypes) will characterize medicine in the future. We can no longer rely on a single physician's brain to interpret all the data about the patient including *in vitro* genetic and *in vivo* molecular imaging images. Nuclear medicine has the opportunity to face this challenge.

Is it beyond our wildest dreams to be able to bring this about? Technological advances usually take more than 20 years to reach widespread use. By the end of the decade, every American will have an Electronic Health Record (EHR). Our challenge is to make sure that by 2020, the EHR will contain genetic and molecular imaging information. It is time for a paradigm shift in the practice of medicine. Every person needs to be more closely involved in his or her health care. The principle that "only the doctor knows best" is on the way out.

Until now, the practice of medicine has been managed primarily by physicians, hospital administrators, and insurance companies. The IHMD will give patients an equal role in the operation of the health care system. Patients will be increasingly conscious of the quality and cost of their health care. They will have greater access to information that will permit them to make informed choices. Patients will continue to look to their doctors to advise them and guide them through the complexities of an illness, but the patients will be partners in their care rather than clients. Patients will play an essential role in maintaining their health and taking care of themselves, with their physician's help when they are sick.

Today, over 70 million persons in the United States get health information from the Internet. We have come a long way since the time of "horse-and-buggy" medicine. The horizon is becoming clearer. Nuclear medicine was the first medical specialty to use computers in the everyday practice of medicine. It it can play a key role in helping create whole new model of health care in the 21st century.

We are witnessing a renaissance in radionuclide therapy of cancer. We have "magic bullets" far beyond radioactive iodine and radioactive phosphorus, the patron saints of radioisotope therapy. With the demonstration of the effectiveness of radionuclide therapy in treating non-Hodgkin lymphoma, oncologists will have to become trained in these procedures, just as cardiologists are now trained in nuclear cardiology. Within a decade, radionuclide therapy will have achieved a role equal to surgery in the care of patients with cancer. Not only will there be "molecular diagnosis" of cancer; there will be molecular treatment and monitoring of the effectiveness of therapy.

What the future holds is as uncertain as ever. Looking back, despite the obstacles that they faced, the pioneers of the field of nuclear medicine had a lot of fun. Many were unforgettable personalities, willing to venture beyond well-accepted pathways to profes-

Figure 187 Professor Karl Oeff, former Head of Nuclear Medicine at the Free University of West Berlin. His technologist, Ursula Scheffel, emigrated to Hopkins nuclear medicine and rose to the rank of Associate Professor.

Figure 188 Henry N. Wagner, Jr., during a visit to the Max Planck Institute in Berlin in 1961.

Figure 189 Anne Wagner during a visit to Moscow and Leningrad in 1961, during which she read a lecture of mine on liver scanning in the Russian language.

sional advancement in academic medicine, and face the risk that the dramatic new technology might not achieve their expectations and never become accepted as a new way to define and treat disease. They rejected the admonitions of their superiors who said "think as you are told to think. Do as others do. Learn the rules." Many of the pioneers in nuclear medicine are no longer living, but those who remain can be very proud.

Bibliography

1. Wagner HN Jr: Objective testing of vision with use of the galvanic skin response. Arch Ophthalmol 43: 529–536, 1950.
2. Wagner HN Jr, Woods JW: Interruption of bulbocapnine catalepsy in rats by environmental stress. Arch Neurol Psychiatry 64: 720–725, 1950.
3. Wagner HN Jr: Electrical skin resistance studies in two persons with congenital absence of sweat glands. Arch Dermatol Syphil 65: 543–548, 1952.
4. Wagner HN Jr: Experimental and clinical results with intramuscular teramycin. Bulletin of The Johns Hopkins Hospital 91: 75, 1952.
5. Fisher AM, Wagner HN Jr, Ross RS: Staphylococcal endocarditis. Arch Intern Med 95: 427–437, 1955.
6. Wagner HN Jr, Bennett IL Jr, Lasagna L, Cluff LE, Rosenthal MB, Mirick GS: The effect of hydrocortisone upon the course of pneumococcal pneumonia treated with penicillin. Bulletin of The Johns Hopkins Hospital 98: 197–215, 1956.
7. Wagner HN Jr, Braunwald E: The pressor effect of the antidiuretic principle of the posterior pituitary in orthostatic hypotension. J Clin Invest 35: 1412–1418, 1956.
8. Wagner HN Jr: The influence of autonomic vasoregulatory reflexes on the rate of sodium and water excretion in man. J Clin Invest 36: 1319–1327, 1957.
9. McDonald RK, Wagner HN Jr, Weise VK: Relationship between endogenous antidiuretic hormone activity and ACTH release in man. Proc Soc Exp Biol Med 96: 652–655, 1957.
10. Orloff J, Wagner HN Jr, Davidson DG: The effect of variations in solute excretion and vasopressin dosage on the excretion of water in the dog. J Clin Invest 37: 458–464, 1958.
11. MacGregor GA, Wagner HN Jr: The influence of age on excretion of radioactive iodine. Lancet 2: 612–614, 1958.
12. Wagner HN Jr: An outline of the use of iodine in endemic goiter. Arch Intern Med 103: 484–488, 1959.
13. Wagner HN Jr: Orthostatic hypotension. Bulletin of The Johns Hopkins Hospital 105: 322–359, 1959.
14. Wagner HN Jr, McAfee JG, Mozley JM: Medical radioisotope scanning. JAMA 174: 162–165, 1960.
15. Winkelman JW, Wagner HN Jr, McAfee JG, Mozley JM: Visualization of the spleen in man by radioisotope scanning. Radiology 75: 465–466, 1960.
16. McAfee JG, Wagner HN Jr: Visualization of renal parenchyma by scintiscanning with Hg-203 neohydrin. Radiology 75: 820–821, 1960.
17. Wagner HN Jr, McAfee JG, Mozley JM: Diagnosis of liver disease by radioisotope scanning. Trans Assoc Am Physicians 73: 247–258, 1960.
18. Wagner HN Jr, McAfee JG, Mozley JM: Diagnosis of liver disease by radioisotope scanning. Arch Intern Med 107: 324–334, 1961.
19. Wagner HN Jr, Nelp WB, Watts JC: Clinical and experimental evaluation of the triiodothyronine (T3) red cell uptake test. Bulletin of The Johns Hopkins Hospital 108: 161–170, 1961.
20. Wagner HN Jr, Nelp WB, Dowling JH: Use of neutron activation analysis for studying stable iodide uptake by the thyroid. J Clin Invest 40: 1984–1992, 1961.
21. Razzak MA, Wagner HN Jr: Measurement of hepatic blood flow by colloidal gold clearance. J Appl Physiol 16: 1133–1138, 1961.
22. Wagner HN Jr, McAfee JG, Mozley JM: Diagnosis of pericardial effusion by radioisotope scanning. Arch Intern Med 108: 679–685, 1961.
23. Mason DR, Frye RL, Wagner HN Jr: Radioisotope scanning of the precordial distribution of iodide in patients with myocardial infarction. Circulation 24: 1338–1341, 1961.
24. Razzak MA, Wagner HN Jr: Cardiac photoscanning. Bull Egyptian Soc Cardiol 3: 44–50, 1962.

25. Wagner HN Jr, Razzak MA, Gaertner RA, Caine WP, Feagin OT: Removal of erythrocytes from the circulation. Arch Intern Med 110: 90–97, 1962.
26. Reba RC, Wagner HN Jr, McAfee JG: Measurement of Hg-203 chlormerodrin accumulation by the kidneys for detection of unilateral renal disease. Radiology 79: 134–135, 1962.
27. Wagner HN Jr, McAfee JG, Winkelman JW: Splenic disease diagnosis by radioisotope scanning. Arch Intern Med 109: 679–684, 1962.
28. Winkelman J, McAfee JG, Wagner HN Jr, Long RG: The synthesis of Co-57 tetraphenylporphinesulfonate and its use in the scintillation scanning of neoplasms. J Nucl Med 3: 249–253, 1962.
29. Wagner HN Jr: The current status of nuclear medicine. JAMA 183: 500–503, 1963.
30. Iio M, Wagner HN Jr: Studies of the reticuloendothelial system (RES). I. Measurement of the phagocytic capacity of the RES in man and dog. J Clin Invest 42: 417–426, 1963.
31. Wagner HN Jr, Iio M, Hornick RB: Studies of the reticuloendothelial system (RES). II. Changes in the phagocytic capacity of the RES in patients with certain infections. J Clin Invest 42: 427–434, 1963.
32. Nelp WB, McAfee JG, Wagner HN Jr: Single measurement of plasma radioactive vitamin B-12 as a test for pernicious anemia. J Lab Clin Med 61: 158–165, 1963.
33. Wagner HN Jr: Radioactive pharmaceuticals. Clin Pharmacol Ther 4: 351–370, 1963.
34. Reba RC, McAfee JG, Wagner HN Jr: Radiomercury labeled chlormerodrin for in vivo uptake studies and scintillation scanning of unilateral renal lesions associated with hypertension. Medicine (Baltimore) 42: 269–296, 1963.
35. Nelp WB, Wagner HN Jr, Schwartz FJ: Total body radioactivity and the lack of excretion of Co-60 vitamin B-12 by the aglomerular oyster toadfish Opsanus tau. Chesapeake Science 4: 192–194, 1963.
36. Wolzman GB, Wagner HN Jr, Iio M, Rabinowitz D, Zierler KL: Measurement of muscle blood flow in the human forearm with radioactive krypton and xenon. Circulation 30: 27–34, 1964.
37. Wagner HN Jr: An outline of the use of radioisotope techniques in medical diagnosis. Am J Med Sci 247: 123–154, 1964.
38. Wagner HN Jr: Pulmonary scanning. Northwest Med 63: 857–864, 1964.
39. Shah KD, Neill CA, Wagner HN Jr, Taussig HB: Radioisotope scanning of the liver and spleen in dextrocardia and in situs inversus with levocardia. Circulation 24: 231–241, 1964.
40. Greisman SE, Wagner HN Jr, Iio M, Hornick RB: Mechanisms of endotoxin tolerance. II. Relationship between endotoxin tolerance and reticuloendothelial system phagocytic activity in man. J Exp Med 119: 241–264, 1964.
41. Wagner HN Jr, Sabiston DC Jr, Iio M, McAfee JG, Meyer JK, Langan JK: Regional pulmonary blood flow in man by radioisotope scanning. JAMA 187: 601–603, 1964.
42. Nelp WB, Wagner HN Jr, Reba RC: Renal excretion of vitamin B-12 and its use in measurement of glomerular filtration rate in man. J Lab Clin Med 63: 480–491, 1964.
43. Wagner HN Jr, Sabiston DC Jr, McAfee JG, Tow D, Stern HS: Diagnosis of massive pulmonary embolism in man by radioisotope scanning. Bowman Gray Scanning Symposium 1964. N Engl J Med 271: 377–384, 1964.
44. Wagner HN Jr: Pharmacological principles in the development of radiopharmaceuticals for radioisotope scanning. Bowman Gray Scanning Symposium 1964.
45. Wagner HN Jr, Weiner IM, McAfee JG, Martinez J: 1-Mercuri-2-Hydroxypropane (MHP): A new radiopharmaceutical for visualization of the spleen by radioisotope scanning. Arch Intern Med 113: 696–701, 1964.
46. Holzman GB, Wagner HN Jr, Iio M, Rabinowitz D, Zierler KL: Measurement of muscle blood flow in the human forearm with radioactive krypton and xenon. Circulation 30: 27–34, 1964.
47. Wagner HN Jr, Iio M: Studies of the reticuloendothelial system (RES) II. Blockade of the RES in man. J Clin Invest 43: 1525–1532, 1964.
48. Wagner HN Jr, Sabiston DC Jr, McAfee JG, Tow DE, Stern HS: Diagnosis of massive pulmonary embolism in man by radioisotope scanning. N Engl J Med 271: 377–384, 1964.
49. Bourne HR, Wagner HN Jr, Iio M, Jude JR, Knickerbocker GG: Cerebral blood flow during external cardiac massage. J Nucl Med 5: 738–745, 1964.
50. Lopez-Majano V, Chernick V, Wagner HN Jr, Dutton RE: Comparison of radioisotope scanning and differential oxygen uptake of the lungs. Radiology 83: 696–698, 1964.
51. McAfee JG, Fueger GF, Stern HS, Wagner HN Jr, Migita T: Tc-99m pertechnetate for brain scanning. J Nucl Med 5: 811–827, 1964.
52. Wagner HN Jr: Splenic scanning. Northwest Med 63: 767–770, 1964.
53. Sabiston DC Jr, Wagner HN Jr: The diagnosis of pulmonary embolism by radioisotope scanning. Ann Surg 160: 575–588, 1964.
54. Smith EM, Mozley JM, Wagner HN Jr: Determination of proteinbound iodine (PBI) in human plasma by thermal neutron activation analysis. J Nucl Med 5: 828–839, 1964.

55. Wagner HN Jr, Lopez-Majano V, Langan JK: Clearance of particulate matter from the tracheo-bronchial tree in patients with tuberculosis. Nature 205: 252–254, 1965.

56. Wagner HN Jr, Lopez-Majano V, Tow D, Langan JK: Radioisotope scanning of lungs in the early diagnosis of bronchogenic carcinoma. Lancet 1: 344, 1965.

57. Wagner HN Jr, Jones E, Tow DE, Langan JK: A method for the study of the peripheral circulation in man. J Nucl Med 6: 150–154, 1965.

58. Wagner HN Jr, Jones RH. Massive pulmonary embolism. Physiology for Physicians 3. 1–3, 1965.

59. McAfee JG, Ause RG, Wagner HN Jr: Diagnostic value of scintillation scanning of the liver. Arch Intern Med 116: 95–110, 1965.

60. Proctor DF, Wagner HN Jr: Clearance of particles from the human nose. Arch Environ Health 11: 366–371, 1965.

61. Sabiston DC Jr, Wagner HN Jr: The pathophysiology of pulmonary embolism: relationships to accurate diagnosis and choice of therapy. J Thorac Cardiovas Surg 50: 339–356, 1965.

62. Jones EL, Wagner HN Jr, Zuidema GD: New method for studying peripheral circulation in man. Arch Surg 91: 725–734, 1965.

63. Lopez-Majano V, Wagner HN Jr, Tow DE, Chernick V: Radioisotope scanning of the lungs in pulmonary tuberculosis. JAMA 194: 1053–1058, 1965.

64. Wagner HN Jr: Radioisotope methodology in the study of respiratory disease. Jpn Heart J 6: 87–98, 1965.

65. Bryan AC, MacNamara J, Simpson J, Wagner HN Jr: Effect of acceleration on the distribution of pulmonary blood flow. J Appl Physiol 20: 1129–1132, 1965.

66. Chernick V, Lopez-Majano V, Wagner HN Jr, Dutton RE Jr: Estimation of differential pulmonary blood flow by bronchospirometry and radioisotope scanning during rest and exercise. Amer Rev Respir Dis 92: 958–962, 1965.

67. Wagner HN Jr: Radioisotope techniques in the diagnosis of diseases of the central nervous system. Radiol Clin North Am 1: 203–215, 1966.

68. Baker RR, Wagner HN Jr: Pulmonary embolectomy in the treatment of massive pulmonary embolism. Surg Gynecol Obstet 122: 513–516, 1966.

69. Wagner HN Jr, Tow DE, Lopez-Majano V, Chernick V, Twining R: Factors influencing regional pulmonary blood in man. Scand J Respir Dis, Suppl. 62: 59–72, 1966.

70. Wagner HN Jr, Migita T, Solomon N: Effect of age on reticuloendothelial function in man. J Gerontol 21: 57–62, 1966.

71. Lopez-Majano V, Tow D, Wagner HN Jr: Regional distribution of pulmonary arterial blood flow in emphysema. JAMA 197: 81–84, 1966.

72. Tow DE, Wagner HN Jr, Lopez-Majano V, Smith EM, Migita T: Validity of measuring regional pulmonary arterial blood flow with macroaggregates of human serum albumin. Am J Roentgenol Radium Ther Nucl Med 96: 664–676, 1966.

73. Silk M, Wagner HN Jr: Measurement of glomerular filtration rate in man by external monitoring of the clearance of radioactive vitamin B-12. Invest Urol 3: 609–613, 1966.

74. Rhodes BA, Wagner HN Jr: Are iodotyrosines normal constitutents of plasma? Nature 210: 647–648, 1966.

75. Haller JA, Wagner HN Jr, Jackson DP: Conjoint clinic on thrombophlebitis. J Chronic Dis 19: 785–798, 1966.

76. Holmes RA, Silbiger ML, Karmen A, Wagner HN Jr, Stern HS: Cardiac scanning with technetium-99m labeled albumin. JAMA 198: 67–72, 1966.

77. Wagner HN Jr: Radiomercurials in the study of renovascular hypertension. Postgrad Med 40: 314–319, 1966.

78. Ter-Pogossian MM, Wagner HN Jr: A new look at the cyclotron for making short-lived isotopes. Nucleonics 24: 50–56, 1966.

79. Stern HS, Goodwin D, Wagner HN Jr, Kramer HH: In-113m—a short-lived isotope for lung scanning. Nucleonics 24: 57–59, 1966.

80. Wagner HN Jr: Radiopharmaceuticals—their use in nuclear medicine. Nucleonics 24: 62–66, 84–85, 1966.

81. Goodwin DA, Stern HS, Wagner HN Jr, Kramer HH: In-113m colloid: a new radiopharmaceutical for liver scanning. Nucleonics 24(11): 65, 1966.

82. McIntyre PA, Wagner HN Jr: Comparison of the urinary excretion and eight hour plasma tests for vitamin B-12 absorption. J Lab Clin Med 68: 966–971, 1966.

83. Lopez-Majano V, Wagner HN Jr, Twining RH, Tow DE, Chernick V: Effect of regional hypoxia on the distribution of pulmonary blood flow in man. Circ Res 18: 550–557, 1966.

84. Wagner HN Jr: Nuclear medicine: present and future. Radiology 86: 601–614, 1966.

85. Wagner HN Jr, Bardfeld PA: Evaluation of structure and function of spleen with radioactive tracers. JAMA 199: 202–206, 1967.
86. Mishkin F, Wagner HN Jr: Regional abnormalities in pulmonary arterial blood flow during acute asthmatic attacks. Radiology 88: 142–144, 1967.
87. Kaihara S, Kandel GE, Wagner HN Jr: Continuous measurement of partition of pulmonary blood flow between right and left lung. J Appl Physiol 23: 976–978, 1967.
88. Baker RR, Migita T, Wagner HN Jr: A quantitative method of measuring liver regeneration in the dog. J Surg Res 7: 578–582, 1967.
89. Wagner HN Jr: Communications: Videotape in the teaching of medical history-taking. J Med Educ 42: 1055–1058, 1967.
90. Sasahara AA, Potchen EJ, Thomas DP, Wagner HN Jr, Davis WC: Diagnostic requirements and therapeutic decisions in pulmonary embolism. JAMA 202: 553, 1967.
91. Power GG, Longo LD, Wagner HN Jr, Kuhl DE, Forster RE II: Uneven distribution of maternal and fetal placental blood flow as demonstrated using macroaggregates and its response to hypoxia. J Clin Invest 46: 2053–2063, 1967.
92. Stern HS, Goodwin DA, Scheffel U, Wagner HN Jr, Kramer HH: In-113m for blood-pool, brain scanning. Nucleonics 25: 62–66, 1967.
93. Tow DE, Wagner HN Jr: Scanning for tumors of brain and bone. JAMA 199: 610–614, 1967.
94. Tow DE, Wagner HN Jr: Recovery of pulmonary arterial blood flow in patients with pulmonary embolism. N Engl J Med 276: 1053–1059, 1967.
95. Tow DE, Wagner HN Jr: Lung scanning in pulmonary diseases. NY State J Med 67: 2089–2095, 1967.
96. Rhodes BA, Wagner HN Jr, Gerrard M: Iodine-123: development and usefulness of a new radiopharmaceutical. Isotopes Rad Tech 4: 275–280, 1967.
97. Tow DE, Wagner HN Jr, Holmes RA: Urokinase in pulmonary embolism. N Engl J Med 277: 1161–1167, 1967.
98. Holmes RA, Herron CS, Wagner HN Jr: A modified vertex view in brain scanning. Radiology 88: 498–503, 1967.
99. Gopala Rao UV, Wagner HN Jr: Effect of an analog ratemeter on the modulation transfer function in radioisotope scanning. Radiology 88: 504–508, 1967.
100. Kaihara S, Kandel GE, Wagner HN Jr: Continuous measurement partition of pulmonary blood flow between right and left lung. J Appl Physiol 23: 976–978, 1967.
101. Bardfeld PA, Lopez-Majano V, Wagner HN Jr: Measurement of the regional distribution of arterial blood flow in the human forearm and hand. J Nucl Med 8: 542–550, 1967.
102. Lopez-Majano V, Twining RH, Goodwin DA, Wagner HN Jr: Reproducibility of lung scans. Invest Radiol 2: 410–418, 1967.
103. Lopez-Majano V, Wagner HN Jr, Twining RH, Chernick V: Time factor in the shifting of blood produced by unilateral hypoxia. Am Rev Respir Dis 96: 1190–1198, 1967.
104. Wagner HN Jr, Tow DE: Radioisotope scanning in the study of pulmonary circulation. Progress in Cardiovascular Disease 9: 382–399, 1967.
105. Wagner HN Jr: Foreword, Korean J Nucl Med Vol. I, 1967.
106. Kaihara S, Wagner HN Jr: Measurement of intestinal fat absorption with carbon-14 labeled tracers. J Lab Clin Med 71: 400–411, 1968.
107. Reba RC, Hosain F, Wagner HN Jr: Indium-113m diethylene-triamine-pentaacetic acid (DTPA): A new radiopharmaceutical for study of the kidneys. Radiology 90: 147–149, 1968.
108. Wagner HN Jr: New directions in pulmonary embolism. Hosp Pract 3: 36–42, 1968.
109. Renda F, Holmes RA, North WA, Wagner HN Jr: Characteristics of thyroid scans in normal persons, hyperthyroidism and nodular goiter. J Nucl Med 9: 156–159, 1968.
110. Van Heerden PD, Wagner HN Jr, Kaihara S: Intestinal blood flow during perfusion of the jejunum with hypertonic glucose in dogs. Am J Physiol 215: 30–33, 1968.
111. Mishkin FS, Wagner HN Jr: Regional distribution of pulmonary arterial blood flow in acute asthma. JAMA 203: 1019–1021, 1968.
112. Twining RH, Lopez-Majano V, Wagner HN Jr, Chernick V, Dutton RE: Effect of regional hypercapnia on the distribution of pulmonary blood flow in man. Johns Hopkins Med J 123: 95–103, 1968.
113. Quinlan MF, Wagner HN Jr: Lung imaging with the pinhole Anger camera. J Nucl Med 9: 497–498, 1968.
114. Carulli N, Kaihara S, Wagner HN Jr: Radioisotopic assay of arginase activity. Anal Biochem 24: 515–522, 1968.
115. Clements JP, Wagner HN Jr, Stern HS, Goodwin DA: Indium-113m diethyltriaminopentaacetic acid (DTPA): a new radiopharmaceutical for brain scanning. Am J Roentgenol Radium Ther Nucl Med 104: 139–144, 1968.
116. Lopez-Majano V, Wagner HN Jr: Clinical application of lung scanning. Dis Chest 54: 46–51, 1968.

117. Mishkin FS, Wagner HN Jr, Two DE: Regional distribution of pulmonary arterial blood flow in acute asthma. JAMA 203: 1019–1021, 1968.
118. Wagner HN Jr: Current status of lung scanning (editorial). Radiology 91: 1235–1237, 1968.
119. Goodwin DA, Stern HS, Wagner HN Jr: Ferric hydroxide particles labeled with indium In-113m for lung scanning. JAMA 206: 339–343, 1968.
120. Rutherford RB, Kaihara S, Schwentker EP, Wagner HN Jr: Regional blood flow in hemorrhagic shock by the distribution of labeled microspheres. Surg Forum 19: 14–15, 1968
121. Poulose K, Reba RC, Wagner HN Jr: Characterization of the shape and location of perfusion defects in certain pulmonary diseases. N Engl J Med 279: 1020–1025, 1968.
122. DeLand FH, Wagner HN Jr: Regeneration of the liver after hepatectomy. J Nucl Med 9: 587–589, 1968.
123. Wagner HN Jr, Lopez-Majano V, Langan JK, Joshi RC: Radioactive xenon in the differential diagnosis of pulmonary embolism. Radiology 91: 1168–1174, 1968.
124. Hosain F, Reba RC, Wagner HN Jr: Ytterbium-169 diethylenetriaminepentaacetic acid complex: a new radiopharma ceutical for brain scanning. Radiology 91: 1199–1203, 1968.
125. Wagner HN Jr: The new challenge. Clin Res 16: 411–413, 1968.
126. Carulli N, Kaihara S, Wagner HN Jr: A simple system for continuous recording of 14C02 production rate for in vitro metabolic studies. Anal Biochem 26: 334–337, 1968.
127. Kaihara S, Van Heerden PD, Migita T, Wagner HN Jr: Measurement of distribution of cardiac output. J Appl Physiol 25: 696–700, 1968.
128. Quinlan MF, Salman SD, Swift DL, Wagner HN Jr, Proctor DF: Measurement of mucociliary function in man. Am Rev Respir Dis 99: 13–23, 1969.
129. DeLand FH, Wagner HN Jr: Brain scanning as a diagnostic aid in the detection of eighth nerve tumors. Radiology 92: 571–575, 1969.
130. Greisman SE, Hornick RB, Wagner HN Jr, Woodward WE, Woodward TE: The role of endotoxin during typhoid fever and tularemia in man. J Clin Invest 48: 613–629, 1969.
131. Anghileri LJ, Reba RC, Wagner HN Jr: Uptake of radioarsenic labeled chromic B-glycerophosphate by experimental brain tumors. Invest Radiol 4: 91–96, 1969.
132. Rivlin RS, Wagner HN Jr: Anemia in hyperthyroidism. Ann Intern Med 70: 507–516, 1969.
133. Hosain F, McIntyre PA, Poulose KP, Stern HS, Wagner HN Jr: Binding of trace amounts of ionic indium-113m to plasma transferrin. Clin Chim Acta 24: 69–75, 1969.
134. Lopez-Majano V, Kieffer RF Jr, Marine DN, Garcia DA, Wagner HN Jr: Pulmonary resection in bullous disease. Am Rev Respir Dis 99: 554–564, 1969.
135. Lopez-Majano V, Kandel GE, Wagner HN Jr: Inequality between ventilation and perfusion in chronic obstructive lung disease. Bull Physiopathol Respir (Nancy) 5: 39–58, 1969.
136. Ohlsson EG, Rutherford RB, Haslebos M, Wagner HN Jr, Zuidema GD: The distribution of portal blood flow before and after hepatic resection in dogs. J Surg Res 9: 657–663, 1969.
137. Strauss HW, Hurley PJ, Rhodes BA, Wagner HN Jr: Quantification of right-to-left transpulmonary shunts in man. J Lab Clin Med 74: 597–607, 1969.
138. Castronovo FP Jr, Reba RC, Wagner HN Jr: System for sustained intravenous infusion of a sterile solution of 137m Ba ethylenediaminetetraacetic acid (EDTA). J Nucl Med 10: 242–245, 1969.
139. Haroutunian LM, Neill C, Wagner HN Jr: Radioisotope scanning of the lung in cyanotic congenital heart disease. Am J Cardiol 23: 387–395, 1969.
140. Castronovo FP Jr, Reba RC, Wagner HN Jr: System for sustained intravenous infusion of a sterile solution of 137mBa-ethylenediaminetetraacetic acid (EDTA). J Nucl Med 10: 242–245, 1969.
141. Kaihara S, Carulli N, Wagner HN Jr: Comparison of radio isotopic and column chromatographic assay of serum thyroxine. J Nucl Med 10: 281–283, 1969.
142. Buchanan JW, Rhodes BA, Wagner HN Jr: Labeling albumin microspheres with 113mIn. J Nucl Med 10: 487–490, 1969.
143. Wagner HN Jr, Hosain F, Rhodes BA: Recently developed radiopharmaceuticals: Ytterbium-169 DTPA and technetium-99m micro spheres. Radiol Clin North Am 7: 233–241, 1969.
144. Rhodes BA, Zolle I, Buchanan JW, Wagner HN Jr: Radioactive albumin microspheres for studies of the pulmonary circulation. Radiology 92: 1453–1460, 1969.
145. Rutherford RB, Rhodes BA, Wagner HN Jr: The distribution of extremity blood flow before and after vagectomy in a patient with hypertrophic pulmonary osteoarthropathy. Dis Chest 56: 19–23, 1969.
146. Kaihara S, Rutherford RB, Schwentker EP, Wagner HN Jr: Distribution of cardiac output in experimental hemorrhagic shock in dogs. J Appl Physiol 27: 218–222, 1969.
147. DeLand FH, Wagner HN Jr: Early detection of bacterial growth, with carbon-14 labeled glucose. Radiology 92: 154–155, 1969.
148. Tow DE, Wagner HN Jr, DeLand FH, North WA: Brain scanning in cerebral vascular disease. JAMA 207: 105–108, 1969.

149. Hosain P, Hosain F, Iqbal QM, Carulli N, Wagner HN Jr: Measurements of plasma volume using 99mTc and 113mIn labeled proteins. Br J Radiol 42: 627–630, 1969.
150. Hosain F, Reba RC, Wagner HN Jr: Measurement of glomerular filtration rate using chelated ytterbium-169. Int J Appl Radiat Isot 2Cl 517–521, 1969.
151. Lopez-Majano V, Rhodes BA, Wagner HN Jr: Arteriovenous shunting in extremities. J Appl Physiol 27: 782–786, 1969.
152. Natarajan TK, Wagner HN Jr: A new image display and analysis system (IDA) for radionuclide imaging. Radiology 93: 823–827, 1969.
153. Gilday DL, Reba RC, Hosain F, Longo R, Wagner HN Jr: Evaluation of Ytterbium-169 diethylenetriamine-pentaacetic acid as a brain-scanning agent. Radiology 93: 1129–1134, 1969.
154. Kaihara S, Natarajan TK, Maynard CD, Wagner HN Jr: Construction of a functional image from spatially localized rate constants obtained from serial camera and rectilinear scanner data. Radiology 93: 1345–1348, 1969.
155. Wagner HN Jr, Rhodes BA, Sasaki Y, Ryan JP: Studies of the circulation with radioactive microspheres. Invest Radiol 4: 374–386, 1969.
156. Langan JK, Wagner HN Jr: System for displaying, reproducing and storing scintillation-camera images. J Nucl Med 10(June): 460, 1969.
157. McIntyre PA, Wagner HN Jr: Current procedures for scanning of the spleen. Ann Int Med 73: 995–1001, 1970.
158. Roth JA, Greenfield AJ, Kaihara S, Wagner HN Jr: Total and regional cerebral blood flow in unanesthetized dogs. Am J Physiol 219: 96–101, 1970.
159. Zolle I, Rhodes BA, Wagner HN Jr: Preparation of metabolizable radioactive human serum albumin microspheres for studies of the circulation. Int J Appl Radiat Isot 21: 155–167, 1970.
160. Wagner HN Jr, Hosain F, DeLand FH, Som P: A new radio-pharmaceutical for cisternography: chelated ytterbium-169. Radiology 95: 121–125, 1970.
161. Tow DE, Wagner HN Jr: Lung scanning in diseases of the cardio-pulmonary circulation. Therapeutische Umschau/Revue Therapeutique 27: 44–50, 1970.
162. Adelstein SJ, Parker R, Wagner HN Jr: First phase in objective evaluation of new diagnostic tests. Invest Radiol 5: 154–163, 1970.
163. Tow DE, Wagner HN Jr: Effect of pleural fluid on the appearance of the lung scan. J Nucl Med 11: 138–319, 1970.
164. Wagner HN Jr: Lung scanning in pulmonary embolism. Bull Physiopathol Respir (Nancy) 6: 65–96, 1970.
165. Poulose KP, Reba RC, Gilday DL, DeLand FH, Wagner HN Jr: Diagnosis of pulmonary embolism. A correlative study of the clinical scan, and angiographic findings. Br Med J 3: 67–71, 1970.
166. DeLand FH, Wagner HN Jr: Automated radiometric detection of bacterial growth in blood cultures. J Lab Clin Med 75: 529–534, 1970.
167. Wagner HN Jr: Current research in nuclear medicine: Hickey Lecture 1970. Am J Roentgenol Radium Ther Nucl Med 109: 669–675, 1970.
168. Strauss HW, Hurley PJ, Wagner HN Jr: Advantages of 99mTc pertechnetate for thyroid scanning in patients with decreased radioiodine uptake. Radiology 97: 307–310, 1970.
169. James AE Jr, Wagner HN Jr: Role of nuclear medicine in cancer. Postgrad Med 48: 88–94, 1970.
170. Blackmon JR, Sautter RD, Wagner HN Jr, and other members of the Uro kinase Pulmonary Embolism Trial Study Group: Urokinase Pulmonary Embolism Trial, Phase 1 Results, A Cooperative Study. JAMA 214: 2163–2172, 1970.
171. Sasaki Y, Iio M, Kameda H, Ueda H, Ayoagi T, Christopher NL, Bayless TM, Wagner HN Jr: Measurement of the 14C-Lactose absorption in the diagnosis of lactase deficiency. J Lab Clin Med 76: 824–835, 1970.
172. James AE Jr, DeLand FH, Hodges FJ III, Wagner HN Jr: Cerebrospinal fluid (CSF) scanning: cisternography. Am J Roentgenol Radium Ther Nucl Med 110: 74–87, 1970.
173. James AE Jr, DeLand FH, Hodges FJ III, Wagner HN Jr: Radionuclide imaging in the detection and differential diagnosis of cranio-pharyngiomas. Am J Roentgenol Radium Ther Nucl Med 109: 692–700, 1970.
174. Hurley PJ, Strauss HW, Wagner HN Jr: Radionuclide angiocardiography in cyanotic congenital heart disease. Johns Hopkins Med J 127: 46–54, 1970.
175. Strauss HW, Natarajan TK, Sziklas JJ, Poulose KP, Fukushima T, Wagner HN Jr: Computer assistance in the interpretation and quantification of lung scans. Radiology 97: 277–281, 1970.
176. James AE Jr, DeLand FH, Hodges FJ III, Wagner HN Jr: Normal-Pressure Hydrocephalus: Role of cisternography in diagnosis. JAMA 213: 1615–1622, 1970.
177. Rothfeld B, Carulli N, Kaihara S, Reus I, Wagner HN Jr: Serum arginase concentrations in acute alcoholism. Biochem Med 4: 36–42, 1970.

178. Dibos PE, Spaulding MB, Judisch JM, Wagner HN Jr, McIntyre PA: Estudio del sistema reticuloendotelial en padecimientos hematologicos. Revista de Biologia y Medicina Nuclear 3: 1, 1971.

179. Chamroenngan S, Langan JK, Trattner JM, Muehllehner G, Wagner HN Jr: Tomographic imaging with the scintillation camera in the detection and characterization of brain lesions. Radiology 98: 445–448, 1971.

180. Wagner HN Jr: Remarks at dedication of the biomedical cyclotron at UCLA, June 30, 1971. SNM Newsletter, 1974, page 4.

181. Castronovo FP, Wagner HN Jr: Factors affecting the toxicity of the element indium. Brit J Exp Path 52: 543, 1971.

182. Sziklas JJ, Hosain F, Reba RC, Wagner HN Jr: Comparison of 169Yb-DTPA, 113mIn-DTPA, 14C-inulin and endogeneous creatinine to estimate glomerular filtration. J Nucl Biol Med 15: 122, 1971.

183. Hosain F, Wagner HN Jr: Measurement of extracellular fluid volume with 169Yb-diethylenetriaminepentaacetate. J Lab Clin Med 77: 699–704, 1971.

184. Hosain F, Syed IB, Wagner HN Jr, Poggenburg JK: Ionic barium-135m: a new agent for bone scanning. Radiology 98: 684–686, 1971.

185. James AE Jr, Cooper M, White RI Jr, Wagner HN Jr: Perfusion changes on lung scans in patients with congestive heart failure. Radiology 100: 99–106, 1971.

186. James AE Jr, Hosain F, DeLand FH, Reba RC, Wagner HN Jr, North WA: 169Ytterbium diethylenetriaminepentaacetic acid (169Yb-DTPA)—a versatile radiopharmaceutical. Journal de L'Association Canadienne des Radiologistes 22: 1971.

187. Hurley PJ, Cooper M, Reba RC, Poggenburg KJ, Wagner HN Jr: 43KCl: A new radiopharmaceutical for imaging the heart. J Nucl Med 12: 516–519, 1971.

188. Sasaki Y, Wagner HN Jr: Measurement of the distribution of cardiac output in unanesthetized rats. J Appl Physiol 30: 879–884, 1971.

189. James AE Jr, Cooper M, White RI Jr, Wagner HN Jr: Perfusion changes on lung scans in patients with congestive heart failure. Radiology 100: 99–106, 1971.

190. Buchanan JW, Rhodes BA, Wagner HN Jr: Labeling iron-free albumin microspheres with 113mIn. J Nucl Med 12: 616–619, 1971.

191. James AE Jr, White RI Jr, Cooper M, Wagner HN Jr: Pretreatment diagnostic evaluation with reference to pulmonary scans in lung cancer. J Thorac Cardiovas Surg 61: 530–540, 1971.

192. Sherr HP, Sasaki Y, Newman A, Banwell JG, Wagner HN Jr, Hendrix TR: Detection of bacterial deconjugation of bile salts by a convenient breath-analysis technic. N Engl J Med 285: 656–661, 1971.

193. Hurley PJ, Strauss HW, Pavoni P, Langan JK, Wagner HN Jr: The scintillation camera with pinhole collimator in thyroid imaging. Radiology 101: 133–138, 1971.

194. Rhodes BA, Stern HS, Buchanan JW, Zolle I, Wagner HN Jr: Lung scanning with 99mTc-microspheres. Radiology 99: 613–621, 1971.

195. Cooper JF, Levin J, Wagner HN Jr: Quantitative comparison of in-vitro and in-vivo methods for the detection of endotoxin. J Lab Clin Med 78: 138–148, 1971.

196. Hosain F, Pavoni P, Wagner HN Jr: Measurement of blood flow in skeletal muscle with chelated Ytterbium-169 and electrical stimulation. J Nucl Biol Med 15: 21, 1971.

197. DeLand FH, James AE Jr, Wagner HN Jr, Hosain F: Cisternography with 169Yb-DTPA. J Nucl Med 12: 683–689, 1971.

198. Pavoni P, Moen T, Rhodes BA, Wagner HN Jr: Changes in 133Xe clearances resulting from electrically induced muscle contractions. I. Study in dog. J Nucl Biol Med 15: 16–18, 1971.

199. Pavoni P, Moen T, Rhodes BA, Wagner HN Jr: Changes in 133Xe clearances resulting from electrically induced muscle contractions. II. Study in man. J Nucl Biol Med 15: 19–20, 1971.

200. DeBlanc HJ Jr, DeLand FH, Wagner HN Jr: Automated radiometric detection of bacteria in 2,967 blood cultures. Appl Microbiol 22: 846–849, 1971.

201. Rutherford RB, Reddy CM, Walker AG, Wagner HN Jr: A new quantitative method of assessing the functional status of the leg vein. Am J Surg 122: 594–602, 1971.

202. Castronovo FP, Wagner HN Jr: Factors affecting the toxicity of the element indium. Br J Exp Path 52: 543, 1971.

203. McIntyre PA, Laleli YR, Hodkinson BA, Wagner HN Jr: Evidence for anti-leukocyte antibodies as a mechanism for drug-induced agranulocytosis. Trans Assoc Am Physicians 84: 217–228, 1971.

204. Wesselhoeft H, Hurley PJ, Wagner HN Jr, Rowe RD: Nuclear angiocardiography in the diagnosis of congenital heart disease in infants. Circulation 45: 77–91, 1972.

205. Wagner HN Jr, Natarajan TK: Computers in nuclear medicine. Hosp Pract 7: 121–131, 1972.

206. Spence RJ, Rhodes BA, Wagner HN Jr: Regulation of arterio venous anastomotic and capillary blood flow in the dog leg. Am J Physiol 222: 326–332, 1972.

207. Moses DC, James AE Jr, Strauss HW, Wagner HN Jr: Regional cerebral blood flow estimation in the diagnosis of cerebro vascular disease. J Nucl Med 13: 135–141, 1972.

208. Hurley PJ, Maisey MN, Natarajan TK, Wagner HN Jr: A computerized system for rapid evaluation of thyroid function. J Clin Endocrinol Metab 34: 354–360, 1972.

209. Dibos PE, Judisch JM, Spaulding MB, Wagner HN Jr, McIntyre PA: Scanning the reticuloendothelial system (RES) in hematological disease. Johns Hopkins Med J 130: 68–82, 1972.

210. DeLand FH, James AE Jr, Muehllehner G, Wagner HN Jr: The value of tomography in liver scanning. Radiology 102: 429–432, 1972.

211. Som P, Hosain F, Wagner HN Jr, Scheffel U: Cisternography with chelated complex of 99mTc. J Nucl Med 13: 551–553, 1972.

212. Scheffel U, Rhodes BA, Natarajan TK, Wagner HN Jr: Albumin microspheres for study of the reticulo-endothelial system. J Nucl Med 13: 498–503, 1972.

213. Lomas F, Dibos PE, Wagner HN Jr: Increased specificity of liver scanning with the use of Gallium-67 citrate. N Engl J Med 286: 1323–1329, 1972.

214. Rhodes BA, Turaihi KS, Bell WR, Wagner HN Jr: Radioactive urokinase for blood clot scanning. J Nucl Med 13: 646–648, 1972.

215. Dibos PE, Muhletaler CA, Natarajan TK, Wagner HN Jr: Intravenous radionuclide arteriography in peripheral occlusive arterial disease. Radiology 102: 181–183, 1972.

216. Evdokimoff V, Wagner HN Jr: Hepatic phagocytosis as a mechanism for increasing heavy-metal toxicity. J Reticuloendothel Soc 11: 148–153, 1972.

217. Wagner HN Jr: Radioactive tracers and the circulation. Circulation 45: 8–10, 1972.

218. Rao UVG, Wagner HN Jr: Normal weights of human organs. Radiology 102: 337–339, 1972.

219. Rhodes BA, Rutherford RB, Lopez-Majano V, Greyson ND, Wagner HN Jr: Arteriovenous shunt reasurements in extremities. J Nucl Med 13: 357–362, 1972.

220. Greyson ND, Rhodes BA, Buchanan JW, Wagner HN Jr: Local increases in reticuloendothelial function during healing. J Reticuloendothel Soc 11: 293–299, 1972.

221. Siegel ME, Malmud LS, Rhodes BA, Bell WR, Wagner HN Jr: Scanning of thromboemboli with 131I-streptokinase. Radiology 103: 695–696, 1972.

222. Wagner HN Jr, Rhodes BA: Cardiovascular Diseases—Radioactive tracers in diagnosis of cardiovascular disease. J Cardiovas Dis 15: 1–24, 1972.

223. Hurley PJ, Wagner HN Jr: Diagnostic Value of brain scanning in children. JAMA 221: 877–881, 1972.

224. Wagner HN Jr: Nuclear medicine in cardiovascular disease. Hosp Pract 7: 108–116, 1972.

225. Rhodes BA, Greyson ND, Hamilton CR Jr, White RI Jr, Giargiana FA Jr, Wagner HN Jr: Absence of anatomic arteriovenous shunts in Paget's disease of bone. N Engl J Med 287: 686–689, 1972.

226. Moses DC, Davis LE, Wagner HN Jr: Brain scanning with 99mTcO4- in multiple sclerosis. J Nucl Med 13: 847–848, 1972.

227. DeBlanc HJ Jr, Charache P, Wagner HN Jr: Automatic radiometric measurerent of antibiotic effect on bacterial growth. Antimicrob Agents Chemother 2: 360–366, 1972.

228. Som P, Hosain F, Wagner HN Jr: Accelerated clearance of radioactive chelate from cerebrospinal fluid in experimental meningitis. Concise Communication—J Nucl Med 13: 942–944, 1972.

229. Lomas F, Wagner HN Jr: Accumulation of ionic 67Ga in empyema of the gallbladder. Work in Progress—Radiology 105: 689–692, 1972.

230. Murphy BEP, Wagner HN Jr: Nuclear In-vitro procedures. I: Competitive protein-binding radioassays. II: Functional radioassays and other techniques. Hosp Pract 7: 65–72, 1972.

231. Strauss HW, James AE Jr, Hurley PJ, DeLand FH, Moses DC, Wagner HN Jr: Nuclear cerebral angiography—usefulness in the differential diagnosis of cerebrovascular disease and tumor. Arch Intern Med 131: 211–216, 1973.

232. Maisey MN, Natarajan TK, Hurley PJ, Wagner HN Jr: Validation of a rapid computerized method of measuring 99mTc pertechnetate uptake for routine assessment of thyroid structure and function. J Clin Endocrinol Metab 36: 317–322, 1973.

233. McKusick KA, Soin JS, Ghiladi A, Wagner HN Jr: Gallium-67 accumulation in pulmonary sarcoidosis. JAMA 22: 688 Letters, 1973.

234. Maisey MN, Moses DC, Hurley PJ, Wagner HN Jr: Improved methods for thyroid scanning. JAMA 223: 761–763, 1973.

235. McKusick KA, Malmud LS, Kirchner PT, Wagner HN Jr: An interesting artifact in radionuclide imaging of the kidneys. Case Report J Nucl Med 14: 113–114, 1973.

236. Moses DC, Natarajan TK, Preziosi, Udvarhelyi GB, Wagner HN Jr: Quantitative cerebral circulation studies with sodium pertechnetate. J Nucl Med 14: 142–148, 1973.

237. Briedis D, McIntyre PA, Judisch J, Wagner HN Jr: An evaluation of a dual-isotope method for the measurement of vitamin B-12 absorption. J Nucl Med 14: 135–141, 1973.

238. Wagner HN Jr: Nuclear tracer studies of the cerebral circulation. Hosp Pract 8: 152–159, 1973.

239. Soin JS, McKusick KA, Wagner HN Jr: Regional lung-function abnormalities in narcotic addicts. JAMA 224: 13, 1717–1720, 1973.

240. Wagner HN Jr: Estado Actual de la Medicina Nuclear. El Hospital 21-23, Agosto/Septiembre 1973.

241. Larson SM, Charache P, Chen M, Wagner HN Jr: Automated detection of haemophilus influenzae. Appl Microbiol 25: 6, 1011–1012, 1973.

242. Castronova FP, Wagner HN Jr: Comparative toxicity and pharmaco-dynamics of ionic indium chloride and hydrated indium oxide. J Nucl Med 14: 9, 677–682, 1973.

243. Siegel ME, Giargiana FA, Rhodes BA, White RI, Wagner HN Jr: Effect of Reactive hyperemia on the distribution of radioactive microspheres in patients with peripheral vascular disease. Am J Roentgenol Radium Ther Nucl Med 118: 814–819, 1973.

244. Soin JS, James AE Jr, Wagner HN Jr: Detection of pulmonary hypertension by perfusion lung scan. Am J Roentgenol Radium Ther Nucl Med 118: 792–800, 1973.

245. Rhodes BA, Greyson ND, Siegel ME, Giargiana FA, White RI, Williams GM, Wagner HN Jr: The distribution of radioactive microspheres after intra-arterial injection in the legs of patients with peripheral vascular disease. Am J Roentgenol Radium Ther Nucl Med 118: 820–826, 1973.

246. Greyson ND, Rhodes BA, Williams GM, Wagner HN Jr: Radiometric detection of venous function and disease. Surg Gynecol Obstet 137: 220–226, 1973.

247. Poulose KP, Reba RC, Cameron JL, Wagner HN Jr: The value and limitations of liver scanning for the detection of hepatic metastases in patients with cancer. J Indian Med Assoc 61: 199–205, 1973.

248. McKusick KA, Malmud LS, Kordela PA, Wagner HN Jr: Radionuclide cisternography: Normal values for nasal secretion of intrathecally injected 111In-DTPA. J Nucl Med 14: 933–934, 1973.

249. Syed IB, Hosain F, Dugal P, Wagner HN Jr: Bone scanning agents: Present status and choice. Indian J Cancer 10: 280–284, 1973.

250. Wagner HN Jr, Strauss HW: A new approach to coronary heart disease. Circulation 48: 229–231, 1973.

251. Langan JK, Wagner HN Jr: A system to record, view, store and distribute nuclear medicine images and records. J Nucl Med 14: 588–590, 1973.

252. Giargiana FA, Siegel ME, James AE Jr, Rhodes BA, Wagner HN Jr, White RI: A preliminary report on the complementary roles of arteriography and perfusion scanning in assessment of peripheral vascular disease. Radiology 108: 619–627, 1973.

253. Wagner HN Jr: Nuclear medicine in cardiovascular diseases. La Ricerca Clin Lab 4: 209–226, 1974.

254. Rhodes BA, Lopez-Majano, Wagner HN Jr: Letters to the Editor, Arteriovenous shunting after sympathectomy. Surgery 75: 153–154, 1974.

255. Larson SM, Charache P, Chen M, Wagner HN Jr: Inhibition of the metabolism of streptococci and salmonella by specific antisera. Appl Microbiol 27: 351–355, 1974.

256. Rhodes BA, Wagner HN Jr: Letters to the editor, Adverse reaction to radiopharmaceuticals. J Nucl Med 15: 213–214, 1974.

257. McKusick KA, Wagner HN Jr, Soin JS, Benjamin JJ, Cooper M, Ball WC, Jr: Measurement of regional lung function in the early detection of chronic obstructive pulmonary disease. Scand J Resp Dis Suppl 85: 61–63, 1974.

258. Myers WG, Wagner HN Jr: Nuclear Medicine: How it began. Hosp Pract 9: 103–113, 1974.

259. Wagner HN Jr: It all depends on whose ox is gored. Medical Laboratory Observer, May–June, 83–84, 1974.

260. Wagner HN Jr: Color in nuclear medicine: contribution or camouflage. Hosp Pract July, 87–91, 1974.

261. Siegel ME, Giargiana FA Jr, Wagner HN Jr: Verification and quantification of anatomic arteriovenous shunting in a hypernephroma. J Urol 112: 16–18, 1974.

262. Siegel ME, Friedman BH, Wagner HN Jr: A new approach to breast cancer, Breast uptake of 99m Tc-HEDSPA. JAMA 229: 1769–1771, 1974.

263. Russell CD, DeBlanc HJ Jr, Wagner HN Jr: Components of variance in laboratory quality control. Johns Hopkins Med J 135: 344–357, 1974.

264. Rhodes BA, Kamanetz GS, Wagner HN Jr: The use of limulus testing to reduce the incidence of adverse reactions to cisternographic agents. Neurology 24: 9, 1974.

265. Mer T, Malmud L, McKusick K, Wagner HN Jr: The mechanism of 67GA association with lymphocytes. Cancer Research 34: 2495–2499, 1974.

266. Lovegrove F, Langan J, Wagner HN Jr: Quality control in nuclear medicine procedures. J Nucl Med Tech 2: 44–51, 1974.

267. Wagner HN Jr, Kashima HK, McKusick KA, Malmud LS: Gallium-67 scanning in patients with head and neck cancer. Laryngoscope 84: 1078–1089, 1974.

268. Hill JH, Wagner HN Jr: 67Ga-uptake in the regenerating rat liver. J Nucl Med 15: 818–820, 1974.

269. Heshiki A, Schatz SL, McKusick KA, Bowersox DW, Soin JS, Wagner HN Jr: Gallium 67 citrate scanning in patients with pulmonary sarcoidosis. Am J Roentgenol Radium Ther Nucl Med 122: 744–749, 1974.

270. Friedman BH, Lovegrove FT, Wagner HN Jr: An unusual variant in cerebral circulation studies. J Nucl Med 15: 5, 1974.

271. Chen M, Rhodes BA, Larson SM, Wagner HN Jr: Sterility testing of radiopharmaceuticals. J Nucl Med 15: 12, 1974.

272. Camargo EE, Larson SM, Tepper BS, Wagner HN Jr: Radiometric measurement of metabolic activity of mycobacterium lepraemurium. Appl Microbiol 28: 3, 1974.

273. Amrein PC, Larson SM, Wagner HN Jr: A rapid automated system for measurement of antibody titers. J Nucl Med 15: 12, 1974.

274. Amrein PC, Larson SM, Wagner HN Jr: An automated system for leukocyte metabolism. J Nucl Med 15: 5, 1974.

275. Prokop EK, Strauss HW, Shaw J, Pitt B, Wagner HN Jr: Comparison of 5egional myocardial perfusion determined by ionic potassium-43 to that determined by microspheres. Circulation 50: 978–984, 1974.

276. Fee HJ Jr, Prokop EK, Cameron JL, Wagner HN Jr: Liver scanning in patients with suspected abdominal tumor. JAMA 230: 1675–1677, 1974.

277. Prokop EK, Buddemeyer EU, Strauss HW, Wagner HN Jr: Detection and localization of an occult vesicoenteric fistula. Am J Roentgenol Radium Ther Nucl Med 121: 811–818, 1974.

278. Giargiana FA Jr, White RI Jr, Greyson ND, Rhodes BA, Siegel ME, Wagner HN Jr, James AE Jr: Absence of arteriovenous shunting in peripheral arterial disease. Invest Radiol 9: 4, 1974.

279. Wagner HN Jr: Nuclear medicine and coronary heart disease. Trans Assoc Life Ins Med Dir Am 58: 173–179, 1974.

280. Larson SM, Chen M, Charache P, Wagner HN Jr: Radiometric identification of Streptococcus Group A in throat cultures. J Nucl Med 16: 1085–1086, 1975.

281. Soin JS, Wagner HN Jr, Thomashaw D, Brown TC: Increased sensitivity of regional measurements in early detection of narcotic lung disease. Chest 67: 325–330, 1975.

282. Eikman EA, Cameron JL, Colman M, Natarajan TK, Dugal P, Wagner HN Jr: A test for patency of the cystic duct in acute cholecystitis. Ann Intern Med 82: 318–322, 1975.

283. Oster SH, Larson SM, Strauss HW, Wagner HN Jr: Analysis of liver scanning in a general hospital. J Nucl Med 16, 6: 450–453, 1975.

284. Kirchner PT, James AE Jr, Reba RC, Wagner HN Jr: Patterns of excretion of radioactive chelates in obstructive uropathy. Radiology 114: 655–661, 1975.

285. Wagner HN Jr, Strauss HW: Radioactive tracers in cardiac diagnosis. Cardiovasc Clin 6: 319–336, 1975.

286. Siegel ME, Giargiana FA Jr, Rhodes BA, Williams GM, Wagner HN Jr: Perfusion of ischemic ulcers of the extremity. Arch Surg 10: 265–268, 1975.

287. Wagner HN Jr: Nuclear medicine pioneer citation—1975—George V. Taplin, M.D. J Nucl Med 16: 504–507, 1975.

288. Adelstein SJ, Jansen C, Wagner HN Jr: Report of the intersociety commission for heart disease resources. Optimal resource guidelines for radioactive tracer studies of the heart and circulation. Circulation 52: A9–A22, 1975.

289. Siegel ME, Williams GM, Giargiana FA, Wagner HN Jr: A useful, objective criterion for determining the healing potential of an ischemic ulcer. J Nucl Med 16: 993–995, 1975.

290. Klingensmith WC III, Eikman EA, Maumenee L, Wagner HN Jr: Widespread abnormalities of radiocolloid distribution in patients with mucopolysaccharidoses. J Nucl Med 16: 1002–1006, 1975.

291. Cummings DM, Ristroph D, Camargo EE, Larson SM, Wagner HN Jr: Radiometric detection of the metabolic activity of Mycobacterium tuberculosis. J Nucl Med 16: 1189–1191, 1975.

292. Ross JF, Wagner HN Jr, Hutchens TT, Tauxe WN, Adelstein SJ, Bender MA, Fish MB, Gottschalk A, Kriss JP, Kuhl DE, McAfee JG, Peterson RE: The American Board of Nuclear Medicine's Information, Policies, and Procedures, 1975. J Nucl Med 16: 691–696, 1975.

293. Hill JH, Merz T, Wagner HN Jr: Iron-induced enhancement of 67Ga uptake in a model human leukocyte culture system. J Nucl Med 12: 1183–1186, 1975.

294. Siegel ME, Giargiana FA Jr, White RI Jr, Friedman BH, Wagner HN Jr: Peripheral vascular perfusion scanning, correlation with the arteriogram and clinical assessment in the patient with peripheral vascular disease. Am J Roentgenol Radium Ther Nucl Med 125: 628–633, 1975.

295. Camargo EE, Larson SM, Tepper BS, Wagner HN Jr: A radiometric method for predicting effectiveness of chemotherapeutic agents in murine leprosy. Int J Leprosy 43: 234–238, 1975.

296. Wagner HN Jr: State of the art—the use of radioisotope techniques for the evaluation of patients with pulmonary disease. Am Rev Respir Dis 113: 203–218, 1976.

297. Brody KR, Hosain P, Spencer RP, Hosain F, Wagner HN Jr: Technetium-99m labeled imidodiphosphate: an improved bone-scanning radiopharmaceutical. Br J Radiol 49: 267–269, 1976.

298. Oster AH, Larson SM, Wagner HN Jr: Possible enhancement of 67Ga-citrate imaging by iron dextran. J Nucl Med 17: 356–358, 1976.

299. Tsan MF, Chen WY, Newman B, Wagner HN Jr, McIntyre PA: Effects of mitogens on glucose oxidation by lymphocytes from normal individuals and patients with chronic leukemia. Johns Hopkins Med J 138: 113–118, 1976.

300. Camargo EE, Larson SM, Tepper BS, Wagner HN Jr: Radiometric studies of Mycobacterium lepr-aemurium. Int J Leprosy 44: 294, 1976.

301. Siegel ME, Wagner HN Jr: Radioactive tracers in peripheral disease. Semin Nucl Med 6: 253–278, 1976.

302. Kim HR, Buchanan JW, D'Antonio R, Larson SM, Morgan RP, Thorell JI, McIntyre PA, Wagner HN Jr: Toadfish serum as a binder for in vitro assay of vitamin B12. J Nucl Med 17: 737–739, 1976.

303. Wagner HN Jr: Cardiovascular nuclear medicine: a progress report. Hosp Pract, July, 77–83, 1976.

304. D'Antonio N, Tsan MF, Charache P, Larson S, Wagner HN Jr: Simple radiometric technique for rapid detection of herpes simplex virus Type I in WI-38 cell culture. J Nucl Med 17: 503–507, 1976.

305. Klingensmith WC, Tsan MF, Hsu CK, Wagner HN Jr: Intravascular phagocytic activity of the lung during varying levels of circulating monocytes and neutrophils. J Reticuloendothel Soc 19: 375–381, 1976.

306. Klingensmith WC III, Tsan MF, Wagner HN Jr: Factors affecting the uptake of 99mTc-sulfur colloid by the lung and kidney. J Nucl Med 17: 681–684, 1976.

307. Adachi H, Strauss HW, Ochi H, Wagner HN Jr: The effect of hypoxia on the regional distribution of cardiac output in the dog. Circ Res 39: 314–319, 1976.

308. Cook DJ, Bailey I, Strauss HW, Rouleau J, Wagner HN Jr, Pitt B: Thallium-201 for myocardial imaginq: appearance of the normal heart. J Nucl Med 17: 583–589, 1976.

309. Klingensmith WC III, Danish EH, Dover GJ, Wagner HN Jr: Delineation of peripheral bone infarcts in a child with a rare hemoglobinopathy (SOArab) and purpura fulminans: case report. J Nucl Med 17: 1062–1064, 1976.

310. Wagner HN Jr (Guest Editor): Introduction. Symposium on Advances in Cardiovascular Nuclear Medi-cine. Am J Cardiol 38: 709–710, 1976.

311. Wagner HN Jr, Wake R, Nickoloff E, Natarajan TK: The nuclear stethoscope: a simple device for genera-tion of left ventricular volume curves. Am J Cardiol 38: 747–750, 1976.

312. Gray HW, Tsan MF, Wagner HN Jr: A quantitative study of leukocyte cohesion: effects of divalent cations and pH. J Nucl Med 18: 147–150, 1977.

313. Wagner HN Jr: Nuclear medicine in motion. J Nucl Med 18: 2–4, 1977.

314. Wagner HN Jr, Knowles LG: Nuclear cardiology: A new medical specialty. APL Tech Digest 16: 2–10, 1977.

315. Buchanan JW, McIntyre PA, Scheffel U, Wagner HN Jr: Comparison of toadfish-serum competitive binding and microbiologic assays of vitamin B12. J Nucl Med 18: 394–398, 1977.

316. Chen MF, McIntyre PA, Wagner HN Jr: A radiometric microbiologic method for vitamin B12 assay. J Nucl Med 18: 388–393, 1977.

317. Wagner HN Jr: The diagnostic process after 25 years. Johns Hopkins Med J 141: 177–181, 1977.

318. Thomashow O, Summer WR, Soin J, Wagner HN, Brown TC: Lung disease in reformed drug addicts: diagnostic and physiologic correlations. Johns Hopkins Med J 141: 1–8, 1977.

319. Holder LE, Martire JR, Holmes ER, Wagner HN Jr: Testicular radionuclide angiography and static imaging: anatomy, scintigraphic interpretation, and clinical indications. Radiology 125: 739–752, 1977.

320. Holmes ER III, Klingensmith WC III, Kirchner PT, Wagner HN: Phantom kidney in technetium-99m DTPA studies of renal blood flow: case report. J Nucl Med 18: 702–705, 1977.

321. Klingensmith WC III, Lotter MG, Knowles LG, Motazedi A, Wagner HN Jr: Physiological interpretation of time-activity curves from cerebral flow studies. Medical Imaging 2: 43–44, 1977.

322. Tsan Min-Fu, Chen WY, Scheffel U, Wagner HN Jr: Studies on gallium accumulation in inflammatory lesions: I. Gallium uptake by human polymorphonuclear leukocytes. J Nucl Med 19: 36–43, 1978.

323. Menon S, Wagner HN Jr, Tsan Min-Fu: Studies on gallium accumulation in inflammatory lesions: II. Uptake by Staphylococcus aureus: Concise Communication. J Nucl Med 19: 44–47, 1978.

324. Tran N, Wagner HN Jr: Liquid scintillation vial for radiometric assay of lymphocyte carbohydrate metabolism in response to mitogens. J Nucl Med 19: 61–63, 1978.

325. Wagner HN Jr, Lotter MG, Douglass KH, Alderson PO, Knowles LG: Cinematic display of regional func-tion in nuclear imaging, Johns Hopkins Med J 142: 61–66, 1978.

326. D'Antonio NL, Tsan MF, Griffin DE, Charache PA, Wagner HN Jr: Radiometric detection of herpes simplex viruses. J Nucl Med 19: 185–190, 1978.

327. Harrison K, Wagner HN Jr: Biodistribution of intravenously injected [14C] doxorubicin and [14C] daunorubicin in mice: concise communication. J Nucl Med 19: 84–86, 1978.

328. Hurlburt EM, Ki PF, Wagner HN Jr: Effect of cytomegalovirus infection on metabolism of WI-38 cell cultures: concise communication. J Nucl Med 19: 191–194, 1978.

329. Kertcher JA, Chen MF, Charache P, Hwangbo CC, Camargo EE, McIntyre PA, Wagner HN Jr: Rapid radiometric susceptibility testing of mycobacterium tuberculosis. Am Rev Resp Dis 117: 631–637, 1978.

330. Klingensmith WC III, Yang SL, Wagner HN Jr: Lung uptake of Tc-99m sulfur colloid in liver and spleen imaging. J Nucl Med 19: 31–35, 1978.

331. Qureshi S, Wagner HN Jr, Alderson PO, Housholder DF, Douglass KH, Lotter MG, Nickoloff EL, Tanabe M, Knowles LG: Evaluation of left-ventricular function in normal persons and patients with heart disease. J Nucl Med 19: 135–141, 1978.

332. Alderson PO, Wagner HN Jr, Gomez-Moeiras JJ, Rehn TG, Becker LC, Douglass KH, Manspeaker HF, Scindledecker GR: Simultaneous detection of myocardial perfusion and wall motion abnormalities by cinematic 201T1 imaging. Radiology 127: 531–533, 1978.

333. Wagner HN: Images of the future. J Nucl Med 19: 599–605, 1978.

334. Ward K, Klingensmith WC III, Sterioff S, Wagner HN Jr: The origin of lymphoceles following renal transplantation. Transplantation 25: 346–347, 1978.

335. Wagner HN Jr: Lewis A. Conner Memorial Lecture, "Nuclear Cardiology: 1978". Abstracts of the 51st Scientific Sessions, American Heart Association. Circulation 57 & 58, Suppl II: II-1–II-3, 1978.

336. Wagner HN Jr: Time for a FASORS? Letter to the Editor. J Nucl Med 20: 581–582, 1979.

337. Yang SL, Alderson O, Kaizer HA, Wagner HN Jr: Serial Ga-67 citrate imaging in children with neoplastic disease: concise communication. J Nucl Med 20: 210–214, 1979.

338. Wagner HN Jr, Rigo P, Baxter RH, Alderson PO, Douglass KH, Housholder DF: Monitoring ventricular function at rest and during exercise with a nonimaging nuclear detector. Am J Cardiol 43: 975–979, 1979.

339. Douglass KH, Wagner HN Jr, Shindledecker JG: Use of color display for selection of left ventricular regions of interest. (Technical Notes) Radiology 131: 249–250, 1979.

340. Wagner HN Jr, Rigo P, Baxter RH, Alderson PO: Scintillation probe detector in the assessment of cardio-vascular disease. Medical Instrumentation 13: 152–155, 1979.

341. Buchanan JW, Wagner HN Jr: Teamwork in cardiovascular nuclear medicine. J Nucl Med 20: 377–378, 1979.

342. Camargo EE, Wagner HN Jr, Tsan MF: Studies of gallium accumulation in inflammatory lesions. IV. Kinetics of accumulation and role of polymorphonuclear leukocytes in the distribution of gallium in experimental inflammatory exudates. Nucl Med (Stutt) 18: 147–150, 1979.

343. Wagner HN Jr: Invasive vs noninvasive testing in possible coronary artery disease (Questions and Answers). JAMA 242: 765–766, 1979.

344. Camargo EE, Kertcher, JA, Larson SM, Tepper BS, Wagner HN Jr: Radiometric measurement of differen-tial metabolism of fatty acids by Mycobacterium lepraemurium. Int J Leprosy 47: 126–132, 1979.

345. Alderson PO, Vieras F, Housholder DF, Mendenhall KG, Wagner HN Jr: Gated and cinematic perfusion lung imaging in dogs with experimental pulmonary embolism. J Nucl Med 20: 407–412, 1979.

346. Rigo P, Becker LC, Griffith LSC, Alderson PO, Bailey IK, Pitt B, Burow RD, Wagner HN Jr: Influence of coronary collateral vessels on the results of thallium-201 myocardial stress imaging. Am J Cardiol 44: 452–458, 1979.

347. Rigo P, Alderson PO, Robertson RM, Becker LC, Wagner HN Jr: Measurement of aortic and mitral regur-gitation by gated cardiac blood pool scans. Circulation 60: 306–312, 1979.

348. Alderson PO, Lee H, Summer WR, Motazedi A, Wagner HN Jr: Comparison of SE-133 washout and single-breath imaging for the detection of ventilation abnormalities. J Nucl Med 20: 917–922, 1979.

349. Langan JD, Wagner HN Jr, Buchanan JW: Design concepts of a nuclear medicine department. J Nucl Med 20: 1093–1094, 1979.

350. Lipson A, Nickoloff EL, Hsu TH, Kasecamp WR, Drew HM, Shakir R, Wagner HN Jr: A study of age-dependent changes in thyroid function tests in adults. J Nucl Med 20: 1124–1130, 1979.

351. Alderson PO, Douglass KH, Mendenhall KG, Guadiani VA, Watson DC, Links JM, Wagner HN Jr: Decon-volution analysis in radionuclide quantitation of left-to-right cardiac shunts. J Nucl Med 20: 502–506, 1979.

352. Katz RD, Alderson PO, Rosenshein NB, Bowerman JW, Wagner HN Jr: Utility of bone scanning in detect-ing occult skeletal metastases from cervical carcinoma. Radiology 133: 469–472, 1979.

353. Smaldone GC, Itoh H, Swift DL, Wagner HN Jr: Effect of flow-limiting segments and cough on particle deposition and mucociliary clearance in the lung. Am Rev Resp Dis 120: 747–758, 1979.

354. Bourguignon MH, Wagner HN Jr: Noninvasive measurement of ventricular pressure throughout systole. Am J Cardiol 44: 466–471, 1979.

355. Wagner HN Jr, Bourguignon MN: Noninvasive measurement of ventricular pressure—I. Am J Cardiol 46: 715, 1980.

356. Bourguignon MH, Wagner HN Jr: Reply—Noninvasive measurement of ventricular pressure—II. Am J Cardiol 46: 716, 1980.

357. Baxter RH, Becker LW, Alderson PO, Rigo P, Wagner HN Jr, Weisfeldt ML: Quantification of aortic valvular regurgitation in dogs by nuclear imaging. Circulation 61: 404–410, 1980.

358. Boonyaprapa S, Alderson PO, Garfinkel DJ, Chipps BE, Wagner HN Jr: Detection of pulmonary aspiration in infants and children with respiratory disease. J Nucl Med 21: 314–318, 1980.

359. Buchanan JW, Wagner HN Jr: Nuclear medicine and the patient with cancer. Surg Rounds 3: 50–62, 1980.
360. Rigo P, Bailey IK, Griffith LSC, Pitt B, Burow RD, Wagner HN Jr, Becker LC: Value and limitations of segmental analysis of stress thallium myocardial imaging for localization of coronary artery disease. Circulation 61: 973–981, 1980.
361. Tzen KY, Oster ZH, Wagner HN Jr, Tsan MF: Role of iron-binding proteins and enhanced capillary permeability on the accumulation of gallium-67. J Nucl Med 21: 31–35, 1980.
362. van Aswegen A, Alderson PO, Nickoloff EL, Housholder DF, Wagner HN Jr: Temporal resolution requirements for left ventricular time-activity curves. Radiology 135: 165–170, 1980.
363. Wagner HN Jr: Nuclear imaging: new developments. Hosp Pract, April, 117–126, 1980.
364. Wagner HN Jr: A new phase of nuclear medicine. Isotope News, No. 309, page 1, 1980.
365. Wagner HN Jr: University-hospital cyclotron. Japanese Society of Nuclear Medicine 16 (Supplementum): 165–167, 1980.
366. Wagner HN Jr: President's viewpoint. BCMS Newsletter 9(1): 4, 1980 (Jan).
367. Wagner HN Jr: President's viewpoint. BCMS Newsletter 9(2): 1980 (Feb).
368. Wagner HN Jr: President's viewpoint. BCMS Newsletter 9(2): 1980 (Mar).
369. Wagner HN Jr: President's viewpoint. BCMS Newsletter, 1980 (July).
370. Wagner HN Jr: President's viewpoint. BCMS Newsletter 9 (8): 4, 1980 (Sept).
371. Wagner HN Jr: President's viewpoint. BCMS Newsletter, 1980.
372. Wagner, HN: Chongqing Yiyao (2): 18–24, 1980.
373. ••, Wagner, HN, Jr: Medical Information, 7: 112–113, 1980. (Printed in Chinese language).
374. Wagner HN Jr, Buchanan JW: Radioactive tracers and the heart. Symposium on Noninvasive Cardiac Diagnosis I. Med Clin North Am 64: 83–98, 1980.
375. Wagner HN Jr, Buchanan JW: Radioactive tracers and the heart. Japanese Society of Nuclear Medicine 16 (Supplementum): 184–191, 1980.
376. Wagner HN Jr, Buchanan JW: Nuclear medicine: Get the most from lung scans. Diagnosis 3: 25–47, 1980.
377. Camargo EE, Harrison KS, Wagner HN Jr, Bourguignon MH, Reid PR, Alderson PO, Baxter RH: Noninvasive beat to beat monitoring of left ventricular function by a nonimaging nuclear detector during premature ventricular contractions. Am J Cardiol 45: 1219–1224, 1980.
378. Burns HD, Marzilli LG, Dannals RF, Dannals TE, Tragesor TC, Conti P, Wagner HN Jr: 125I-4-iodophenyltrimethylammonium ion, an iodinated acetylcholinesterase inhibitor with potential as a myocardial imaging agent. J Nucl Med 21: 875–879, 1980.
379. Liu Xiujie, Wagner HN Jr: Evaluation of left ventricular function by using nuclear stethoscope and 113mIndium. Chinese J Cardiol, July, 90–94, 1980.
380. Links JM, Douglass KH, Wagner HN Jr: Patterns of ventricular emptying by Fourier analysis of gated blood-pool studies. J Nucl Med 21: 978–982, 1980.
381. Wagner HN Jr: Nuclear imaging: past, present and future. Diagnostic Imaging, June, 2, 3, 14, 1980.
382. Leitl GP, Buchanan JW, Wagner HN Jr: Monitoring cardiac function with nuclear techniques. Am J Cardiol 46: 1125–1132, 1980.
383. Conklin JJ, Camargo EE, Wagner HN Jr: Bone scan detection of peripheral periosteal leiomyoma. J Nucl Med 22: 97, 1981.
384. Bourguignon MH, Douglass KH, Links JM, Wagner HN Jr: Fully automated data acquisition, processing, and display in equilibrium radio-ventriculography. Eur J Nucl Med 6: 343–347, 1981.
385. Bourguignon MH, Schindeldecker JG, Carey GA, Douglass KH, Burow RD, Camargo EE, Becker L, Wagner HN Jr: Quantification of left ventricular volume in gated equilibrium radioventriculography. Eur J Nucl Med 6: 349–353, 1981.
386. LaFrance ND, Wagner HN Jr, Whitehouse P, Corley E, Duelfer T: Decreased accumulation of isopropyl-iodoamphetamine (I-123) in brain. J Nucl Med 22: 1081–1083, 1981.
387. Brown JF, Buchanan JW, Wagner HN Jr: Pitfalls in technetium-99m HIDA biliary imaging: duodenal diverticulum simulating the gallbladder. J Nucl Med 22: 747–748, 1981.
388. Maurer AH, Chen DCP, Camargo EE, Wong DF, Wagner HN Jr, Alderson PO: Utility of three-phase skeletal scintigraphy in suspected osteomyelitis. J Nucl Med 22: 941–949, 1981.
389. Bourguignon MH, Links JM, Douglass KH, Alderson PO, Roland JM, Wagner HN Jr: Quantification of left to right cardiac shunts by multiple deconvolution analysis. Am J Cardiol 48: 1086–1091, 1981.
390. Wagner HN Jr: Hevesy Nuclear Medicine Pioneer Lecture. J Nucl Med 22: 573–576, 1981.
391. Wagner HN Jr: Use of the nuclear stethoscopeTM to monitor ventricular function. Practical Cardiology 7: 113–129, 1981.
392. Rigo P, Bailey IK, Griffith LSC, Pitt B, Wagner HN Jr, Becker LC: Stress thallium-201 myocardial scintigraphy for the detection of individual coronary arterial lesions in patients with and without previous myocardial infarction. Am J Cardiol 48: 209–216, 1981.

393. Espinola D, Rupani HD, Camargo EE, Wagner HN Jr: Ventilation perfusion imaging in pulmonary papillomatosis. J Nucl Med 22: 975–977, 1981.
394. ••, Wagner HN Jr: Assessment of left ventricular function with the nuclear stethoscope and 113mIn—A report of 130 case studies. Chinese J Nucl Med 3-6, 1981. (Printed in Chinese language)
395. Wagner HN Jr: Bayes' theorem: an idea whose time has come. Editorial. Am J Cardiol 49: 875–877, 1982.
396. Wagner HN Jr: Letter to the Editor (re Patholoqy and Probabilities). N Engl J Med 306: 305, 1982.
397. Links JM, Becker LC, Shindledecker JG, Guzman P, Burow RD, Nickoloff EL, Alderson PO, Wagner HN: Measurement of absolute left ventricular volume from gated blood pool studies. Circulation 65: 82–91, 1982.
398. D'Antonio RC, Camargo EE, Gedra T, Wagner HN Jr, Charache P: Rapid radiometric serum test for antibiotic activity. Antimicrob Agents Chemother 21: 236–240, 1982.
399. Wagner HN Jr: The nuclear cardiac probe. Hosp Pract, April, 163–165, 169–177, 1982.
400. Strauss HW, Boucher CA, Ritchie JL, Holman BL, Wagner HN Jr: Nuclear cardiology: Introduction. Am J Cardiol 49: 1337–1341, 1982.
401. Wagner HN Jr: The future of nuclear cardiology. Am J Cardiol 49: 1355–1361, 1982.
402. Harrison KS, Liu X, Han ST, Camargo EE, Wagner HN Jr: Evaluation of a miniature CdTe detector for monitoring left ventricular function. Eur J Nucl Med 7: 204–206, 1982.
403. Liu XJ, Harrison KS, Wagner HN Jr: Measurement of left ventricular ejection fraction with ionic 113mIn and a cardiac probe. Eur J Nucl Med 7: 410–412, 1982.
404. Wagner HN Jr: Advances in chemistry highlight SNM's 1982 meeting. SNM Newsline 7: 5, 10, 1982 (September).
405. Wagner HN Jr: World Congress of Nuclear Medicine and Biology A Smashing Success. SNM Newsline: 1, 6, 1982 (December).
406. Wagner HN Jr: Nuclear techniques in ischemic heart disease. Am Heart J 103: 681–688, 1982.
407. Douglass KH, Tibbits P, Kasecamp W, Han ST, Koller D, Links JM, Wagner HN Jr: Performance of a fully automated program for measurement of left ventricular ejection fraction. Eur J Nucl Med 7: 564–566, 1982.
408. Links JM, Wagner HN Jr: Specification of performance of positron emission tomography scanners. Letter to the Editor. J Nucl Med 23: 82, 1982.
409. Jiang MS, Corley EG, Wagner HN Jr, Tsan MF: Localization of abscess with an iodinated synthetic chemotactic peptide. Nucl Med (Stutt) 21: 110–113, 1982.
410. Guilarte TR, Burns HD, Dannals RF, Wagner HN Jr: A simple radiometric in vitro assay for acetylcholinesterase inhibitors. J Pharm Sci 72: 90–92, 1983.
411. Wagner HN Jr: Radiation: the risks and the benefits. Am J Roentgenol 140: 595–603, 1983.
412. DeLand FH, Wagner HN Jr: Nuclear medicine in hepatic mass lesions. Semin Roentgenol 18: 106–113, 1983.
413. Liu XJ, Wagner HN Jr, Tao S: Measurement of effects of the Chinese herbal medicine higenamine on left ventricular function using a cardiac probe. Eur J Nucl Med 8: 233–236, 1983.
414. Wagner HN Jr: Chemistry continues dominant role in nuclear medicine. SNM Newsline, 1983, pp 6–8 (September).
415. Wagner HN Jr, Cooper J: Horseshoe crab provides alternative to rabbit bioassay. The Johns Hopkins Center for Alternatives to Animal Testing 2: 1–2, 1983.
416. Wagner HN Jr, Burns HD, Dannals RF, Wong DF, Langstrom B, Duelfer T, Frost JJ, Ravert HT, Links JM, Rosenbloom S, Lukas SE, Kramer AV, Kuhar MJ: Assessment of dopamine receptor activity in the human brain with carbon-11 N-methyl spiperone: Dopamine receptors have been imaged in baboon and human brain by positron tomography. Science 221: 1264–1266, 1983.
417. Liu XJ, Wagner HN Jr, Ehrlich W, Harrison KS: Nuclear stethoscope. Chinese J Nucl Med 3: 151–154, 1983. (Article published in Chinese)
418. Liu, XJ, Wagner HN Jr, Harrison KS, Douglass KH, Camargo EE: Left ventricular ejection fraction. Chinese J Nucl Med 3: 155–158, 1983. (Article published in Chinese)
419. Wagner HN Jr: Two positron emission tomography meetings: Seventh Nobel Conference on the Metabolism of the Human Brain Studied with Positron Emission Tomography, Saltsjobaden, Stockholm, Sweden, May 17–20, 1983, and Research Issues in Positron Emission Tomography, Bethesda, Maryland, July 16–17, 1983. J Comput Assist Tomogr 7: 1128–1131, 1983.
420. Conklin JJ, Alderson PO, Zizic TM, Hungerford DS, Densereaux JY, Gober A, Wagner HN: Comparison of bone scan and radiograph sensitivity in the detection of steroid-induced ischemic necrosis of bone. Radiology 147: 221–226, 1983.
421. Al-Eid MA, Tutschka PJ, Wagner HN Jr, Santos GW, Tsan MF: Functional asplenia in patients with chronic graft-versus-host disease: concise communication. J Nucl Med 24: 1123–1126, 1983.

422.   Waud JM, Chan DW, Drew HM, Oropeza MJ, Sucupira MS, Scheinin B, Garrison GM, Mayo M, Taylor E, Stem J, Graham D, Coyle JT, Niebyl J, Wagner HN Jr: Clinical evaluation of two direct procedures for free thyroxin, and of free thyroxin index determined nonisotopically and by measuring thyroxin-binding globulin. Clin Chem 29: 1908–1911, 1983.

423.   Wong DF, Espinola D, Camargo EE, Douglass KH, Koller DW, Wagner HN Jr: Sequential computer-assisted hepatobiliary scintigraphy in the evaluation of conjoined twins. Am J Roentgenol 142: 479–481, 1984.

424.   Rosenbloom S, Lukas SE, Kramer AV, Kuhar MJ: Assessment of dopamine receptor activity in the human brain with carbon-11 labeled N-methylspiperone. Ann Neurol 15: S79–S84, 1984.

425.   Frost JJ, Dannals RF, Duelfer T, Burns HD, Ravert HT, Langstrom B, Balasubramanian V, Wagner HN Jr: In vivo studies of opiate receptors. Ann Neurol 15: S85–S92, 1984.

426.   Wagner HN Jr, Frost JJ: Summary of discussion: Normal physiology as defined by positron emission tomography. Ann Neurol 15: S110–S111, 1984.

427.   Frost JJ, Wagner HN Jr: Research Reports. Kinetics of binding to opiate receptors in vivo predicted from in vitro parameters. Brain Res 305: 1–11, 1984.

428.   Wagner HN Jr: Imaging CNS receptors: the dopaminergic system. Hosp Pract, June 187–202, 1984.

429.   Wagner HN Jr, Burns HD, Dannals RF, Wong DF, Langstrom B, Duelfer T, Frost JJ, Ravert HT, Links JM, Rosenbloom S, Lukas SE, Kramer AV, Kuhar MJ: Assessment of dopamine receptor activity in the human brain with carbon-11 N-methylspiperone. In: Proceedings from the Third Symposium on the Medical Application of Cyclotrons, Turku, Finland, June 13–16, 1983. Ann Univ Turkvensis D 17: 263–268, 1984.

430.   Nuclear Medicine Study Group (Adelstein SJ, Holman BL, Wagner HN Jr, Zaret BL, members): Optimal resources for radioactive tracer studies of the heart and circulation. Circulation 70: 525A–536A, 1984. (This is one of the 40 guideline reports of the Inter-Society Commission for Heart Disease Resources, published in Circulation.)

431.   Scheffel U, Wagner HN Jr, Frazier JM, Tsan MF: Gallium-67 uptake by the liver: studies using isolated rat hepatocytes and perfused livers. J Nucl Med 25: 1094–1100, 1984.

432.   Burns HD, Dannals RF, Langstrom B, Ravert HT, Zemyan SE, Duelfer T, Wong DF, Frost JJ, Kuhar MJ, Wagner HN Jr: (3-N[11C]methyl)-spiperone, a ligand binding to dopamine receptors: radiochemical synthesis and biodistribution studies in mice. J Nucl Med 2: 1222–1227, 1984.

433.   Wong DF, Wagner HN Jr, Dannals RF, Links JM, Frost JJ, Ravert HT, Wilson AA, Rosenbaum AE, Gjedde A, Douglass KH, Petronis JD, Folstein MF, Toung JKT, Burns HD, Kuhar MJ: Effects of age on dopamine and serotonin receptors measured by positron tomography in the living human brain. Science 226: 1393–1396, 1984.

434.   Wagner HN Jr, Dannals RF, Frost JJ, Wong DF, Ravert HT, Wilson AA, Links JM, Burns HD, Kuhar MJ, Snyder SH: Imaging neuroreceptors in the living human brain. Korean J Nucl Med 18: 95–101, 1984.

435.   Scheffel U, Wagner HN Jr, Frazier JM, Tsan MF: Gallium-67 uptake by the liver: studies using isolated rat hepatocytes and perfused livers. J Nucl Med 25: 1094–1100, 1984.

436.   Wagner HN Jr: Probing the chemistry of the mind. Editorial. N Engl J Med 312: 44–46, 1985.

437.   Inoue Y, Wagner HN Jr, Wong DF, Links JM, Frost JJ, Dannals RF, Rosenbaum AE, Takeda K, DiChiro G, Kuhar MJ: Atlas of dopamine receptor imagings (PET) of the human brain. J Comput Assist Tomogr 9: 129–140, 1985.

438.   Frost JJ, Wagner HN Jr, Dannals RF, Ravert HT, Links JM, Wilson AA, Burns HD, Wong DF, McPherson RW, Rosenbaum AE, Kuhar MJ, Snyder SH: Imaging opiate receptors in the human brain by positron tomography. J Comput Assist Tomogr 9: 231–236, 1985.

439.   Links JM, Raichlen JS, Wagner HN Jr, Reid PR: Assessment of the site of ventricular activation by Fourier analysis of gated blood-pool studies. J Nucl Med 26: 27–32, 1985.

440.   Dannals RF, Ravert HT, Frost JJ, Wilson AA, Burns HD, Wagner HN Jr: Radiosynthesis of an opiate receptor binding radiotracer: [11C]carfentanil. Int J Appl Radiat Isot 36: 303–306, 1985.

441.   Wagner HN Jr, Dannals RF, Frost JJ, Wong DF, Ravert HT, Wilson AA, Links JM, Burns HD, Kuhar MJ, Snyder SH: Imaging neuroreceptors in the human brain in health and disease. Radioisotopes 34: 103–107, 1985.

442.   Hartig PR, Scheffel U, Frost JJ, Wagner HN Jr: In vivo binding of 125I-LSD to serotonin 5-HT2 receptors in mouse brain. Life Sci 36: 657–664, 1985.

443.   Itoh H, Smaldone GC, Swift DL, Wagner HN Jr: Quantitative evaluation of aerosol deposition in constricted tubes. J Aerosol Sci 16: 167–174, 1985.

444.   Wagner HN Jr: Nuclear medicine in the 1990's: The challenge of change. SNM Scientific Meeting Highlights. SNM Newsline. J Nucl Med 26: 679–686, 1985.

445.   Wagner HN Jr: Has the time arrived for clinical PET imaging? Diagnostic Imaging 7: 138–145, 1985.

446. Wagner HN Jr: Radiolabeled drugs as probes of central nervous system neurons. Clin Chem 31: 1521–1524, 1985.
447. Scheffel U, Wagner HN Jr, Klein JL, Tsan MF: Gallium-67 uptake by hepatoma: studies in cell cultures, perfused livers and intact rats. J Nucl Med 26: 1438–1444, 1985.
448. Lever SZ, Burns HD, Kervitsky TM, Goldfarb HW, Woo DV, Wong DF, Epps LA, Kramer AV, Wagner HN Jr: Design, preparation, and biodistribution of a technetium-99m triaminedithiol complex to assess regional cerebral blood flow. J Nucl Med 26: 1287–1294, 1985.
449. Bice AN, Links JM, Wong DF, Wagner HN Jr: Absorbed fractions for dose calculations of neuroreceptor PET studies. Eur J Nucl Med 11: 127–131, 1985.
450. Saenger EL, Buncher CR, Specker BL, McDevitt RA (The Society of Nuclear Medicine Committee on Public Health and Efficacy, HN Wagner Jr, member): Determination of clinical efficacy: nuclear medicine as applied to lung scanning. J Nucl Med 26: 793–806, 1985.
451. Clausen M, Bice AN, Wagner HN Jr: Resolution of line sources in SPECT with 180° sampling, A technical note. Nuc Compact 16: 449–454, 1985.
452. Wong DF, Wagner HN Jr, Pearlson G, Dannals RF, Links JM, Ravert HT, Wilson AA, Suneja S, Bjorvinssen E, Kuhar MJ, Tune L: Dopamine receptors binding of C-11-3N-methylspiperone in the caudate in schizophrenia and bipolar disorder: a preliminary report. Psychopharm Bull 2: 595–598, 1985.
453. Wagner HN Jr: Commentary: Nuclear medicine–icons or ideas. SNM Newsline. J Nucl Med: 338, 1985.
454. Wong DF, Wagner HN Jr, Pearlson G, Dannals RF, Links JM, Ravert HT, Wilson AA, Suneja S, Bjorvinsson E, Kuhar MJ, Tune L: Dopamine receptors binding of C-11 3-N-methylspiperone in the caudate in schizophrenia and bipolar disorder: A preliminary report. Psychopharmacol Bull 21: 595–598, 1985.
455. Alexander EL, Firestein GS, Weiss JL, Heuser RR, Leitl G, Wagner HN Jr, Brinker JA, Ciuffo AA, Becker LC: Reversible cold-induced abnormalities in myocardial perfusion and function in systemic sclerosis. Ann Intern Med 105: 661–668, 1986.
456. Bice AN, Clausen M, Loncaric S, Wagner HN Jr: Resolution and attenuation effects in SPECT with a rotating scintillation camera. Med Rev 18: 24–29, 1986.
457. Bice AN, Clausen M, Loncaric S, Zeeberg B, Wagner HN Jr: Distanceweighted backprojection: a SPECT reconstruction technique. (Letter to the Editor) Radiology 161: 852–853, 1986.
458. Bice AN, Clausen M, Wagner HN Jr: Comparison of short axis circumferential profiles in 2-D-echocardiography and TL-201 myocardial SPECT. Medical Review 19: 46–52, 1986.
459. Bice AN, Wagner HN Jr: Estimation of bladder wall absorbed dose. [Letter] J Nucl Med 27: 567–568, 1986.
460. Bice AN, Wagner HN Jr, Frost JJ, Natarajan TK, Lee MC, Wong DF, Dannals RF, Ravert HT, Wilson AA, Links JM: Simplified detection system for neuroreceptor studies in the human brain. J Nucl Med 27: 184–191, 1986.
461. Bice AN, Wagner HN Jr, Lee MC, Frost JJ: A simple instrument for biochemical studies of the living human brain. J Med Instrum 20: 244–247, 1986.
462. Bok BD, Scheffel U, Goldfarb HW, Burns HD, Lever SZ, Wong DF, Wagner HN Jr: Comparative pharmacokinetics of technetium-99M-labeled radiopharmaceuticals with cerebral tropism. J Biophys Biomed 10: 5–7, 1986.
463. Clausen M, Bice AN, Civelek AC, Hutchins GM, Wagner HN Jr: Circumferential wall thickness measurements of the human left ventricle: reference data for thallium-201 single-photon emission computed tomography. Am J Cardiol 58: 827–831, 1986.
464. Dannals RF, Ravert HT, Wilson AA, Wagner HN Jr: An improved synthesis of (3-N-^{11}C]methyl)spiperone. Int J Rad Appl Instrum Part A Appl Radiat Isot 37: 433–434, 1986.
465. Fowler JS, Arnett CD, Wolf AP, Shiue C-Y, MacGregor RR, Halldin C, Laangstrom B, Wagner HN Jr: A direct comparison of the brain uptake and plasma clearance of N-11C]methylspiroperidol and [18F]N-methyl-spiroperidol in baboon using PET. Intl J Rad Appl Instrum Part B Nucl Med Biol 13: 281–284, 1986.
466. Frost JJ, Smith AC, Wagner HN Jr: 3H-Diprenorphine is selective for mu opiate receptors in vivo. Life Sci 38: 1597–1606, 1986.
467. Frost JJ, Wagner HN Jr, Dannals RF, Ravert HT, Wilson AA, Links JM, Rosenbaum AE, Trifiletti RR, Snyder SH: Imaging benzodiazepine receptors in man with C-11 suriclone by positron emission tomography. Eur J Pharm 122: 381–383, 1986.
468. Gjedde A, Wong DF, Wagner HN Jr: Transient analysis of irreversible and reversible tracer binding in human brain in vivo. Neurol Neurobiol 21: 223–235, 1986.
469. Guilarte TR, Wagner HN Jr: 3-Hydroxykynurenine as a possible mechanism of epileptic seizures associated with neonatal vitamin B6 deficiency. Trans Assoc Am Physicians 99: 73–77, 1986.
470. Harris JC, Wong DF, Wagner HN Jr, Rett A, Naidu S, Dannals RF, Links JM, Batshaw ML, Moser HW: Positron emission tomographic study of D2 dopamine receptor binding and CSF biogenic amine metabolites in Rett syndrome. Am J Med Genet Suppl 1: 201–210, 1986.

471. Laube BL, Swift DL, Wagner HN Jr, Norman PS, Adams GK III: The effect of bronchial obstruction on central airway deposition of a saline aerosol in patients with asthma. Am Rev Respir Dis 133: 740–743, 1986.

472. Lyon RA, Titeler M, Frost JJ, Whitehouse PJ, Wong DF, Wagner HN Jr, Dannals RF, Links JM, Kuhar MJ: 3H-3-N-Methylspiperone labels D2 dopamine receptors in basal ganglia and S2 serotonin receptors in cerebral cortex. J Neurosci 6: 2941–2949, 1986.

473. Scheffel U, Wagner HN Jr, Tsan MF: Enhancement of hepatic gallium-67 uptake by asialo-transferrin. J Nucl Med 27: 395–398, 1986.

474. Wagner HN Jr: Rett syndrome: positron emission tomography (PET) studies. Am J Med Genet Suppl 1 211–224, 1986.

475. Wagner HN Jr: Quantitative imaging of neuroreceptors in the living human brain. Semin Nucl Med 16: 51–62, 1986.

476. Wagner HN Jr: Quantitative imaging of dopamine, serotonin, and opiate receptors in the living human brain. Neurol Neurobiol 19: 233–254, 1986.

477. Wagner HN Jr: PET and SPECT advances herald new era in human biochemistry: The Society of Nuclear Medicine scientific meeting highlights. J Nucl Med 27: 1227–1238, 1986.

478. Wagner HN Jr: Clinical PET opens gates to in vivo biochemistry. Diag Imag 82–91, 1986.

479. Wagner HN: Positron emission tomography (PET) imaging and the neurobiological revolution. New Developments Med 1: 3–17, 1986.

480. Wagner HN Jr: Nuclear medicine. JAMA 256: 2096–2097, 1986.

481. Wagner HN Jr: Radiology after Chernobyl. Administrative Radiology 4: 39–44, 1986.

482. Wagner HN Jr: Images of the brain: Past as prologue. J Nucl Med 27: 1929–1937, 1986.

483. Wagner HN Jr: Imaging brain chemistry with positron emission tomography. Liver Aging, 1986, Proc Tokyo Symp 3rd 279–290, 1986.

484. Wagner HN Jr: Chemical neutransmission in man. Curr Concepts Diag Nucl Med 14–18, 1986.

485. Wagner HN Jr: PET imaging of the chemistry of the brain Proc Spie-Int Soc Opt Eng 626: 32–38, 1986.

486. Wilson AA, Dannals RF, Ravert HT, Burns HD, Wagner HN Jr: Iodine-125 and Iodine-123 labelled iodo-benzyl bromide, a useful alkylating agent for radiolabelling biologically important molecules. J Labelled Compd Radiopharm 23: 83–93, 1986.

487. Wong DF, Gjedde A, Wagner HN Jr: Quantification of neuroreceptors in the living human brain. I. Irreversible binding of ligands. J Cereb Blood Flow Metab 6: 137–146, 1986.

488. Wong DF, Gjedde A, Wagner HN Jr, Dannals RF, Douglass KH, Links JM, Kuhar MJ: Quantification of neuroreceptors in the living human brain. II. Inhibition studies of receptor density and affinity. J Cereb Blood Flow Metab 6: 147–153, 1986.

489. Wong DF, Wagner HN Jr, Dannals RF, Links JM, Kuhar MJ, Gjedde A: Human brain receptor distribution. Reply to comments. Science 232: 1270–1271, 1986.

490. Wong DF, Wagner HN Jr, Tune LE, Dannals RF, Pearlson GD, Links JM, Tamminga CA, Broussolle EP, Ravert HT, Wilson AA, Toung JKT, Malat J, Williams JA, O'Tuama LA, Snyder SH, Kuhar MJ, Gjedde A: Positron emission tomography reveals elevated D2 dopamine receptors in drug-naive schizophrenics. Science 234: 1558–1563, 1986. [published erratum appears in Science 6:235(4789):623, 1987.]

491. Bice AN, Clausen M, Loncaric S, Wagner HN Jr: Comparison of transaxial resolution in 180° and 360° SPECT with a rotating scintillation camera. Eur J Nucl Med 13: 7–11, 1987.

492. Bice AN, Clausen M, Wagner HN Jr: Comparison of short axis circumferential profiles in 2D-echocardiography and TL-201 myocardial SPECT. Medical Review 19: 46–52, 1987.

493. Bice AN, Wong DF, Wagner HN Jr: On estimating the loss of quantification in PET due to finite detector resolution. Eur J Nucl Med 13: 1–6, 1987.

494. Bok BD, Scheffel U, Goldfarb HW, Burns HD, Lever SZ, Wong DF, Bice A, Wagner HN Jr: Comparison of 99mTc complexes (NEP-DADT, ME-NEP-DADT and HMPAO) with 123IAMP for brain SPECT imaging in dogs. Nucl Med Commun 8: 631–641, 1987.

495. Bok BD, Bice AN, Clausen M, Wong DF, Wagner HN Jr: Artifacts in camera based single photon emission tomography due to time activity variation. Eur J Nucl Med 13: 439–442, 1987.

496. Camargo EE, Kertcher JA, Chen MF, Charache P, Wagner HN Jr: Radiometric studies of Mycobacterium tuberculosis. Rev Inst Med Trop Sao Paulo 29: 18–25, 1987.

497. Camargo EE, Kopajtic TM, Hopkins GK, Cannon NP, Wagner HN Jr: Radiometric studies on the oxidation of (U-14C)L-amino acids by drug-susceptible and drug-resistant mycobacteria. Rev Inst Med Trop Sao Paulo 29: 312–316, 1987.

498. Camargo EE, Kopajtic TM, Hopkins GK, Cannon NP, Wagner HN Jr: Radiometric studies on the oxidation of [1–14C]fatty acids by drug-susceptible and drug-resistant mycobacteria. Rev Inst Med Trop Sao Paulo 29: 9–17, 1987.

499. Camargo EE, Wagner HN Jr: Radiometric studies on the oxidation of [1–14C]fatty acids and [U-14C]-amino acids by mycobacteria. Int J Rad Appl Instrum Part B Nucl Med Biol 14: 43–49, 1987.

500. Douglass KH, Links JM, Chen DC, Wong DF, Wagner HN Jr: Linear discriminant analysis of regional ejection fractions in the diagnosis of coronary artery disease. Eur J Nucl Med 12: 602–604, 1987.

501. Epps LA, Burns HD, Lever SZ, Goldfarb HW, Wagner HN Jr: Brain imaging agents: synthesis and characterization of (N-piperidinylethyl hexamethyl diaminodithiolate) oxo technetium (V) complexes. Intl J Radiat Appl Instrumen, Part A, Appl Radiat Isot 38: 661–664, 1987.

502. Frost JJ, Smith AC, Kuhar MJ, Dannals RF, Wagner HN Jr: In vivo binding of 3H-N-methylspiperone to dopamine and serotonin receptors. Life Sci 40: 987–995, 1987.

503. Guilarte TR, Wagner HN Jr: Increased concentrations of 3-hydroxykynurenine in vitamin B6 deficient neonatal rat brain. J Neurochem 49: 1918–1926, 1987.

504. Guilarte TR, Wagner HN Jr, Frost JJ: Effects of perinatal vitamin B6 deficiency on dopaminergic neurochemistry. J Neurochem 48: 432–439, 1987. [Published erratum appears in J Neurochem 49(1): 328, 1987.]

505. Lever JR, Dannals RF, Wilson AA, Ravert HT, Wagner HN Jr: Synthesis of carbon-11 labeled diprenorphine: a radioligand for positron emission tomographic studies of opiate receptors. Tetrahedron Lett 28: 4015–4018, 1987.

506. Liu XJ, Liu YZ, He ZX, Zhou BG, Liu LS, Wagner HN Jr: Monitoring effects of alpha-human atrial natriuretic polypeptide (alpha-hANP) on left ventricular function in patients with essential hypertension with a cardiac probe. Eur J Nucl Med 13: 335–337, 1987.

507. Natarajan TK, Wise RA, Karam M, Permutt S, Wagner HN Jr: Immediate effect of expiratory loading on left ventricular stroke volume. Circulation 75: 139–145, 1987.

508. Ravert HT, Wilson AA, Dannals RF, Wong DF, Wagner HN Jr: Radiosynthesis of a selective dopamine D-1 receptor antagonist: R(+)-7-chloro-8-hydroxy-3-[11C]methyl-1-phenol-2-,3,4,5-tetrahydro-1H-3-benzazepine ([11C]SCH 23390). Int J Rad Appl Instrum Part A Appl Radiat Isot 38: 305–306, 1987.

509. Wagner HN Jr: Scientific highlights: "Slices of life." (The Society of Nuclear Medicine 34th Annual Meeting 1987) J Nucl Med 28: 1235–1246, 1987.

510. Wagner HN Jr: Lung imaging. Past as prologue. Chest 92: 353–359, 1987.

511. Wagner HN Jr: Contempo '86: Nuclear Medicine (JAMA 256(15): 2096–2097) Translated and published in Japanese JAMA edition, June 1987.

512. Wagner HN Jr: After Chernobyl. Administrative Radiology 5(5): 18–21, 1987.

513. Wagner HN Jr: Radiation and women. Administrative Radiology 6(9): 12–15, 1987.

514. Wong DF, Lever JR, Hartig PR, Dannals RF, Villemagne V, Hoffman BJ, Wilson AA, Ravert HT, Links JM, Scheffel U, Wagner HN Jr: Localization of serotonin 5-HT2 receptors in living human brain by positron emission tomography using N1-([11C]-Methyl)-2-Br-LSD. Synapse 1: 393–398, 1987.

515. Zeeberg BR, Bice AN, Wagner HN Jr: Concerning strategies for in vivo measurement of receptor binding using positron emission tomography. J Cereb Blood Flow Metab 7: 252–255, 1987.

516. Zeeberg BR, Bice AN, Wagner HN Jr: Concerning strategies for in vivo measurement of receptor binding using positron emission tomography. II [letter] J Cereb Blood Flow Metab 7: 818–820, 1987.

517. Zeeberg BR, Wagner HN Jr: Analysis of three- and four-compartment models for in vivo radioligand-neuroreceptor interaction. Bull Math Biol 49: 469–486, 1987.

518. ACNP/SNM Task Force on Clinical PET (members: Kuhl DE, Wagner HN, Alabi A, Coleman RE, Gould KL, Larson SM, Mintun MA, Siegel BA, Strudier PK): Positron emission tomography: Clinical status in the United States in 1987. J Nucl Med 29: 1136–1143, 1988.

519. Bourguignon MH, Valette H, Le Guludec D, Oddou C, Merlet P, Buchanan JW, Raynaud C, Syrota A, Wagner HN Jr: Non-invasive measurement of pulmonary arterial pressure: I. A haemodynamic modelling approach. Phys Med Biol 33: 205–214, 1988.

520. Bourguignon MH, Le Guludec D, Valette H, Davy J, Motte G, Oddou C, Buchanan JW, Raynaud C, Syrota A, Wagner HN Jr: Non-invasive measurement of pulmonary arterial pressure: II. A radionuclide method. Phys Med Biol 33: 215–225, 1988.

521. Chen JJ, LaFrance ND, Rippin R, Allo MD, Wagner HN Jr: Iodine-123 SPECT of the thyroid in multinodular goiter. J Nucl Med 29: 110–113, 1988.

522. Clausen M, Civelek AC, Bice AN, Petronis J, Koller D, Loncaric S, Weiss JL, Wagner HN Jr: Short-axis circumferential profiles of the heart in healthy subjects: Comparison of TL-201 SPECT and two-dimensional echocardiography. Radiology 168: 723–726, 1988.

523. Dannals RF, Langstrom B, Ravert HT, Wilson AA, Wagner HN Jr: Synthesis of radiotracers for studying muscarinic cholinergic receptors in the living human brain using positron emission tomography: [11C]dexetimide and [11C]levetimide. Int J Rad Appl Instrum Part A Appl Radiat Isot 39: 291–295, 1988.

524. Dannals RF, Ravert HT, Wilson AA, Wagner HN Jr: Synthesis of high specific activity carbon-11 labeled tracers for neuroreceptor studies. Synthesis and Applications of Isotopically Labelled Compounds, Proceedings of the Third International Symposium, Innsbruck, Austria, July 17–21, 1988, pp. 457–464.

525. Frost JJ, Mayberg HS, Fisher RS, Douglass KH, Dannals RF, Links JM, Wilson AA, Ravert HT, Rosenbaum AE, Snyder SH, Wagner HN Jr: Mu-opiate receptors measured by positron emission tomography are increased in temporal lobe epilepsy. Ann Neurol 23: 231–237, 1988.

526. Guilarte TR, Block LD, Wagner HN Jr: The putative endogenous convulsant 3-hydroxykynurenine decreases benzodiazepine receptor binding affinity: implications to seizures associated with neonatal vitamin B-6 deficiency. Pharmacol Biochem Behav 30: 665–668, 1988.

527. Laube BL, Links JM, Wagner HN Jr, Norman PS, Koller DW, LaFrance ND, Adams GK 3d: Simplified assessment of fine aerosol distribution in human airways. J Nucl Med 29: 1057–1065, 1988.

528. Lee MC, Wagner HN Jr, Tanada S, Frost JJ, Bice AN, Dannals RF: Duration of occupancy of opiate receptors by naltrexone. J Nucl Med 29: 1207–1211, 1988.

529. Li QS, Frank TL, Franceschi D, Wagner HN Jr, Becker LC: Technetium-99m methoxyisobutyl isonitrile (RP-30) for quantification of myocardial ischemia and reperfusion in dogs. J Nucl Med 29: 1539–1548, 1988.

530. Mayberg HS, Robinson RG, Wong DF, Parikh R, Bolduc P, Starkstein SE, Price T, Dannals RF, Links JM, Wilson AA, Ravert HT, Wagner HN Jr: PET imaging of cortical S2 serotonin receptors after stroke: Lateralized changes and relationship to depression. Am J Psychiatry 145: 937–943, 1988.

531. Mukai T, Links JM, Douglass KH, Wagner HN Jr: Scatter correction in SPECT using non-uniform attenuation data. Phys Med Biol 33: 1129–1140, 1988.

532. O'Tuama LA, Guilarte TR, Douglass KH, Wagner HN Jr, Wong DF, Dannals RF, Ravert HT, Wilson AA, LaFrance ND, Bice AN, Links JM: Assessment of [11C]-L-methionine transport into the human brain. J Cereb Blood Flow Metab 8: 341–345, 1988.

533. Scheffel U, Goldfarb HW, Lever SZ, Gungon RL, Burns DH, Wagner HN Jr: Comparison of technetium-99m aminoalkyl diaminodithiol (DADT) analogs as potential brain blood flow imaging agents. J Nucl Med 29: 73–82, 1988.

534. Vallette H, Bourguignon MH, Apoil E, Moyse D, Wise RA, Buchanon JW, Wagner HN Jr, Syrota A: Accuracy of gated equilibrium radioventriculography in measuring left ventricular function in dogs. Nucl Med Commun 9: 999–1004, 1988.

535. Vallette H, Bourguignon M, Douglass K, del Buono A, Merlet P, Buchanan JW, Raynaud C, Syrota A, Wagner HN: Automatic drawing of the left epicardial region of interest on thallium 201 scintigraphic images. Eur J Nucl Med 14: 485–488, 1988.

536. Wagner HN Jr: Positron emission tomography in assessment of regional stereospecificity of drugs. Biochem Pharmacol 37: 51–59, 1988.

537. Wagner HN Jr: Scientific Highlights 1988: "The Future is Now". J Nucl Med 29: 1329–1337, 1988.

538. Wagner HN Jr: Science, politics, and radiation. JAMA 260: 697–698, 1988.

539. Wagner HN Jr: PET and the chemistry of mental illness. Proceedings of the Chinese Academy of Medical Sciences and the Peking Union Medical College 3 supplement I: 6–8, 1988.

540. Wagner HN: Positron emission tomography and the chemistry of mental illness. Alumni Bulletin Int Med 19: 93–101, 1988.

541. Wagner HN Jr: Building a bridge to in vivo chemistry. Administrative Radiology 7(5): 52–53, 1988.

542. Wagner HN Jr: Imaging dopamine receptors in the human brain. Am J Physiologic Imaging 3: 63, 1988.

543. Wagner HN Jr, Weinberger DR, Kleinman JE, Casanova MF, Gibbs CJ Jr, Gur RE, Hornykiewicz O, Kuhar MJ, Pettegrew JW, Seeman P: Neuroimaging and neuropathology. Schizophr Bull 14: 383–397, 1988.

544. Wong DF, Broussolle EP, Wand G, Villemagne V, Dannals RF, Links JM, Zacur HA, Harris J, Naidu S, Braestrup C, Wagner HN Jr, Gjedde A: In vivo measurement of dopamine receptors in human brain by positron emission tomography—Age and sex differences. In: Joseph JA, ed. Central Determinants of Age-Related Declines in Motor Function. Annals of the New York Academy of Sciences 515: 203–214, 1988.

545. Yanai K, Dannals RF, Wilson AA, Ravert HT, Scheffel U, Tanada S, Wagner HN Jr: (N-methyl-[11C])pyrilamine, a radiotracer for histamine H-1 receptors: radiochemical synthesis and biodistribution study in mice. Int J Radiat Appl Instrum, Part B, Nucl Med Biol 15: 605–610, 1988.

546. Frost JJ, Douglass KH, Mayberg HS, Dannals RF, Links JM, Wilson AA, Ravert HT, Crozier WC, Wagner HN Jr: Multicompartmental analysis of [11C]-carfentanil binding to opiate receptors in humans measured by positron emission tomography. J Cereb Blood Flow Metab 9(3): 398–409, 1989.

547. Humayun MS, Presty SK, LaFrance ND, Holcomb HH, Loats H, Long DM, Wagner HN, Gordon B: Local cerebral glucose abnormalities in mild closed head injured patients with cognitive impairments. Nuclear Medicine Communications 10(5): 335–344, 1989.

548. Laube BL, Links JM, LaFrance ND, Wagner HN Jr, Rosenstein BJ: Homogeneity of bronchopulmonary distribution of 99mTc aerosol in normal subjects and in cystic fibrosis patients. Chest 95: 822–830, 1989.

549. Lever JR, Dannals RF, Wilson AA, Ravert HT, Scheffel U, Hoffman BJ, Hartig PR, Wong DF, Wagner HN Jr: Synthesis and in vivo characterization of D-(+)-N1-[11C]methyl)-2-Br-LSD: a radioligand for posi-

tron emission tomographic studies of serotonin 5-HT2 receptors. Int J Radiat Appl Instrum Part B Nucl Med Biol 16(7): 697–704, 1989.

550. Li Q-S, Solot G, Frank TL, Wagner HN Jr, Becker LC: Serial rest and dipyridamole tomographic myocardial perfusion studies with the rapidly clearing technetium-99m agent SQ 30217. Nuklearmedizin Suppl (Stuttgart) 25: 243–246, 1989.

551. Li Q-S, Solot G, Franceschi D, Wagner HN Jr, Becker LC: Comparison of RP-30 and SQ 32014: Myocardial redistribution in dogs with temporary coronary artery occlusion. Nuklearmedizin Suppl (Stuttgart) 25: 240–242, 1989.

552. London ED, Margolin RA, Wong DF, Links JM, LaFrance ND, Cascella NG, Broussolle EPM, Wagner HN, Snyder FR, Jasinski DR: Cerebral glucose utilization in human heroin addicts: Case reports from a positron emission tomographic study. Res Comm Sub Abuse 10(2): 141–144, 1989.

553. Sadzot B, Frost JJ, Wagner HN Jr: In vivo labeling of central benzodiazepine receptors with the partial inverse agonist [H-3]RO 15–4513. Brain Res 491(1): 128–135, 1989.

554. Sanchez-Roa PM, Grigoriadis DE, Wilson AA, Sharkey J, Dannals RF, Villemagne VL, Wong DF, Wagner HN Jr, Kuhar MJ: [125I]-Spectramide: a novel benzamide displaying potent and selective effects at the D2 dopamine receptor. Life Sci 45(19): 1821–1829, 1989.

555. Wagner HN Jr: SNM highlights-1989: "Why not?" J Nucl Med 30(8): 1283–1295, 1989.

556. Wagner HN Jr: Peace through mind/brain science. JAMA 262(5): 625–626, 1989.

557. Wagner HN Jr: Nuclear medicine. JAMA 261(19): 2860–2862, 1989.

558. Wagner HN: Better living through brain chemistry—Reply. JAMA 262(19): 2682–2682, 1989.

559. Wilson AA, Dannals RF, Ravert HT, Frost JJ, Wagner HN Jr: Synthesis and biological evaluation of [125I]- and [123I]-4-iododexetimide, a potent muscarinic cholinergic receptor antagonist. J Medicinal Chem 32(5): 1057–1062, 1989.

560. Wilson AA, Dannals RF, Ravert HT, Wagner HN Jr: Preparation of [11C]- and [125I]IMB: A dopamine D-2 receptor antagonist. Int J Radiat Appl Instrum, part A, Appl Radiat Isot 40(5): 369–373, 1989.

561. Wilson AA, Dannals RF, Ravert HT, Burns HD, Lever SZ, Wagner HN Jr: Radiosynthesis of [11C]nifedipine and [11C]nicardipine. J Labelled Compd Radiopharm 27(5): 589–598, 1989.

562. Wilson AA, Grigoriadis DE, Dannals RF, Ravert HT, Wagner HN: A one-pot radiosynthesis of [I-125]iodo-azido photoaffinity labels. J Labelled Compd Radiopharm 27(11): 1299–1305, 1989.

563. Yanai K, Dannals RF, Wilson AA, Ravert HT, Scheffel U, Tanada S, Wagner HN Jr: Biodistribution and radiation absorbed dose of [N-methyl]C-11]pyrilamine—a histamine H-1 receptor radiotracer. Int J Rad Appl Instrum Part B Nucl Med Biol 16(4): 361–363, 1989.

564. Brandt J, Folstein SE, Wong DF, Links J, dannals RF, McDonnell-Sill A, Starkstein S, Anders P, Strauss ME, Tune LE, Wagner HN Jr, Folstein MF: D2 Receptors in Huntington's disease: Positron emission tomography findings and clinical correlates. J Neuropsychiatry Clin Neurosci 2: 20–27, 1990.

565. Dannals RF, Ravert HT, Wilson AA, Wagner HN Jr: Synthesis of a selective serotonin uptake inhibitor: [11C]citalopram. Int J Rad Appl Instrum A Appl Radiat Isot 41: 541–543, 1990.

566. Frost JJ, Mayberg HS, Sadzot B, Dannals RF, Lever JR, Ravert HT, Wilson AA, Wagner JH Jr, Links JM: Comparison of 11C-diprenorphine and 11C-carfentanil binding to opiate receptors in man by positron emission tomography. J Cereb Blood Flow Metab 10: 484–492, 1990.

567. Lever JR, Mazza SM, Dannals RF, Ravert HT, Wilson AA, Wagner HN: Facile synthesis of [11C]buprenorphine for positron emission tomographic studies of opioid receptors. Appl Radiat Isot Int J Rad Appl Instrum, Part A 41: 745–752, 1990.

568. Lever JR, Scheffel UA, Stathis M, Musachio JL, Wagner HN Jr: In vitro and in vivo binding of (E)-and (Z)-N-(iodoallyl)spiperone to dopamine D2 and serotonin 5-HT2 neuroreceptors. Life Sci 46: 1967–1976, 1990.

569. Li Q-S, Solot G, Frank TL, Wagner HN Jr, Becker LC: Myocardial redistribution of technetium 99m-methoxyisobutyl isonitrile (SESTAMBI). J Nucl Med 31: 1069–1076, 1990.

570. London ED, Broussolle EPM, Links JM, Wong DF, Cascella NG, Dannals RF, Sano M, Herning R, Snyder FR, Rippetoe LR, Toung TJK, Jaffe JH, Wagner HN: Morphine-induced metabolic changes in human brain: Studies with positron emission tomography and fluorodeoxyglucose. Arch Gen Psychiatry 47(1): 73–81, 1990.

571. London ED, Cascella NG, Wong DF, Phillips RL, Dannals RF, Links JM, Herning R, Grayson R, Jaffe JH, Wagner HN Jr: Cocaine-induced reduction of glucose utilization in human brain. A study using positron emission tomography and [fluorine 18]fluorodeoxyglucose. Arch Gen Psychiatry 47: 567–574, 1990.

572. Mayberg HS, Starkstein SE, Sadzot B, Preziosi T, Andrezejewski PL, Dannals RF, Wagner HN, Robinson RG: Selective hypometabolism in the inferior frontal lobe in depressed patients with Parkinson's disease. Ann Neurol 28: 57–64, 1990.

573. Meltzer CC, Bryan RN, Holcomb H, Kimball AW, Mayberg HS, Sadzot B, Leal JP, Wagner HN Jr, Frost JJ: Anatomical localization for PET using MR imaging. J Comput Assist Tomogr 14: 418–426, 1990.

574. Meltzer CC, Leal JP, Mayberg HS, Wagner HN Jr, Frost JJ: Correction of PET data for partial volume effects in human cerebral cortex by MR imaging. J Comput Assist Tomogr 14: 561–570, 1990.
575. O'Tuama LA, Phillips PC, Strauss LC, Carson BC, Uno Y, Smith QR, Dannals RF, Wilson AA, Ravert HT, Loats S, Loats HA, LaFrance ND, Wagner HN Jr: Two-phase [11C]L-methionine PET in childhood brain tumors. Pediatr Neurol 6: 163–170, 1990.
576. Sadzot B, Sheldon J, Dannals RF, Ravert HT, Wagner HN Jr, Frost JJ: Localization of peripheral cholecystokinin receptors in vivo using the cholecystokinin antagonist [3H](±)-MK-329. Eur J Pharmacol 185: 195–201, 1990.
577. Starkstein SE, Mayberg HS, Berthier ML, Fedoroff P, Price TR, Dannals RF, Wagner HN, Leiguarda R, Robinson RG: Mania after brain injury: neuroradiological and metabolic findings. Ann Neurol 27: 652–659, 1990.
578. Suehiro M, Dannals RF, Scheffel U, Stathis M, Wilson AA, Ravert HT, Villemagne VL, Sanchez-Roa PM, Wagner HN Jr: In vivo labeling of the dopamine D2 receptor with N-11C-methylbenperidol. J Nucl Med 31: 2015–2021, 1990.
579. Wagner HN: Letter from IPOH—Reply. JAMA 263(5): 662–662, 1990.
580. Wagner HN Jr: Clinical applications of positron emission tomography (PET). Neurology and Neurosurgery Update Series Vol. 8 (22), pp 2–8, 1990.
581. Wagner HN Jr: Scientific highlights 1990: The universe within. J Nucl Med 31: 17A–26A, 1990.
582. Wagner HN Jr: Highlights of Nuclear Medicine. Indian J Nucl Med 5(1): 24–36, 1990.
583. Wagner HN Jr: Merrill C. Sosman lecture Drugs, behavior and brain chemistry. Am J Roentgenol 155: 925–931, 1990.
584. Wilson AA, Dannals RF, Ravert HT, Wagner HN Jr: Reductive amination of [18F]fluorobenzaldehydes: Radiosynthesis of [2-18F] and [4-18F]fluorodexetimides. J Labelled Compd Radiopharm 28(10): 1189–1199, 1990.
585. Camargo EE, Sostre S, Sadzot B, Shafique I, Szabo Z, Links JM, Dannals RF, Wagner HN Jr: Global and regional cerebral metabolic rate of 2-[18F]fluoro-2-deoxy-D-glucose in the presence of ofloxacin, a gamma-aminobutyric acid A receptor antagonist. Antimicrob Agents and Chemother 35: 648–652, 1991.
586. Civelek AC, Durski K, Shafique I, Matsumura K, Sostre S, Wagner HN Jr, Ladenson PW: Failure of perchlorate to inhibit Tc-99m isonitrile binding by the thyroid during myocardial perfusion studies. Clinical Nuclear Medicine 16: 358–361, 1991.
587. Civelek AC, Shafique I, Brinker JA, Durski K, Weiss JL, Links JM, Natarajan TK, Ozguven MA, Wagner HN Jr: Reduced left ventricular cavitary activity ("black hole sign") in thallium-201 SPECT perfusion images of anteroapical transmural myocardial infarction. Am J Cardiol 68: 1132–1137, 1991.
588. Jeffery PJ, Monsein LH, Szabo Z, Hart J, Fisher RS, Lesser RP, Debrun GM, Gordon B, Wagner HN Jr, Camargo EE: Mapping the distribution of amobarbital sodium in the intracarotid Wada test by use of Tc-99m HMPAO with SPECT. Radiology. 178: 847–850, 1991.
589. Lever JR, Scheffel UA, Musachio JL, Stathis M, Wagner HN Jr: Radioiodinated D-(+)-N1-ethyl-2-iodolysergic acid diethylamide: A ligand for in vitro and in vivo studies of serotonin receptors. Life Sci 48: PL-73–PL-78, 1991.
590. Li Q-S, Solot G, Frank TL, Wagner HN Jr, Becker LC: Tomographic myocardial perfusion imaging with technetium-99m-teboroxime at rest and after dipyridamole. J Nucl Med 32: 1968–1976, 1991.
591. Matsumura K, Uno Y, Scheffel U, Wilson AA, Dannals RF, Wagner Hn JR: In vitro and in vivo characterization of 4-[125I]Iododexetimide binding to muscarinic cholinergic receptors in the rat heart. J Nucl Med 32: 76–80, 1991.
592. Mayberg HS, Sadzot B, Meltzer CC, Fisher RS, Lesser RP, Dannals RF, Lever JR, Wilson AA, Ravert HT, Wagner HN Jr, Bryan RN, Cromwell CC, Frost JJ: Quantification of mu and non-mu opiate receptors in temporal lobe epilepsy using positron emission tomography. Ann Neurol 30: 3–11, 1991.
593. O'Tuama LA, Philips PC, Smith QR, Uno Y, Dannals RF, Wilson AA, Ravert HT, Loats S, Loats HA, Wagner HN Jr: L-Methionine uptake by human cerebral cortex: maturation from infancy to old age. J Nucl Med 32: 16–22, 1991.
594. Ravert HT, Dannals RF, Wilson AA, Wong DF, Wagner HN Jr: Synthesis of 18F-labeled reduced haloperidol and 11C-labeled reduced 3-N-methylspiperone. J Labelled Compd Radiopharm 29: (3) 337–343, 1991.
595. Sadzot B, Price JC, Mayberg HS, Douglass KH, Dannals RF, Lever JR, Ravert HT, Wilson AA, Wagner HN Jr: Quantification of human opiate receptor concentration and affinity using high and low specific activity[11C]diprenorphine and positron emission tomography. J Cereb Blood Flow Metab 11: 204–219, 1991.
596. Suehiro M, Wilson AA, Scheffel U, Dannals RF, Ravert HT, Wagner HN Jr: Radiosynthesis and evaluation of N-(3-[18F]fluoropropyl) paroxetine as a radiotracer for in vivo labeling of serotonin uptake sites by PET. Int J Radiat Appl Instrum Part B Nucl Med Biol 18: 791–796, 1991.

597. Takeda K, LaFrance ND, Weisman HF, Wagner HN Jr, Becker LC: Comparison of indium-111 antimyosin antibody and technetium-99m pyrophosphate localization in reperfused and nonreperfused myocardial infarction. J Am Coll Cardiol 17: 519–526, 1991.
598. Takeda K, Ueda K, Scheffel U, Ravert H, LaFrance ND, Baumgartner WA, Reitz BA, Herskowitz A, Wagner HN Jr: Indium-111 myosin-specific antibodies and technetium-99m pyrophosphate in the detection of acute cardiac rejection of transplanted hearts: studies in a heterotopic rat heart model. Eur J Nucl Med 18: 461–466, 1991.
599. Tune L, Brandt J, Frost JJ, Harris G, Mayberg H, Steele, Burns A, Sapp J, Folstein MF, Wagner HN et al.: Physotigmine in Alzheimer's disease: effects on cognitive functioning, cerebral glucose metabolism analyzed by positron emission tomography and cerebral blood flow analyzed by single photon emission tomography. Acta Psychiatr Scand Suppl 366: 61–65, 1991.
600. Uno Y, Matsumura K, Scheffel U, Wilson AA, Dannals RF, Wagner HN Jr: Effects of atropine treatment on in vitro and in vivo binding of 4-[125I]-dexetimide to central and myocardial muscarinic receptors. Eur J Nucl Med 18: 447–452, 1991.
601. Villemagne VL, Dannals RF, Sanchez-Roa PM, Ravert HT, Vazquez S, Wilson AA, Natarajan TK, Wong DF, Yanai K, Wagner HN Jr: Imaging histamine H1 receptors in the living human brain with carbon-11-pyrilamine. J Nucl Med 32: 308–311, 1991.
602. Wagner HN: Drugs, behaviour and the brain chemistry. Def Sci J 41: 137–141, 1991.
603. Wagner HN Jr: Clinical PET: its time has come. J Nucl Med 32: 567–564, 1991.
604. Wagner HN Jr: Molecular medicine: from science to service [news] J. Nucl Med 32: 11N–12N, 14N–16N, 17N–23N, 1991.
605. Wagner HN Jr: Henry N. Wagner Jr. M.D.: winner of the American Medical Association's Scientific Achievement Award [interview by Michelle Burke and Betsy Newman]. Maryland Medical Journal 40: 557–561, 1991.
606. Wagner HN Jr, Conti PS: Advances in medical imaging for cancer diagnosis and treatment. Cancer 67: 1121–1128, 1991.
607. Wilson AA, Conti PS, Dannals RF, Ravert HT, Wagner HN Jr: Radiosynthesis of [11C]-N-methylacyclovir. J Labelled Compd Radiopharm 29: 765–768, 1991.
608. Wilson AA, Dannals RF, Ravert HT, Sonders MS, Weber E, Wagner HN Jr: Radiosynthesis of sigma receptor ligands for positron emission tomography: 11C-and 18F-labeled guanidines. J Medicinal Chem 34: 1867–1870, 1991.
609. Wilson AA, Scheffel UA, Dannals RF, Stathis M, Ravert HT, Wagner HN Jr: In vivo biodistribution of two [18F]-labelled muscarinic cholinergic receptor ligands: 2-[18F]-and 4-[18F]-fluorodexetimide. Life Sci 48: 1385–1394, 1991.
610. Young LT, Wong DF, Goldman S, Minkin E, Chen C, Matsumura K, Scheffel U, Wagner HN Jr: Effects of endogenous dopamine on kinetics of [3H]N-methylspiperone and [3H]raclopride binding in the rat brain. Synapse 9: 188–194, 1991.
611. Brust P, Shaya EK, Jeffries KJ, Dannals RF, Ravert HT, Wilson AA, Conti PS, Wagner HN Jr, Gjedde A, Ermish A, Wong DF: Effects of vasopressin on blood-brain transfer of methionine in dogs. J Neurochem 59: 1421–1429, 1992.
612. Camargo EE, Szabo Z, Links JM, Sostre S, Dannals RF, Wagner HN: The influence of biological and technical factors on the variability of global and regional brain metabolism of 2-[18F]fluorodeoxy-D-glucose. J Cereb Blood Flow Metab 12: 281–290, 1992.
613. Civelek AC, Brinker JA, Camargo EE, Links JM, Wagner HN Jr: Rest thallium-201 myocardial perfusion imaging in a patient with leukaemic infiltration of the heart. Eur J Nucl Med 19: 306–308, 1992.
614. Civelek AC, Gozukara I, Durski K, Ozguven MA, Brinker JA, Links JM, Camargo EE, Wagner HN Jr, Flaherty JT: Detection of left anterior descending coronary artery disease in patients with left bundle branch block. Am J Cardiol 70: 1565–1570, 1992.
615. Horti A, Ravert HT, Dannals RF, Wagner HN Jr: Synthesis of [N-11C]-methyl-L-DOPA. J Labelled Compd Radiopharm 31: 1029–1036, 1992.
616. Lever JR, Scheffel U, Kinter CM, Ravert HT, Dannals RF, Wagner HN Jr, and Frost JJ: In vivo binding of N1'-([11C]methyl)naltrindole to delta-opioid receptors in mouse brain. Eur J Pharmacol 216: 459–460, 1992.
617. Links JM, Leal JP, Mueller-Gaertner HW, Wagner HN Jr: Improved positron emission tomography quantification by Fourier-based restoration filtering. Eur J Nucl Med 19: 925–932, 1992.
618. Müller-Gärtner HW, Links JM, Prince JL, Bryan RN, McVeigh E, Leal JP, Davatzikos C, Frost JJ: Measurement of radiotracer concentration in brain gray matter using positron emission tomography: MRI-based correction for partial volume effects. J Cereb Blood Flow Metab 12: 571–583, 1992.
619. Müller-Gärtner HW, Wilson AA, Dannals RF, Wagner HN Jr, Frost JJ: Imaging muscarinic cholinergic receptors in human in vivo with SPECT, [123I]4-Iodo-dexetimide, and [123I]4-Iodolevetimide. J Cereb Blood Flow Metab 12: 562–570, 1992.

620. Noble GD, Dannals RF, Ravert HT, Wilson AA, Wagner HN Jr: Synthesis of a radiotracer for studying kappa-subtype opiate receptors: N-[11C-methyl]-N-(trans-2-pyrrolidinyl-cyclohexyl)-3,4-dichlorophenylacetamide ([11C](+)U-50488H). J Labelled Compd Radiopharm 31: 81–89, 1992.

621. Ravert HT, Dannals RF, Wilson AA, Wagner HN Jr: (N-[11C]-methyl)doxepin: synthesis of a radiotracer for studying the histamine H-1 receptor. J Labelled Compd Radiopharm 31: 403–407, 1992.

622. Shaya EK, Scheffel U, Dannals RF, Ricaurte GA, Carroll FI, Wagner HN Jr, Kuhar MJ, Wong DF: In vivo imaging of dopamine reuptake sites in the primate brain using single photon emission computed tomography (SPECT) and iodine-123 labeled RTI-55. Synapse 10: 169–172, 1992.

623. Singer HS, Wong DF, Brown JE, Brandt J, Ravert L, Shaya E, Dannals RF, Wagner HN Jr: PET/PAIR scan tomography evaluation of D-2 receptors in adults with Tourette syndrome. Adv Neurol 58: 233–239, 1992.

624. Sostre S, Kalloo AN, Spiegler EJ, Camargo EE, Wagner HN Jr: A noninvasive test of spincter of Oddi dysfunction in postcholecystectomy patients: the scintigraphic score. J Nucl Med 33: 1216–1222, 1992.

625. Suehiro M, Ravert HT, Dannals RF, Scheffel U, Wagner HN Jr: Synthesis of a radiotracer for studying serotonin uptake sites with positron emission tomography: [11C]McN-5652-Z. J Labelled Compd Radiopharm 31: 841–848, 1992.

626. Suehiro M, Ravert HT, Wilson AA, Scheffel U, Dannals RF, Wagner HN Jr: Further investigation on the radiosynthesis of alpha-[11C]methyl-tryptophan. J Labelled Compd Radiopharm 31: 151–157, 1992.

627. Suehiro M, Scheffel U, Dannals RF, Wilson AA, Ravert HT, Wagner HN Jr: Synthesis and biodistribution of a new radiotracer for in vivo labeling of serotonin uptake sites by PET, cis-N,N-[11C]dimethyl-3-(2',4'-dichlorophenyl)-indanamine (cis-[11C]DDPI) Int J Rad Appl Instrum Part B Nucl Med Biol 19: 549–553, 1992.

628. Szabo Z, Camargo EE, Sostre S, Shafique I, Sadzot B, Links JM, Dannals RF, Wagner HN Jr: Factor analysis of regional cerebral glucose metabolic rates in healthy men. Eur J Nucl Med 19: 469–475, 1992.

629. Szabo Z, Links JM, Seki C, Rhine J, Wagner HN Jr: Scatter, spatial resolution, and quantitative recovery in high resolution SPECT. J Comput Assist Tomogr 16: 461–467, 1992.

630. Wagner HN Jr: Masahiro Iio memorial lecture: New Horizons for Nuclear Medicine. Ann Nucl Med 6: 1–8, 1992.

631. Wagner HN Jr: Annual Meeting Highlights: Molecules with messages. J Nucl Med 33: 10N–27N, 1992.

632. Wagner HN Jr: '92 SNM Annual Meeting Highlights: Molecules which bring messages. (Translation in Japanese published by Joknawa Medical System). 1992.

633. Wagner HN Jr: Positron emission tomography at the turn of the century: A perspective. Seminars in Nuclear Medicine 22: 285–288, 1992.

634. Wong DF, Wilson AA, Chen C, Minken E, Dannals RF, Ravert HT, Sanchez-Roa P, Villemagne V, Wagner HN Jr: In vivo studies of [125I]iodobenzamide and [11C]iodobenzamide: A ligand suitable for positron emission tomography and single photon emission tomography imaging of cerebral D2 dopamine receptors. Synapse 12: 236–241, 1992.

635. Civelek AC, Sitzmann JV, Chin BB, Venbrux A, Wagner HN Jr, Grochow LB: Misperfusion of the liver during hepatic artery infusion chemotherapy: Value of preoperative angiography and postoperative pump scintigraphy. Am J Roentgenol 160: 865–870, 1993.

636. Dannals RF, Neumeyer JL, Milius RA, Ravert HT, Wilson AA, Wagner HN Jr: Synthesis of a radiotracer for studying dopamine uptake sites in vivo using PET: 2b-(4-fluorophenyl)-[N-11C-methyl]tropane ([11C]CFT or [11C]WIN-35,428). J Labelled Compd Radiopharm 33: 147–152, 1993.

637. Müller-Gärtner H-W, Mayberg HS, Fisher RS, Lesser RP, Wilson AA, Ravert HT, Dannals RF, Wagner HN Jr, Uematsu S. Frost JJ: Decreased hippocampal muscarinic cholinergic receptor binding measured by 123I-iododexetimide and single-photon emission computed tomography in epilepsy. Ann Neurol 34: 235–238, 1993.

638. Sasaki M, Müller-Gärtner H-W, Lever JR, Ravert HT, Dannals RF, Guilarte TR, Wagner HN Jr: Assessment of brain muscarinic acetylcholinergic receptors in living mice using a simple probe, [I-125]-dexetimide, and [I-125]-levetimide. Neuropharmacology 32: 1441–1443, 1993.

639. Sostre S, Osman M, Szabo Z, Drew HH, Rivera-Luna H, Civelek AC, Wagner HN Jr: Estimating renal function from the visual analysis of Tc-99m DTPA images. Clinical Nuclear Medicine 18: 281–285, 1993.

640. Suehiro M, Scheffel U, Dannals RF, Ravert HT, Ricaurte GA, Wagner HN Jr: A PET radiotracer for studying serotonin uptake sites: Carbon-11-McN-5652Z. J Nucl Med 34: 120–127, 1993.

641. Suehiro M, Scheffel U, Ravert HT, Dannals RF, Wagner HN Jr: [11C](+)McN-5652 as a radiotracer for imaging serotonin uptake sites with PET. Life Sciences 53: 883–892, 1993.

642. Szabo Z, Ravert HT, Gözükara I, Geckle W, Seki C, Sostre S, Peller P, Monsein L, Natarajan TK, Links JM, Wong DF, Dannsl RF, Wagner HN Jr: Noncompartmental and compartmental modeling of the kinetics of carbon-11 labeled pyrilamine in the human brain. Synapse 15: 263–275, 1993.

643. Wagner HN Jr: The new molecular medicine. J Nucl Med 33: 165–166, 1993.

644. Wagner HN Jr: Oncology: a new engine for PET/SPECT. Highlights of the Society of Nuclear Medicine annual meeting. J Nucl Med 34: 13N–16N, 19N, 22N–29N, 1993.

645. Wagner HN Jr, Holman BL: The future of nuclear medicine: autonomy or integration? J Nucl Med 34: 27N–27N, 31N, 33N, 1993.

646. Wong DF, Yung B, Dannals RF, Shaya EK, Ravert HT, Chen CA, Chan B, Folio T, Scheffel U, Ricaurte GA, Neumeyer JL, Wagner HN Jr, Kuhar MJ: In vivo imaging of a baboon and human dopamine transporters by positron emission tomography using [11C]WIN 35,428. Synapse 15: 130–142, 1993.

647. Holcomb HH, Cascella NG, Medoff DR, Gastineau EA, Loats H, Thaker GK, Conley RR, Dannals RF, Wagner HN, Jr, Tamminga CA: PET-FDG test-retest reliability during a visual discrimination task in schizophrenia. J Comput Assist Tomogr 17: 704–709, 1993.

648. Lever SZ, Sun SY, Scheffel UA, Kaltovich FA, Baidoo KE, Goldfarb H, Wanger HN Jr: Pulmonary Accumulation of Neutral Diamine Dithiol Complexes of Technetium-99m. Journal of Pharmaceutical Sciences 83: 802–809, 1994.

649. Wagner HN Jr: Annual Meeting Highlights: Disease as Dissonance. J Nucl Med 35: 13N–26N, 1994.

650. Wagner HN, Jr: PET, SPECT, and probes in drug design and development. [Review] [14 refs]. NIDA Res Monogr 138: 1–13, 1994.

651. Wagner HN: Nuclear medicine and health care reform [news]. J Nucl Med 35: 34N, 1994.

652. Wagner HN: Patient care and market share [news]. J Nucl Med 35: 22N–23N, 1994.

653. Villemagne VL, Frost JJ, Dannals RF, Lever JR, Tanada S, Natarajan TK, Wilson AA, Ravert HT, Wagner HN, Jr: Comparison of [11C]diprenorphine and [11C]carfentanil in vivo binding to opiate receptors in man using a dual detector system. Eur J Pharmacol 257: 195–197, 1994.

654. Scheffel U, Dannals RF, Suehiro M, Ricaurte GA, Carroll FI, Kuhar MJ, Wagner HN, Jr: Development of PET/SPECT ligands for the serotonin transporter. [Review]. NIDA Res Monogr 138: 111–130, 1994.

655. Mayberg HS, Lewis PJ, Regenold W, Wagner HN, Jr: Paralimbic hypoperfusion in unipolar depression. J Nucl Med 35: 929–934, 1994.

656. Krause BJ, Szabo Z, Becker LC, Dannals RF, Scheffel U, Seki C, Ravert HT, Dipaola AF, Jr, Wagner HN, Jr: Myocardial perfusion with [11C]methyl triphenyl phosphonium: measurements of the extraction fraction and myocardial uptake. J Nucl Biol Med 38: 521–526, 1994.

657. Civelek AC, Meyerrose GE, Wagner HN, Jr, Blumenthal RS: Atypical presentation of left main coronary artery disease. Am J Card Imaging 8: 316–320, 1994.

658. May CH, Guilarte TR, Wagner HN Jr, Vogel S: Intrastriatal Infusion of Lisuride-a Potential Treatment for Parkinson's Disease? Behavioral and Autoradiographic Studies in 6-OHDA Lesioned Rats. Neurodegeneration 3: 305–313, 1994.

659. Woiciechowsky C, Guilarte TR, May CH, Vesper J, Wagner HN, Jr, Vogel S: Intrastriatal dopamine infusion reverses compensatory increases in D2- dopamine receptors in the 6-OHDA lesioned rat. Neurodegeneration 4: 161–169, 1995.

660. Wagner HN, Jr: The new face of health care [news]. J Nucl Med 36: 22N, 39N, 1995.

661. Wagner HN Jr: Molecular Nuclear Medicine: From Genotype to Phenotype via Chemotype. J Nucl Med 36: 5S, 1995.

662. Wagner, HN Jr: A New Era of Certainty. J Nucl Med 36: 13N–28N, 1995.

663. Wagner HN, Jr: Tribute to Professor Masahiro Iio. Radiat Med 13: 1–3, 1995.

664. Szabo Z, Scheffel U, Suehiro M, Dannals RF, Kim SE, Ravert HT, Ricaurte GA, Wagner HN, Jr. Positron emission tomography of 5-HT transporter sites in the baboon brain with [11C]McN5652. J Cereb Blood Flow Metab 15: 798–805, 1995.

665. Suehiro M, Musachio JL, Dannals RF, Mathews WB, Ravert HT, Scheffel U, Wagner HN, Jr. An improved method for the synthesis of radiolabeled McN5652 via thioester precursors. Nucl Med Biol 22: 543–545, 1995.

666. Pearlson GD, Wong DF, Tune LE, Ross CA, Chase GA, Links JM, Dannals RF, Wilson AA, Ravert HT, Wagner HN, Jr: In vivo D2 dopamine receptor density in psychotic and nonpsychotic patients with bipolar disorder. Arch Gen Psychiatry 52: 471–477, 1995.

667. Civelek AC, Pacheco EM, Natarajan TK, Wagner HN Jr, Iliff NT: Quantitative Measurement of vascularization and vascular ingrowth rate of coralline hydroxyapatite Ocular Implant by Tc-99m MDP Bone Imaging. Clinical Nuclear Medicine 20, 1995.

668. Kao PF, Wagner HN Jr, Lever JR, Ravert HT, Dannals RF: Assessing neuroreceptor occupancy by continuous infusion of carbon-11 labeled radioligands. Eur J Nucl Med 23(2): 1996.

669. Wagner HN Jr: Nuclear Oncology: RSNA Categorical Course in Diagnostic Radiology: Nuclear Medicine. pp 47–50, 1996.

670. May CH, Sing HC, Vogel S, Shaya EK, Wagner HN Jr: A New Method of Monitoring motor activity in baboons: Behavior Research Methods, Instru a& Computers. 28: 23–26, 1996.

671. Wagner HN, Jr: The Future. Seminars in Nuclear Medicine 26: 194–200, 1996.

672. Wagner HN, Jr: 1996 SNM annual meeting: medical problem solving [news]. J Nucl Med 37: 11N–14N, 17N, 26N, 1996.
673. Wagner HN, Jr: Nuclear medicine: 100 years in the making [news]. J Nucl Med 37: 18N, 24N, 37N, 1996.
674. Tune L, Barta P, Wong D, Powers RE, Pearlson G, Tien AY, Wagner HN: Striatal dopamine D2 receptor quantification and superior temporal gyrus: volume determination in 14 chronic schizophrenic subjects. Psychiatry Res 67: 155–158, 1996.
675. Wong DF, Singer HS, Brandt J, Shaya E, Chen C, Brown J, Kimball AW, Gjedde A, Dannals RF, Ravert HT, Wilson PD, Wagner HN: D2-like dopamine receptor density in Tourette syndrome measured by PET. J Nucl Med 38: 1243–1247, 1997.
676. Wagner HN, Jr: The incremental value of diagnostic tests [editorial]. J Nucl Med 38: 241–242, 1997.
677. Wagner HN, Jr: Molecular nuclear medicine—the best kept secret in medicine. J Nucl Med 38: N15 ff, 1997.
678. Wagner HN, Jr: F-18 FDG in Oncology: its time has come. Applied Radiology 26: 29–31, 1997.
679. Wagner HN, Jr: Nuclear medicine faces chronic identity crisis. Diagnostic Imaging, June, 1997.
680. Wagner HN, Jr: A new approach to clinical decision-making in oncology. Resident and Staff Physician 43: 12–22, 1997.
681. Mochizuki T, Villemagne VL, Scheffel U, Liu X, Musachio JL, Dannals RF, Wagner HJ: A simple probe measures the pharmacokinetics of [125I]RTI-55 in mouse brain in vivo. Eur Pharmacol 338(1): 17–23, 1997.
682. Kim S, Wagner HJ, Villemagne VL, Kao PF, Dannals RF, Ravert HT, Joh T, Dixon RB, Civelek AC: Longer occupancy of opioid receptors by nalmefene compared to naloxone as measured in vivo by a dual-detector system. J Nucl Med 38(11): 1726–1731, 1997.
683. Liu X, Musachio JL, Wagner HJ, Mochizuki T, Dannals RF, London ED: External monitoring of cerebral nicotinic acetylcholine receptors in living mice. Synapse 27(4): 378–380, 1997.
684. Villemagne VL, Horti A, Scheffel U, Ravert HT, Finley P, Clough DJ, London ED, Wagner HJ, Dannals RF: Imaging nicotinic acetylcholine receptors with fluorine-18-FPH, an epibatidine analog. J Nucl Med 38(11): 1737–1741, 1997.
685. Wagner HJ. Time for a change? [letter]. J Nucl Med 39(5): 1998.
686. Baidoo KE, Lin KS, Zhan Y, Finley P, Scheffel U, Wagner HJ: Design, synthesis, and initial evaluation of high-affinity technetium bombesin analogues. Bioconjugate Chemistry 9(2): 218–225, 1998.
687. Baidoo KE, Scheffel U, Stathis M, Finley P, Lever SZ, Zhan Y, Wagner HJ: High-affinity no-carrier-added 99mTc-labeled chemotactic peptides for studies of inflammation in vivo. Bioconjugate Chemistry 9(2): 208–217, 1998.
688. Lever JR, Ilgin N, Musachio JL, Scheffel U, Finley PA, Flesher JE, Natarajan TK, Wagner HJ, Frost JJ: Autoradiographic and SPECT imaging of cerebral opioid receptors with an iodine-123 labeled analogue of diprenorphine. Synapse 29(2): 172–182, 1998.
689. Musachio JL, Scheffel U, Finley PA, Zhan Y, Mochizuki T, Wagner HJ, Dannals RF: 5-[I-125/123]iodo-3(2(S)-azetidinylmethoxy)pyridine, a radioiodinated analog of A-85380 for in vivo studies of central nicotinic acetylcholine receptors. Life Sciences 62(22): 351–357, 1998.

Index